Arithmetic Skills Worktext

Second Edition

CALMAN GOOZNER

When ordering this book, please specify
either **R 457 W**
or Arithmetic Skills Worktext: Second Edition

AMSCO SCHOOL PUBLICATIONS, INC.
315 Hudson Street New York, N.Y. 10013

ISBN 0-87720-263-X

PRINTED IN THE UNITED STATES OF AMERICA

PREFACE

The ARITHMETIC SKILLS WORKTEXT: SECOND EDITION offers a complete one-year course of instruction designed to develop the student's ability to perform arithmetic computations. For the student who has experienced difficulty in learning arithmetic, this book should prove especially helpful.

The first part of the book covers the fundamentals of arithmetic, and teaches computational skills with whole numbers, fractions, decimals, and percents. In the latter part of the book, these skills are applied to a variety of topics, such as measurement, ratio and proportion, probability, statistics, personal finance, and the solution of simple algebraic equations.

In ARITHMETIC SKILLS WORKTEXT, the teaching is by example, rather than by exposition. The examples in each unit form a vital part of the text proper.

The student's understanding of each topic is immediately reinforced: first, in exercises requiring demonstration of the mastery of the particular computational skill; next, in application problems stressing the ongoing use of critical thinking and problem-solving skills. Wherever necessary, a sample solution at the beginning of a set of exercises or application problems shows the student how to proceed.

Since the book is designed to permit self-study, the more competent students can proceed on their own, with a minimum of supervision, thus freeing the teacher to work with students who need more help.

Suggestions made by users of the first edition, as well as modifications in mathematics curricula, have prompted this revision. In addition to a complete update of real-life references, the revision contains a number of significant changes:

(1) The single unit on solving word problems has been expanded to several units describing different problem-solving suggestions and strategies. The problem-solving approach previously offered in model examples has been maintained and extended.

(2) A unit on the calculator has been added, and exercises specifically designated for calculator use have been incorporated throughout the text.

(3) The topics of area of a triangle and of a circle have been added to the material on geometric measurements.

(4) Reflecting the increasing use of the metric system in the United States, metric measures are interspersed throughout the book.

(5) Applications of ratio include a new unit on probability. Also, scale drawings are now presented as an application of proportion.

(6) A part on statistical averages has been added to precede the work on statistical graphs.

(7) The number of application problems has been doubled, and the number of computational exercises has been significantly increased. Thus, a wealth of material is available for classwork, homework, and testing.

(8) In addition to the Part Reviews previously offered, there are now four Cumulative Reviews. These latter sets of exercises are extended reviews that follow topics in which the emphasis is on computation rather than application, thus providing greatly needed reinforcement in skills work.

The author has had much experience with students who have difficulty with computation, and has developed original ideas to facilitate learning. Visual aids such as arrows showing how numbers are manipulated or how decimal points are moved, as well as handwritten solutions, clarify techniques step by step. The author's familiarity with sources of student errors has enabled him to anticipate areas of difficulty, and to deal with them clearly and effectively, thus making ARITHMETIC SKILLS WORKTEXT uniquely helpful.

CONTENTS

PART VII. Fractions

PART VIII. Adding and Subtracting Fractions

PART IX. Multiplying Fractions

PART X. Dividing Fractions

PART XI. Decimal Fractions

PART XII. Adding, Subtracting, and Multiplying Decimals

PART XIII. Dividing Decimals

PART XIV. Percents

PART XV. Geometric Measures: Length, Area, and Volume

PART XVI. More About Measures

PART XVII. Ratio and Its Applications

PART XVIII. Statistical Averages

PART XIX. Statistical Graphs

PART XX. Earning Money

PART XXI. Spending Money

PART XXII. Personal Banking

PART XXIII. Algebra

PART I. Whole Numbers

UNIT 1. Understanding Whole Numbers

Words to Know

A **whole number** is a number we use for counting.

All whole numbers are formed from the **digits** 1, 2, 3, 4, 5, 6, 7, 8, 9, 0. For example, the number **57** contains the digits **5** and **7.**

The right-hand digit of a whole number is called the **ones digit.** In the number **57,** the ones digit is **7.**

The digit to the left of the ones digit is called the **tens digit.** In the number **57,** the tens digit is **5.** In the number **608,** the tens digit is **0.**

Numbers that consist of the digit 1 followed by one or more zeros are called **powers of ten.** Examples are 10, 100, 1,000, and 10,000.

All over the world, scientists, businesspeople, and students solve arithmetic problems by using numbers. We will first study **whole numbers.**

We use ten symbols, called **digits,** to represent whole numbers:

1 2 3 4 5 6 7 8 9 0

With only these ten symbols, we can express any number at all, small or large. We can start from the smallest whole number, 0, which means "nothing," and we can count all the way up to the number of people in the world or to the number of miles to the most distant visible star.

When you count, 1, 2, 3, 4, 5, 6, 7, 8, 9, 10, the number 10 is different from the first nine numbers. To make the 10, you take the digits 1 and 0 and combine them to form a *different kind of number.* The numbers from 1 to 9 are known as **ones digits,** because they represent values of one. The number 8 means "8 ones" and the number 9 means "9 ones." The number 10, however, means "1 ten." Similarly, the number 20 means "2 tens" and the number 80 means "8 tens." In any number, the digit to the left of the ones digit is known as the **tens digit.**

In real life, you know that 10 pennies make a dime and that 10 dimes make a dollar. Similarly, ten 1's make 10, ten 10's make 100, and ten 100's make 1,000. Thus, as whole numbers become larger, they are expressed in **powers of ten.**

Now you will learn the names and the values of whole numbers. In reading the value of a number, you must note the *position* of each digit in the number.

The number 10 means:

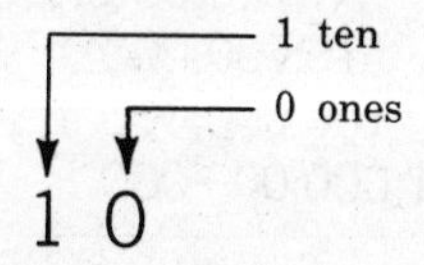

The number 100 means:

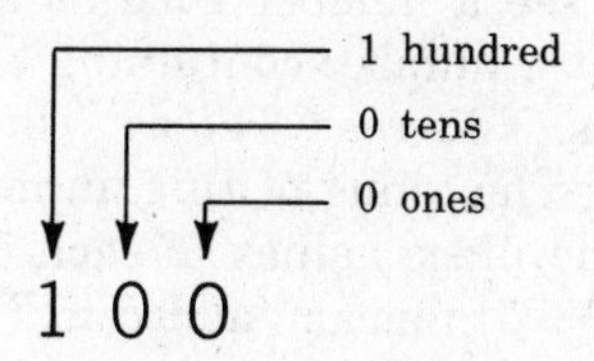

The number 1,000 means:

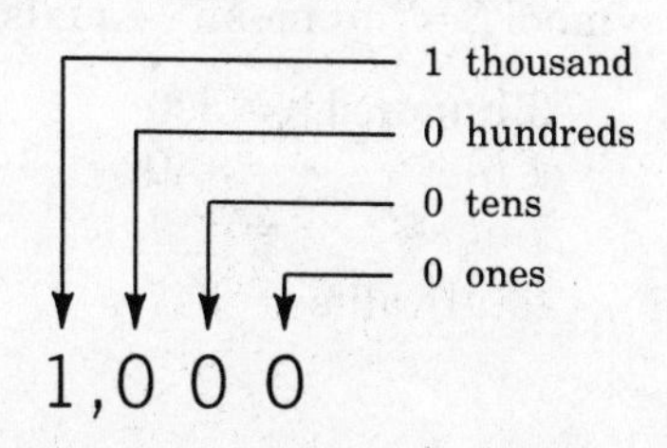

The number 10,000 means:

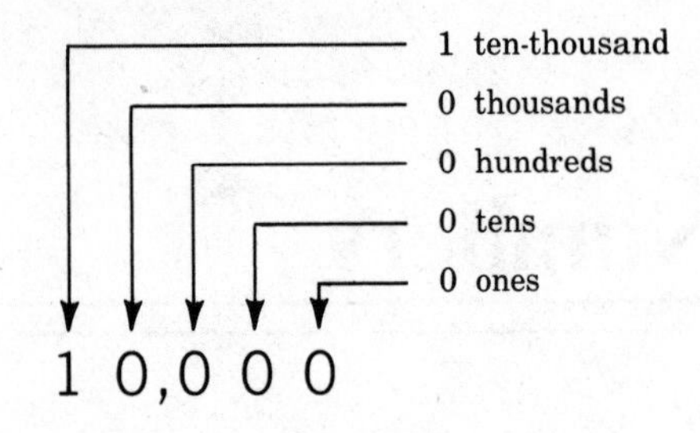

The number 100,000 means:

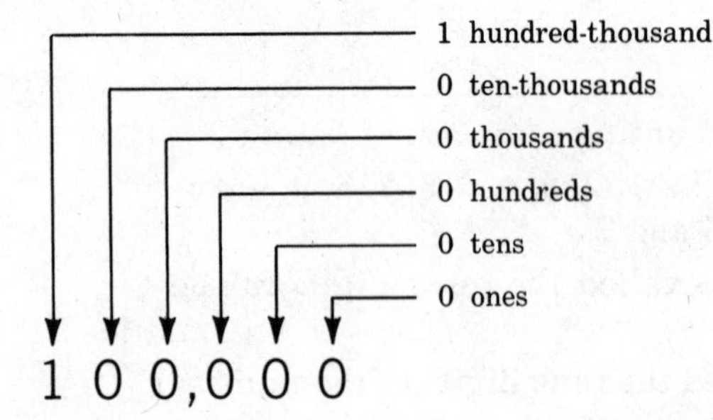

The number 1,000,000 means:

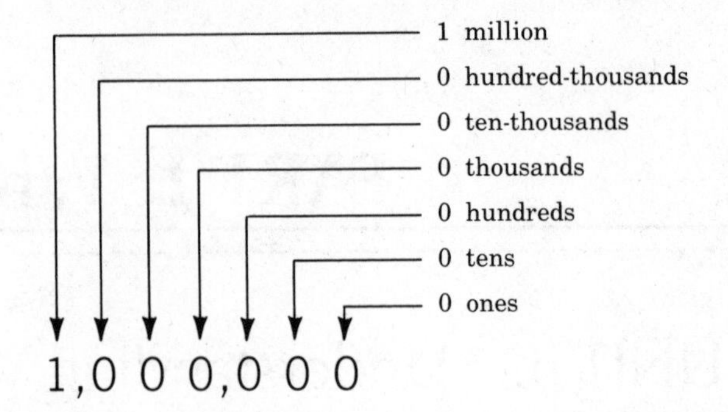

In the preceding six numbers, note that the place value of the digit 1 changes from "ten" to "one million," depending on its *position*.

Table I below shows the possible whole numbers from one to one billion. Also, it tells how many digits are required to write these numbers.

Table I: Whole Numbers

Possible Numbers	How Many Digits	Largest Place Value
1 to 9	1 digit	one
10 to 99	2 digits	ten
100 to 999	3 digits	hundred
1,000 to 9,999	4 digits	thousand
10,000 to 99,999	5 digits	ten-thousand
100,000 to 999,999	6 digits	hundred-thousand
1,000,000 to 9,999,999	7 digits	million
10,000,000 to 99,999,999	8 digits	ten-million
100,000,000 to 999,999,999	9 digits	hundred-million
1,000,000,000	10 digits	billion

When you see a number such as 347, Table I tells you that a number containing 3 digits is in the hundreds. You read 347 as "three hundred forty-seven." The value of this number is found by adding the place values of each digit. Using the symbol "+" to mean "added to," you have

3 hundreds + 4 tens + 7 ones

Using the symbol "=" to mean "equals,"

3 hundreds = 300
4 tens = 40
7 ones = 7
total value = 347

The number 2,465 is read as "two thousand, four hundred sixty-five," and the value is:

2 thousands + 4 hundreds + 6 tens + 5 ones
2,000 + 400 + 60 + 5

It is difficult to read a number such as:

5321642

To make the reading of a large number easier, you must first "punctuate" the number by using commas to separate the number into groups.

To punctuate a number, *place a comma after every three digits of the number, moving from right to left.* Thus:

5,321,642

What you have done is to separate the number into three groups: the *hundreds* group, the *thousands* group, and the *millions* group. (The hundreds group actually consists of hundreds, tens, and ones.)

Remember

You punctuate a number from right to left, but you read a number from left to right.

Let's look at the number again:

5,321,642

The right-hand group of digits means *hundreds, tens,* and *ones.* The three digits to the left of the right-hand group mean *thousands.* Since the number has another digit to the left of the thousands, the highest value of the number is in the *millions.* (Table I shows us that when a number has 7 digits, the value of the largest digit is in the millions.)

How many millions? 5,321,642 — 5 millions

How many thousands? 5,321,642 — 321 thousands

How many hundreds, tens, and ones? 5,321,642 — 6 hundreds, 4 tens, 2 ones

You read the number as "five million, three hundred twenty-one thousand, six hundred forty-two."

EXERCISES

In 1–15, write each number as a word phrase.

SAMPLE SOLUTION

a. 245 *two hundred forty-five* **b.** 5,021 *five thousand, twenty-one*

1. 53 fiftythree

2. 71

3. 95 ninetyfive

4. 17

5. 68 sixtyeight

6. 123 onehundred

7. 467

8. 351

9. 915

10. 673

11. 8,015

12. 29,040

13. 261,451

14. 5,653,000

15. 23,520,825

In 16–25, write the value of each digit in the given number.

SAMPLE SOLUTION

317 = 3 hundreds, 1 tens, 7 ones

16. 14 = ________ tens, ________ ones
17. 37 = ________ tens, ________ ones
18. 89 = ________ tens, ________ ones
19. 335 = ________ hundreds, ________ tens, ________ ones
20. 510 = ________ hundreds, ________ tens, ________ ones
21. 809 = ________ hundreds, ________ tens, ________ ones
22. 1,043 = ________ thousands, ________ hundreds, ________ tens, ________ ones
23. 26,413 = ________ ten-thousands, ________ thousands, ________ hundreds, ________ tens, ________ ones
24. 237,000 = ________ hundred-thousands, ________ ten-thousands, ________ thousands, ________ hundreds, ________ tens, ________ ones
25. 15,865,437 = ______ ten-millions, ______ millions, ______ hundred-thousands, ______ ten-thousands, ______ thousands, ______ hundreds, ______ tens, ______ ones

In 26–35, write each word phrase as a number.

SAMPLE SOLUTION

a. eighty-four 84 **b.** twenty-one thousand, fifteen 21,015

26. twenty-five ________
27. ninety-six ________
28. three hundred five ________
29. eight hundred seventy-three ________
30. four thousand, nineteen ________
31. seven thousand, three hundred twenty-seven ________
32. thirteen thousand, five ________
33. eighty-six thousand, four hundred ________
34. three hundred thousand, four hundred fifty-seven ________
35. nineteen million, four hundred twenty-three thousand, six hundred fifteen ________

APPLICATION PROBLEMS

In 1–10, write a word phrase for the number in each given fact.

SAMPLE SOLUTION

Cynthia bought a car for $8,423.

eight thousand, four hundred twenty-three

1. Frank's total income last year was $18,523.

2. The O'Brien family paid $74,095 for their new home.

3. The new school was completed at a cost of $375,485.

4. The population of Castleton is 13,473,000.

5. A government study reveals that 15,963,400 people are employed as office workers.

6. Alaska, with 589,757 square miles, has the largest area of any state.

7. In the first decade of this century, 8,795,386 immigrants came to the United States.

8. In one year of business trips, computer expert Fran Onassis traveled 205,087 miles.

9. The first time that Abraham Lincoln was elected president, he received 1,866,352 popular votes.

10. Because of their different orbits, the planets Earth and Mercury are not always the same distance apart. The greatest distance between them is about 136,000,000 miles.

UNIT 2. Rounding Off Numbers

WORDS TO KNOW

A number is **rounded off** when its actual value is changed to an approximate value. The population of New York City at one time was 7,895,563. You say that the population of New York City was 8,000,000, rounded off to the nearest million.

In the above example, the number 7,895,563 is **rounded off** to 8,000,000. You round off numbers when you do not need the exact value that the number represents. Rounding off is a way of simplifying a number.

Suppose that 53,968 people go to a baseball game. In the morning newspapers, the attendance figure is reported as 54,000. This value is obtained by rounding off the actual attendance to the nearest thousand.

Before rounding off any number, you must decide on the value you want to round off to. The "value" means the value of the digit in the rounded-off number that comes just before the 0's. In the rounded-off number 8,000,000, the digit 8 has a value of *millions.* In the rounded-off number 54,000, the digit 4 has a value of *thousands.*

Table I (page 2) gives such values from *one* all the way up to *one billion.*

After deciding on the value to be rounded off to, do the following steps:

Step 1: Place parentheses around all digits to the right of this value.

Step 2: A. If the left-hand digit inside the parentheses is 5 or more, add 1 to the part outside the parentheses.

B. If the left-hand digit inside the parentheses is less than 5, do not add 1.

Step 3: Substitute 0's for the digits inside the parentheses.

EXAMPLE 1. Round off 7,465 to the nearest *ten.*

Solution:

Step 1: Place parentheses around the digit to the right of the *tens* value.

7,46(5)

Step 2: Because the digit inside the parentheses is 5, add 1 to the *tens* digit.

$$\begin{array}{r} 7{,}46(5) \\ +1 \\ \hline 7{,}47(5) \end{array}$$

Step 3: Substitute a 0 for the digit inside the parentheses.

7,47(0)

Answer: 7,470

EXAMPLE 2. Round off 7,465 to the nearest *hundred.*

Solution:

Step 1: Place parentheses around the digits to the right of the *hundreds* place.

7,4(65)

Steps 2 and 3: Because the left-hand digit inside the parentheses is 6, which is 5 or more, add 1 to the *hundreds* digit. Then substitute 0's for the digits inside the parentheses.

$$\begin{array}{r} 7{,}4(65) \\ +1 \\ \hline 7{,}5(00) \end{array}$$

Answer: 7,500

EXAMPLE 3. Round off 7,465 to the nearest *thousand.*

Solution:

Step 1: Place parentheses around the digits to the right of the *thousands* place.

7,(465)

Steps 2 and 3: Because the left-hand digit inside the parentheses is 4, which is less than 5, *do not* add 1. (You can think of "do not add 1" as "add zero.") Substitute 0's for all the digits in the parentheses.

$$\begin{array}{r} 7{,}(465) \\ +0 \\ \hline 7{,}(000) \end{array}$$

Answer: 7,000

Sometimes more than one of the digits outside the parentheses are changed.

EXAMPLE 4. Round off 3,796 to the nearest *ten.*

Solution:

Step 1: Place parentheses around the digit to the right of the *tens* place.

3,79(6)

Steps 2 and 3: Because the digit inside the parentheses is 6, which is more than 5, add 1 to the *tens* digit. Adding 1 to the tens digit, the 9, changes *two digits.* Add 1 to 79. Then substitute 0 for the digit inside the parentheses.

$$\begin{array}{r} 3{,}79(6) \\ +1 \\ \hline 3{,}80(0) \end{array}$$

Answer: 3,800

After a little practice, you should be able to perform steps 2 and 3 *mentally.*

EXERCISES

In 1–8, round off each number to the nearest *ten*.

SAMPLE SOLUTION

4(7) 50

1. 68 ______
2. 84 ______
3. 135 ______
4. 324 ______
5. 5,048 ______
6. 8,964 ______
7. 11,475 ______
8. 18,388 ______

In 9–16, round off each number to the nearest *hundred*.

SAMPLE SOLUTION

4(38) 400

9. 362 ______
10. 538 ______
11. 895 ______
12. 2,809 ______
13. 5,243 ______
14. 8,556 ______
15. 15,045 ______
16. 21,099 ______

In 17–24, round off each number to the nearest *thousand*.

SAMPLE SOLUTION

8,(762) 9,000

17. 9,462 ______
18. 6,548 ______
19. 3,817 ______
20. 4,072 ______
21. 21,628 ______
22. 41,314 ______
23. 35,089 ______
24. 61,529 ______

In 25–32, round off each number to the nearest *ten-thousand*.

SAMPLE SOLUTION

1(5,681) 20,000

25. 23,510 ______
26. 15,010 ______
27. 94,215 ______
28. 195,628 ______

29. 507,128 ________ 30. 704,852 ________

31. 5,246,117 ________ 32. 8,375,000 ________

In 33–40, round off each number to the nearest *hundred-thousand.*

SAMPLE SOLUTION

1(21,502) 100,000

33. 158,708 ________ 34. 605,235 ________

35. 820,718 ________ 36. 2,073,210 ________

37. 5,621,187 ________ 38. 37,162,115 ________

39. 65,973,508 ________ 40. 80,053,287 ________

In 41–48, round off each number to the nearest *million.*

SAMPLE SOLUTION

4,(528,116) 5,000,000

41. 3,417,128 ________ 42. 5,617,249 ________

43. 1,500,000 ________ 44. 8,310,000 ________

45. 20,583,117 ________ 46. 25,021,320 ________

47. 61,621,187 ________ 48. 10,421,356 ________

APPLICATION PROBLEMS

1. Round off the numbers in the following sentence to the nearest *thousand:* In a given week, the United States produced 163,478 cars, 143,627 trucks, and 53,067 buses.

   cars: ________

   trucks: ________

   buses: ________

2. Round off, to the nearest *hundred-thousand,* the following numbers that show the populations of ten American cities in a given year. As a sample solution, the population of New York has been rounded off.

   New York: 7,8(95,563) 7,900,000

   Chicago: 3,369,359 ________

   Los Angeles: 2,816,061 ________

   Philadelphia: 1,950,098 ________

   Detroit: 1,512,893 ________

   Houston: 1,232,802 ________

   Baltimore: 905,759 ________

Dallas: 844,401 ______

Washington: 756,510 ______

Cleveland: 750,879 ______

3. Below are the monthly sales, by department, of the Thrifty Supermarket. Round off each amount to the nearest *ten-thousand dollars.*

Grocery: $34,628 ______

Meat: $25,738 ______

Dairy: $19,563 ______

Produce: $8,750 ______

Non-food: $5,620 ______

4. A sales representative traveled the following distances last week. Round off each distance to the nearest *hundred kilometers.*

Monday: 558 kilometers ______

Tuesday: 455 kilometers ______

Wednesday: 443 kilometers ______

Thursday: 548 kilometers ______

Friday: 963 kilometers ______

5. Round off each price to the nearest *ten dollars.*

Blouse: $18 ______

Skirt: $29 ______

Belt: $9 ______

Slacks: $23 ______

Jacket: $37 ______

Hat: $14 ______

Review of Part I (Units 1 and 2)

In 1–5, write each number as a word phrase.

1. 5,125 ______

2. 14,085 ______

3. 321,560 ______

4. 5,321,565 ______

5. 28,671,305 ______

In 6–10, write the value of each digit in the given number.

6. 453: ______ hundreds, ______ tens, ______ ones

7. 905: ______ hundreds, ______ tens, ______ ones

8. 3,438: __________ thousands, __________ hundreds, __________ tens, __________ ones

9. 28,068: __________ ten-thousands, __________ thousands, __________ hundreds, __________ tens, __________ ones

10. 507,234: __________ hundred-thousands, __________ ten-thousands, __________ thousands, __________ hundreds, __________ tens, __________ ones

11. Round off each number to the nearest *ten.*

a. 364 __________ *b.* 495 __________ *c.* 638 __________

12. Round off each number to the nearest *hundred.*

a. 2,658 __________ *b.* 3,085 __________

c. 5,963 __________ *d.* 6,063 __________

13. Round off each number to the nearest *thousand.*

a. 12,485 __________ *b.* 15,862 __________

c. 20,573 __________ *d.* 30,451 __________

14. Round off each number to the nearest *ten-thousand.*

a. 243,670 __________ *b.* 567,210 __________

c. 705,340 __________ *d.* 876,777 __________

15. Round off each number to the nearest *hundred-thousand.*

a. 3,625,138 __________ *b.* 3,063,780 __________

c. 4,965,362 __________ *d.* 7,209,652 __________

16. Round off each number to the nearest *million.*

a. 25,326,120 __________ *b.* 32,725,000 __________

c. 40,860,000 __________ *d.* 84,696,530 __________

PART II. Adding Whole Numbers

UNIT 3. Addition of Numbers

WORDS TO KNOW

When the values of two or more numbers are added to form a larger number, the process is called **addition.**

The larger number that results from addition is called the **sum** or the **total.**

The numbers that are added are called **addends.**

When you perform **addition,** you combine two or more numbers to get one larger number called the **sum** or **total.** The sum of an addition problem must always be larger than any one of the numbers you are adding. Addition is indicated by writing the *plus* symbol "+" with the numbers to be added. To indicate that the number 17 and the number 11 are to be added, write 17 + 11 or

$$\begin{array}{r} 17 \\ +11 \\ \hline \end{array}$$

In setting up the numbers to be added, known as **addends,** you must always be sure to line up your digits correctly: ones digit under ones digit, tens digit under tens digit, etc. A column of ones digits is called the *ones column,* a column of tens digits is called the *tens column,* etc.

EXAMPLE 1. Find the sum of 25 and 7.

Solution: The 5 in the number 25 is the ones digit. The 7 in the number 7 is the ones digit. Therefore, *the 7 must line up directly under the 5.*

$$\begin{array}{r} 25 \\ +7 \\ \hline \end{array}$$

Now, you add the 5 and the 7 (5 ones and 7 ones) and get 12 (1 ten and 2 ones). You write the 2 under the ones column and carry the 1 to the tens column. (The 1 means "1 ten.") Add the 1 to the 2 and write the sum, 3, under the tens column (1 ten + 2 tens = 3 tens).

$$\begin{array}{r} {}^{1} \\ 25 \\ +7 \\ \hline 32 \end{array}$$

Answer: 25 + 7 = 32

EXAMPLE 2. Add: 395 + 37

Solution: Again, line up your digits correctly: ones under ones and tens under tens. Add: 5 plus 7 equals 12.

$$\begin{array}{r} 395 \\ +37 \\ \hline \end{array}$$

Write down the 2 under the ones column and carry the 1 to the tens column. Add: 1 plus 9 plus 3 equals 13. Write the 3 under the tens column and carry the 1 to the hundreds column (this 1 means "1 hundred"). Add: 1 plus 3 equals 4. Write the 4 under the hundreds column.

$$\begin{array}{r} {}^{1\,1} \\ 395 \\ +37 \\ \hline 432 \end{array}$$

Answer: 395 + 37 = 432

EXAMPLE 3. Add: 5,327 + 462 + 578

Solution: Line up your digits correctly: ones under ones, tens under tens, and hundreds under hundreds.

$$\begin{array}{r} 5,327 \\ 462 \\ +578 \\ \hline \end{array}$$

Add: 7 + 2 + 8 = 17. Write the 7 under the ones column and carry the 1 to the tens column.
Add: 1 + 2 + 6 + 7 = 16.
Write the 6 under the tens column and carry the 1 to the hundreds column.
Add: 1 + 3 + 4 + 5 = 13.
Write the 3 under the hundreds column and carry the 1 to the thousands column.
Add: 1 + 5 = 6.
Write the 6 under the thousands column.

$$\begin{array}{r} {}^{1\;\,1\,1} \\ 5,327 \\ 462 \\ +578 \\ \hline 6,367 \end{array}$$

Check: To check an addition problem, add the columns again, but start adding from the bottom up.

$$\begin{array}{r} 5,327 \\ 462 \\ +578 \\ {}^{1\;\,1\,1} \\ \hline 6,367 \end{array}$$

add up

Answer: 5,327 + 462 + 578 = 6,367

EXERCISES

In 1–10, add and check.

1.	2.	3.	4.	5.
234 25	327 68	437 98	368 56	929 82

6.	7.	8.	9.	10.
4,628 407	6,724 368	3,257 845	8,207 636	5,237 568

In 11–18, find the sum and check each answer.

11. 324 + 85 + 7
12. 168 + 67 + 4
13. 763 + 46 + 9
14. 572 + 73 + 8
15. 678 + 59 + 6
16. 368 + 38 + 3
17. 273 + 89
18. 972 + 65 + 7

In 19–26, add and check.

19.	20.	21.	22.
3,428 734 235 58	5,742 837 768 53	7,209 568 47 59	8,796 467 85 93

23.	24.	25.	26.
6,748 264 19 8	5,680 974 63 7	2,471 687 86 3	6,592 749 53 5

In 27–38, find the sum and check each answer.

27. 3,827 + 463 + 75 + 82
28. 5,649 + 329 + 67 + 8
29. 3,558 + 379 + 56 + 7

30. 6,857 + 574 + 38 + 6 = 7475

31. 2,463 + 595 + 63 + 3 = 3124

32. 5,361 + 768 + 56 + 9 = 6194

33. 4,736 + 429 + 38 + 5 = 5208

34. 6,729 + 647 + 58 + 3 = 7437

35. 3,427 + 2,685 + 347 + 22 = 6481

36. 5,692 + 478 + 363 + 42 + 8 = 6583

37. 5,238 + 3,595 + 765 + 528 + 49 = 10175

38. 7,562 + 8,473 + 597 + 98 + 8 = 16738

In 39–42, add and check.

39.	**40.**	**41.**	**42.**
23,478	235,407	478,629	629,319
47,593	346,972	962,523	492,748
39,267	53,863	593,093	68,492
9,432	17,093	468	9,716
827	841	793	345
963	493	658	729
121560	657669	2036164	1201349

APPLICATION PROBLEMS

Sample Solution

Last week, the Council Rock football team gained 84 yards rushing and 121 yards passing. What was the total yardage gained?

84 yd.
+ 121 yd.
205 total yd.

1. A company deposited checks in the following amounts: $235, $76, $865, $245. Find the total amount deposited. 1421 deposited

2. How much will a set of living room furniture cost if a couch costs $1,495, an armchair costs $373, a coffee table costs $485, and an end table costs $299? 2652 costs

3. A grocery store had the following daily sales: Monday, $937; Tuesday, $875; Wednesday, $1,125; Thursday, $1,020; Friday, $1,248; Saturday, $1,368. Find the total sales for the week. $6573 week

4. A school has the following enrollment: 1,627 freshmen; 1,327 sophomores; 1,148 juniors; 953 seniors. What is the total enrollment? 5055 enrollment

5. The local library has 4,389 books of fiction, 3,235 nonfiction books, 358 reference books, and 175 magazines. What is the total number of books and magazines? 8157 books and magazines

6. A salesperson sold the following amounts last week: $896, $968, $1,346, $739, and $1,263. What were the total sales for the week? $5212 week

7. The yearly salaries of five office clerks are: $13,570, $14,750, $15,275, $12,850, and $13,950. How much is the yearly payroll for the five clerks? $70395 five clerks

8. A traveling salesperson had the following expenses last week: hotel, $420; food, $189; gasoline, $79; telephone, $27; other, $83. What were the total expenses? $798. total expenses

9. A meat wholesaler shipped the following quantities of meat: 350 kilograms of steak, 485 kilograms of round roast, 283 kilograms of pork chops, 372 kilograms of veal, and 324 kilograms of chopped meat. What was the total weight of the shipment? 1814 shipment

10. Janet bought the following items: a blouse for $36, a sweater for $29, a coat for $286, and a pair of shoes for $39. How much did Janet spend on her purchases? 393 her purchases

UNIT 4. Addition When the Number of Digits in the Addends Varies

It is often necessary to add a group of numbers that are expressed in dollars. It is not uncommon for amounts of money to vary widely, from a few dollars to thousands of dollars.

To avoid mistakes, you should arrange the column of addends by writing the largest number first, then the next largest, and so on, down to the smallest number. When adding amounts of money, be sure to line up the digits correctly.

Suppose you must add the following amounts:

$23 $1,625 $8
$11,698 $153

Start your column with the largest number, $11,698. Write the number down and then *cross it out* of the original list.

$23 $1,625 $8
~~$11,698~~ $153

$11,698

Select the next largest number, $1,625. Write it down, aligning the digits correctly, and cross it out of the list. (It is not necessary to repeat the dollar sign.)

$23 ~~$1,625~~ $8
~~$11,698~~ $153

$11,698
1,625

Select the next largest number, $153. Write it down and cross it out of the list.

$23 ~~$1,625~~ $8
~~$11,698~~ ~~$153~~

$11,698
1,625
153

Follow this procedure until all the numbers are written down in the descending order of their values and all the numbers have been crossed out of the original list. Complete the addition, bringing down the dollar sign. The completed addition should look like this:

~~$23~~ ~~$1,625~~ ~~$8~~
~~$11,698~~ ~~$153~~

$11,698
1,625
153
23
8
$13,507

The technique of forming a column of numbers in the *descending order of their values* (largest value first) is not limited to the addition of dollars. It is used whenever a group of numbers of varying values must be added. By "varying values," we mean that the number of digits in the addends varies.

EXAMPLE. In the addition problem shown, the answer is wrong. Find the correct sum by forming a column of numbers in the descending order of their values.

4827
123
4
82
50000
wrong ✗232570

Solution: Whoever added the numbers forgot to line up the digits correctly. Here is how the column of numbers should be formed:

~~4827~~	50,000
~~123~~	4,827
~~4~~	123
~~82~~	82
~~50000~~	4
	55,036

Answer: The correct sum is 55,036.

EXERCISES

In 1–14, add the numbers by arranging them in the descending order of their values.

SAMPLE SOLUTION

~~28~~ + ~~5~~ + ~~235~~

235
28
5
268

268

1. 47 + 423 + 8 = 478

2. 2 + 576 + 68 = 646

3. 19 + 486 + 68 + 2,519 = 3092

4. 357 + 8 + 5,694 + 62 = 6121

5. 42 + 634 + 7 + 6,927 + 23 = 7633

6.
248
47
15,620
8
6,537
12
634
23106

7.
6,409
4
254
8,623
34
14,638
29962

8.
36
18,437
7
422
3,294
52
22248

9. $5,246 + $9 + $368 + $84 + $16,472 + $234 + $575 = 22988

10. $38 + $122 + $92 + $1,196 + $7 + $276 = 1731

11. $3 + $2,415 + $47 + $89 + $6 + $342 = 2902

12. 73 + 14,193 + 754 + 8 + 9,634 + 521 + 758 + 38 = 25979

13. 19,627 + 72 + 235,537 + 6 + 9,427 + 135,364 + 47,638 = 447671

14. 94 + 32,763 + 18 + 897 + 7,458 + 347,905 + 8 = 389143

APPLICATION PROBLEMS

SAMPLE SOLUTION

Dave McGraw made a bank deposit of the following items: $2 in pennies, $6 in nickels, $20 in dimes, $14 in one-dollar bills, and a check for $17. How much did he deposit?

$20
17
14
6
2
$59

1. A delivery truck traveled the following distances: 235 miles, 9 miles, 47 miles, 97 miles, and 125 miles. How many miles did the truck travel altogether? 513

2. What is the total seating capacity of the local stadium if there are 893 box seats, 8,329 reserved seats, 19,365 general admission seats, 3,478 bleacher seats, and 8 press seats? 32073

3. Walter Watts made the following purchases: an overcoat for $235, a suit for $149, a tie for $13, a sports jacket for $78, and some socks for $8. How much did he spend? 483

4. Jessica bought a car for $8,378. She also bought the following extras: air conditioning, $675; car speakers, $53; undercoating, $89; power brakes, $368; power steering, $235. Find the total cost. $9798

5. The Jackson family bought a house for $79,378 and made the following improvements: central air conditioning, $1,865; gas barbecue, $195; fence, $978; patio, $825. What was the total cost of the house including the improvements? 83241

6. The attendance at an adult education center varied according to the schedule of courses given each night. The attendance figures for four nights were: Monday, 443; Tuesday, 87; Wednesday, 1,622; Thursday, 2,015. What was the total attendance for the four nights? 4167

7. In the Hopewell High School Social Studies bookroom, there were 35 texts on sociology, 653 texts on world geography, 87 texts on economics, 1,822 texts on American history, and 42 texts on government. What was the total number of all texts? 2639

8. Sandy MacIvor counted all the vegetables she took from her garden in one week. She had 9 beefsteak tomatoes, 23 zucchini squash, 105 cherry tomatoes, 18 green peppers, and 8 carrots. How many vegetables did she have altogether? 163

9. In one recent year, the egg production of several states, in millions of eggs, was reported as follows: Alabama, 3,354; Arizona, 11; Colorado, 464; Connecticut, 1,004; Rhode Island, 84; and Arkansas, 4,153. What was the total egg production, in millions of eggs, of these states? 9070

10. The Starkey children all went to the family doctor for physical examinations. Since their last checkups, Stacey had grown 93 millimeters, Sam 182 millimeters, Stan 7 millimeters, and Susie and Swoozie had grown 47 millimeters each. What was the total height increase for all the children? 329

UNIT 5. Horizontal Addition

There are many times when it is convenient to be able to add a group of numbers that is arranged horizontally. This arrangement may slow down the addition of the numbers unless you practice and develop the skill of adding horizontally.

A good way to develop this skill is to use a *rhythmic pattern of adding.* In other words, add in a steady beat with the eye "bouncing" from one number to the next in regular rhythm.

EXAMPLE 1. Add:

8 + 5 + 7 + 6 + 5 + 9 + 2 + 4

Solution: Add, going from left to right:

8 + 5 + 7 + 6 + 5 + 9 + 2 + 4

Say: 13 20 26 31 40 42 46

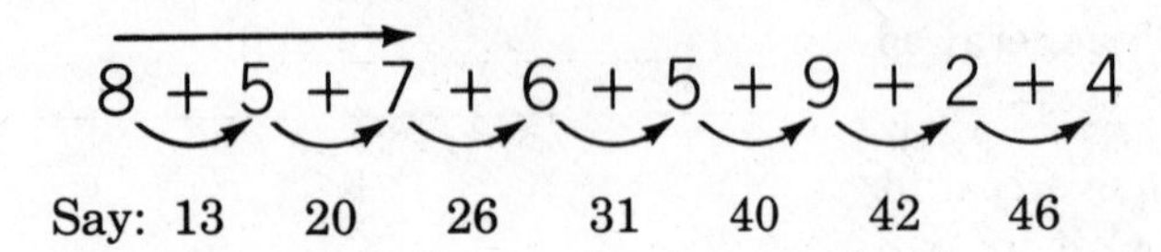

You should also be able to add from right to left:

8 + 5 + 7 + 6 + 5 + 9 + 2 + 4

46 38 33 26 20 15 6

Answer: 46

EXAMPLE 2. Add: 24 + 36 + 83 + 67

Solution: When the addends have more than one digit, add the ones digits first:

24 + 36 + 83 + 67

Say: 10 13 20

Write the 0 of the 20 and carry the 2.

2

__ 0

Now add the tens digits:

24 + 36 + 83 + 67

Say: 5 13 19

Write the 19, and add the 2 you carried.

2

1 9 0

2 1 0

Answer: 210

For addends with more than two digits, continue this procedure.

At times, as on some business forms, you may be required to add groups of figures arranged both vertically and horizontally.

To check the accuracy of such problems, add the totals of the horizontal rows and add the totals of the vertical columns. The *grand totals* (the totals of the totals) should be the same.

EXAMPLE:

15 +	23 +	87 =	125
18 +	52 +	35 =	105
35 +	63 +	52 =	150
68 +	138 +	174 =	380

Since the grand totals both vertically and horizontally are the same (380), you know that there are no arithmetic mistakes in the additions.

EXERCISES

In 1–6, add horizontally.

1. 5 + 7 + 4 + 8 + 5 + 2

2. 9 + 6 + 5 + 3 + 7 + 4 + 8

3. 8 + 5 + 7 + 9 + 3 + 6 + 8 + 2

4. 6 + 9 + 8 + 5 + 3 + 7 + 9 + 5

5. 13 + 23 + 42 + 15

6. 25 + 32 + 47 + 56

In 7–8, add horizontally and vertically.

7. 47 + 68 + 74 =
65 + 46 + 59 =
78 + 96 + 54 = ___
___ + ___ + ___ =

8. 34 + 59 + 72 =
67 + 85 + 98 =
37 + 48 + 54 =
23 + 64 + 82 = ___
___ + ___ + ___ =

APPLICATION PROBLEMS

In these problems, do *not* form columns of numbers. Find the answers by adding horizontally.

1. A salesperson earned the following commissions:

$13 $57 $14 $43 $89

Find the total commission earned. ___

2. A bookstore sold the following numbers of books last week:

MON	TUE	WED	THU	FRI	SAT
58	37	89	68	148	215

What was the total number of books sold? ___

3. Last year, Mrs. Wilson bought the following numbers of gallons of fuel oil:

NOV	DEC	JAN	FEB	MAR
160	210	240	170	180

Find the total number of gallons. ___

4. John delivered the following numbers of newspapers last week:

Monday, 76 Tuesday, 78 Wednesday, 82
Thursday, 69 Friday, 73 Saturday, 87

How many newspapers did he deliver last week? ___

5. A secretary mailed the following numbers of letters:

230 350 280 360 250

How many letters did he send out? ___

6. The kilogram weights of six packages ready for shipment were:

102 34 116 98 56 27

What was the total weight of the shipment? ___

7. Ms. Kurz was trying to get more exercise. On seven consecutive days, she walked the following numbers of blocks:

28 33 24 32 44 36 38

How many blocks did she walk that week? ___

8. As a fund-raiser, the Music Club was selling crates of oranges. The dollar values of the sales made by the most successful members were:

Pete	Joe	Rosa	Meg	Harry
336	224	208	192	176

What was the total amount of their sales? __________

9. Roger was on a diet, and decided to keep track of his weight loss in ounces. He lost the following numbers of ounces:

Sunday, 15 Monday, 22 Tuesday, 16 Wednesday, 4
Thursday, 3 Friday, 18 Saturday, 12

What was the total number of ounces he lost all week? __________

10. The Carey Bus Company had five different routes. On one day, the buses carried the following numbers of passengers:

Route Number	M4	N7	B3	R2	G6
Number of Passengers	218	432	105	337	296

What was the total number of passengers for the day? __________

11. Ted, Anna, Ellen, and Angie went on a fishing trip. They competed each day to see whose catch weighed the most.

	Number of Pounds of Fish Caught					Total Catch for Each Person
	Saturday	Sunday	Monday	Tuesday	Wednesday	
Ted	13	8	26	18	30	
Anna	7	19	22	34	21	
Ellen	16	11	19	23	14	
Angie	24	13	21	28	19	
Daily Totals						
						Grand Total

For *a* and *b*, write the answers in the table.

a. Find the daily totals of fish caught.

b. Find the total for each person for the entire trip.

c. Whose individual total was the highest? __________

d. What was the total weight of all the fish caught? __________

e. Check that the grand totals are the same vertically and horizontally.

UNIT 6. Using a Calculator

A calculator can do arithmetic operations very quickly, but it can do only what you tell it to do. By learning to do the arithmetic yourself, you will be able to work more effectively with a calculator. Also, some problems can be solved more quickly by using your own knowledge of arithmetic than by working them out with a calculator.

In real-life situations that involve computation, you need to know *when* to add, subtract, multiply, and divide. These are decisions *you* must make, since the calculator cannot. In Part VI of this book, we will discuss how to make these decisions.

There are many different models of calculators, some with as many as 40 buttons. A calculator for doing the basic arithmetic operations would look something like this:

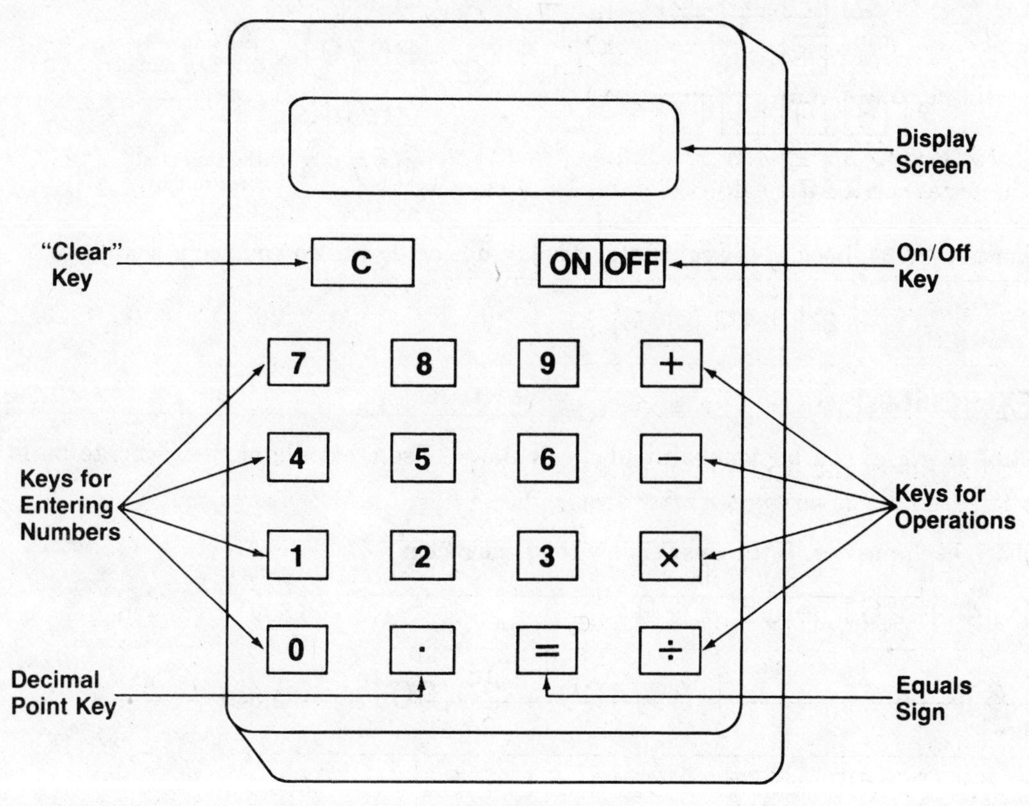

Practice using a calculator. We begin by doing addition. The same methods will apply when doing subtraction, multiplication, or division, just by pressing the [−], [×], or [÷] key, as needed.

Start with the easy sum 2 + 3. Press the following in order, and watch the display screen. The calculator starts each problem with 0.

Press	Display
[ON]	0 (to start)
[2]	2
[+]	2 (no change)
[3]	3
[=]	5 (the answer)

When you press [=], the display shows the answer to your problem.

Press [C] to clear the screen, so that it displays 0 again.

Now, do this harder problem. (It's no harder for the calculator, though.)

Add: 42,016 + 38,775 + 26,943

Press		Display	
[4] [2] [0] [1] [6]	(no commas)	42016	
[+]		42016	
[3] [8] [7] [7] [5]		38775	
[+]		80791	(the sum of the first two numbers)
[2] [6] [9] [4] [3]		26943	
[=]		107734	(the sum of all three numbers)

The answer is 107,734. Look at the calculator display upside down. Do you see a word?

EXERCISES

This set of exercises can be done without a calculator. Use a calculator, however, to build skill in working with it.

In 1–4, check each answer. If the answer is wrong, correct it.

		Answer
1.	3,214 + 261 + 35 + 29	3,539
2.	58,614 + 21,314 + 35,409 + 20,001	135,349
3.	36 + 42 + 31 + 71 + 92 + 61 + 51	384
4.	6,721 + 321 + 4,123 + 29	11,294

In 5–8, add without using the calculator. Check, using the calculator.

5.
```
6,123
   26
   87
   36
-----
```

6.
```
6,192
8,173
4,073
  790
-----
```

7.
```
6,190
2,003
  179
  684
   21
   37
-----
```

8.
```
86,945
79,945
 1,042
   301
------
```

In 9–12, break up each number as shown in the sample solution. Add, using the calculator. The sum should be the same as the number you started with.

SAMPLE SOLUTION

247 = 2 hundreds, 4 tens, 7 ones

200 + 40 + 7 = 247

9. 95 = 9 tens, 5 ones

+ =

10. 5,283 = 5 thousands, 2 hundreds, 8 tens, 3 ones

+ + + =

11. 8,514 = 8 thousands, 5 hundreds, 1 tens, 4 ones

+ + + =

12. 27,635 = 2 ten-thousands, 7 thousands, 6 hundreds, 3 tens, 5 ones

+ + + + =

In 13–14, break up the numbers as shown in the sample solution. Using the calculator, check that the column sums add up to the total sum.

SAMPLE SOLUTION

$$\begin{array}{rcrcr} 38 & = & 30 & + & 8 \\ +25 & = & 20 & + & 5 \\ \hline 63 & = & 50 & + & 13 \end{array}$$

Since 38 + 25 is equal to 63, and 50 + 13 is also equal to 63, the work is correct.

13.
45 = ____ + ____
+37 = ____ + ____
____ = ____ + ____

14.
253 = ____ + ____ + ____
+414 = ____ + ____ + ____
____ = ____ + ____ + ____

In 15–17, use the calculator to find the sum. Then turn the calculator upside down to read a word answer that fits the clue.

	Clue	*Numerical Answer*	*Word Answer*
15. 1,772 + 1,273	Worn on the foot	____	____
16. 289 + 49	Buzzes about	____	____
17. 2,963 + 543 + 1	Fail to win	____	____

18. Perform the calculations without a calculator, and complete the cross-number puzzle. If any of your ACROSS and DOWN entries do not fit together, use the calculator to find errors.

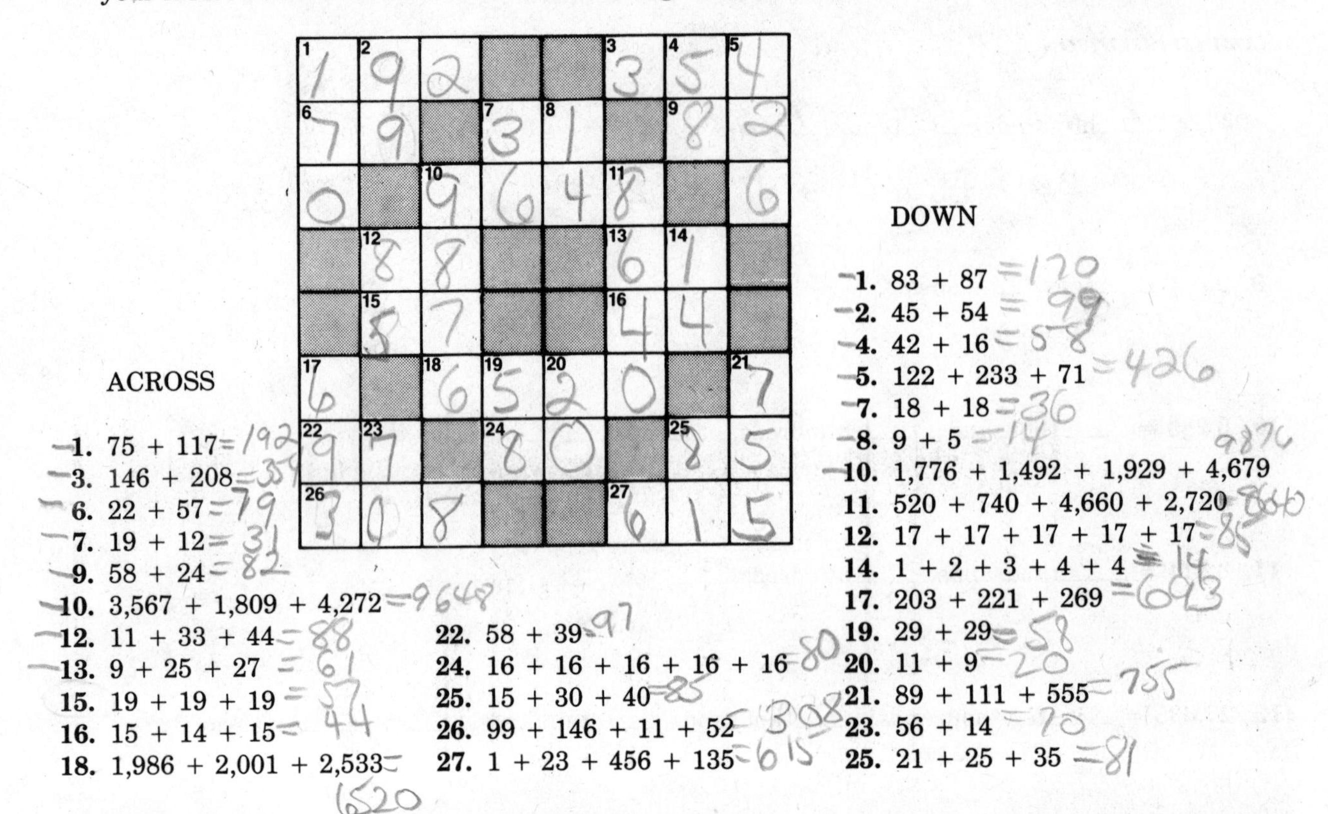

ACROSS

1. 75 + 117
3. 146 + 208
6. 22 + 57
7. 19 + 12
9. 58 + 24
10. 3,567 + 1,809 + 4,272
12. 11 + 33 + 44
13. 9 + 25 + 27
15. 19 + 19 + 19
16. 15 + 14 + 15
18. 1,986 + 2,001 + 2,533
22. 58 + 39
24. 16 + 16 + 16 + 16 + 16
25. 15 + 30 + 40
26. 99 + 146 + 11 + 52
27. 1 + 23 + 456 + 135

DOWN

1. 83 + 87
2. 45 + 54
4. 42 + 16
5. 122 + 233 + 71
7. 18 + 18
8. 9 + 5
10. 1,776 + 1,492 + 1,929 + 4,679
11. 520 + 740 + 4,660 + 2,720
12. 17 + 17 + 17 + 17 + 17
14. 1 + 2 + 3 + 4 + 4
17. 203 + 221 + 269
19. 29 + 29
20. 11 + 9
21. 89 + 111 + 555
23. 56 + 14
25. 21 + 25 + 35

In 19–20, fill in the *magic squares*. In a magic square, every row, every column, and each diagonal must add up to the same sum. (*Hint:* In exercise 19, every sum must be the same as 40 + 28 + 16. Aim for that sum in every other direction, working each time with a row, column, or diagonal that has only one remaining blank.) Use a calculator to check.

19.

40	12	
	28	
24		16

20.

54	12		
	39		30
33	27	24	42
18			

In 21–28, find the missing numbers. Use a calculator to check your guesses.

21.
```
  3□
+ □8
  79
```

22.
```
  □,62□
+ 5,□□7
  7,968
```

23.
```
  15□
  3□1
+ □32
  888
```

24.
```
  1,□23
  2,1□1
+ □,52□
  5,678
```

25.
$$\begin{array}{r} 3\square 6 \\ +\square 4\square \\ \hline 1{,}080 \end{array}$$

26.
$$\begin{array}{r} 4{,}3\square 7 \\ +1{,}\square 0\square \\ \hline 6{,}161 \end{array}$$

27.
$$\begin{array}{r} 45 \\ 8\square \\ +\square 9 \\ \hline 161 \end{array}$$

28.
$$\begin{array}{r} 27\square \\ 5\square 6 \\ +\square 45 \\ \hline 1{,}539 \end{array}$$

Review of Part II (Units 3–6)

In 1–10, add and check.

1.
$$\begin{array}{r} 8{,}368 \\ 9{,}654 \\ \hline \end{array}$$

2.
$$\begin{array}{r} 4{,}763 \\ 2{,}572 \\ \hline \end{array}$$

3.
$$\begin{array}{r} 7{,}248 \\ 5{,}734 \\ \hline \end{array}$$

4.
$$\begin{array}{r} 3{,}873 \\ 9{,}638 \\ \hline \end{array}$$

5.
$$\begin{array}{r} 6{,}973 \\ 4{,}865 \\ \hline \end{array}$$

6.
$$\begin{array}{r} 638 \\ 521 \\ 368 \\ 865 \\ \hline \end{array}$$

7.
$$\begin{array}{r} 374 \\ 463 \\ 985 \\ 742 \\ \hline \end{array}$$

8.
$$\begin{array}{r} 863 \\ 535 \\ 415 \\ 768 \\ \hline \end{array}$$

9.
$$\begin{array}{r} 528 \\ 862 \\ 978 \\ 484 \\ \hline \end{array}$$

10.
$$\begin{array}{r} 338 \\ 569 \\ 342 \\ 894 \\ \hline \end{array}$$

In 11–16, add by arranging the numbers in the descending order of their values.

11. 95 + 1,230 + 5 + 12,562 + 678

12. 28 + 8 + 235,325 + 615 + 18,690

13. 5,385 + 7 + 895 + 35,615 + 38

14. 735 + 215,563 + 25 + 65,426 + 7,325

15. $156 + $2 + $1,500 + $37 + $99

16. $15 + $72 + $162 + $223 + $98

In 17–20, add horizontally and vertically.

17.
$$\begin{array}{rcrcrcl} 324 & + & 563 & + & 235 & = & \\ 473 & + & 248 & + & 563 & = & \\ 254 & + & 537 & + & 364 & = & \\ 637 & + & 432 & + & 764 & = & \underline{\quad\quad} \\ \hline & + & & + & & = & \end{array}$$

18.
$$\begin{array}{rcrcrcl} 243 & + & 416 & + & 425 & = & \\ 354 & + & 625 & + & 867 & = & \\ 678 & + & 345 & + & 568 & = & \\ 278 & + & 465 & + & 753 & = & \underline{\quad\quad} \\ \hline & + & & + & & = & \end{array}$$

19.
$$\begin{array}{rcrcrcl} 563 & + & 428 & + & 754 & = & \\ 425 & + & 634 & + & 325 & = & \\ 653 & + & 719 & + & 466 & = & \\ 327 & + & 532 & + & 845 & = & \underline{\quad\quad} \\ \hline & + & & + & & = & \end{array}$$

20.
$$\begin{array}{rcrcrcl} 235 & + & 464 & + & 532 & = & \\ 528 & + & 427 & + & 852 & = & \\ 735 & + & 563 & + & 895 & = & \\ 215 & + & 472 & + & 368 & = & \underline{\quad\quad} \\ \hline & + & & + & & = & \end{array}$$

In 21–22, use a calculator.

21. Complete the addition table by adding the number at the left of the row to the number at the top of the column. A sample answer is shown.

+	9	55	127
14			
33			160
285			

22. To break the code, add the numbers and write each letter in the space that corresponds to its sum. As a sample, the letter N is written over the number 25.

8 + 17 = N
3 + 89 = L
4 + 36 = O
2 + 59 = A
5 + 63 = R
7 + 46 = H
6 + 98 = S
9 + 23 = E

25 + 380 = Y
42 + 923 = L
67 + 256 = A
38 + 642 = U
19 + 522 = S
88 + 372 = A
55 + 927 = O
71 + 477 = R

234 + 5,678 = C
107 + 2,523 = A
721 + 8,305 = O
563 + 4,872 = C
682 + 5,671 = D
827 + 9,012 = L
331 + 2,897 = T
548 + 1,942 = T

___	___	___	___	___	___	___	___	___	___	___
5,912	460	965	5,435	680	9,839	61	3,228	40	548	541

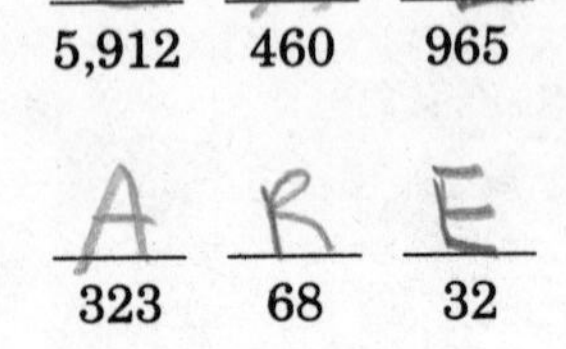

___	___	___	___	___	N	___	___	___	___	___	___	___
323	68	32	53	2,630	25	6,353	405	2,490	982	9,026	92	104

PART III. Subtracting Whole Numbers

UNIT 7. Subtraction of Numbers

Words to Know

When the value of a smaller number is "taken away" from the value of a larger number, the process is called **subtraction.**

The number that results from subtraction is called the **difference.**

The larger number is called the **minuend.**

The smaller number, the number that is subtracted, is called the **subtrahend.**

Subtraction is the opposite of addition. In subtraction, you take away a smaller number (the **subtrahend**) from a larger one (the **minuend**). Your answer is called the **difference.** The difference is *always smaller* than the larger number of the problem. (Only two numbers are involved in a subtraction problem.) Subtraction is indicated by writing the *minus* symbol "−" in front of the number to be subtracted. To indicate that 16 is to be subtracted from 48, write 48 − 16 or

$$\begin{array}{r} 48 \\ -16 \\ \hline \end{array}$$

Setting up a subtraction problem is similar to setting up an addition problem. As in addition, be sure to line up the digits correctly: ones under ones, tens under tens, and so on. Since, in subtraction, you must take the smaller number from the larger number, *always place the larger number above the smaller one.*

Example 1. Subtract: 47 − 23

Solution: Set up the problem with the larger number on top and the digits correctly aligned. Take away 3 from 7, getting 4; write the 4 under the ones column. Take away 2 from 4, getting 2; write the 2 under the tens column.

$$\begin{array}{r} 47 \\ -23 \\ \hline 24 \end{array}$$

Answer: 47 − 23 = 24

Example 2. Subtract 223 from 547.

Solution: Write the minuend on top and correctly line up the subtrahend under it. Subtract: 7 − 3 = 4. Write the 4 under the ones column. Subtract: 4 − 2 = 2. Write the 2 under the tens column. Subtract: 5 − 2 = 3. Write the 3 under the hundreds column.

$$\begin{array}{r} 547 \\ -223 \\ \hline 324 \end{array}$$

Answer: 547 − 223 = 324

You may check your answer in any subtraction problem by *adding* your answer to the number above it. This sum should equal the top number of the problem.

Let us check the preceding example:

$$\begin{array}{rl} 547 & \leftarrow \text{minuend} \\ -223 & \leftarrow \text{subtrahend} \\ \hline 324 & \leftarrow \text{difference} \end{array}$$

Add the answer (the difference) to the number above it (the subtrahend): In the ones column, 4 + 3 = 7. In the tens column, 2 + 2 = 4. In the hundreds column, 3 + 2 = 5. Thus, 324 + 223 = 547. Since the difference plus the subtrahend equals the original minuend of the problem, you know that the arithmetic is correct.

Remember

Difference + Subtrahend = Minuend

EXERCISES

In 1–20, subtract and check.

1. $\begin{array}{r}39\\7\\\hline\end{array}$ **2.** $\begin{array}{r}28\\5\\\hline\end{array}$ **3.** $\begin{array}{r}26\\4\\\hline\end{array}$ **4.** $\begin{array}{r}57\\4\\\hline\end{array}$ **5.** $\begin{array}{r}48\\32\\\hline\end{array}$ **6.** $\begin{array}{r}67\\15\\\hline\end{array}$

7. $\begin{array}{r}49\\26\\\hline\end{array}$ **8.** $\begin{array}{r}27\\13\\\hline\end{array}$ **9.** $\begin{array}{r}576\\32\\\hline\end{array}$ **10.** $\begin{array}{r}769\\47\\\hline\end{array}$ **11.** $\begin{array}{r}687\\327\\\hline\end{array}$ **12.** $\begin{array}{r}6,574\\43\\\hline\end{array}$

13. $\begin{array}{r}7,645\\5,224\\\hline\end{array}$ **14.** $\begin{array}{r}4,857\\1,204\\\hline\end{array}$ **15.** $\begin{array}{r}8,793\\3,670\\\hline\end{array}$ **16.** $\begin{array}{r}15,627\\2,403\\\hline\end{array}$

17. $\begin{array}{r}47,874\\25,521\\\hline\end{array}$ **18.** $\begin{array}{r}84,691\\23,270\\\hline\end{array}$ **19.** $\begin{array}{r}60,507\\20,105\\\hline\end{array}$ **20.** $\begin{array}{r}65,000\\42,000\\\hline\end{array}$

In 21–36, subtract as indicated.

21. 47 − 32 **22.** 85 − 34 **23.** 68 − 26 **24.** 98 − 43

25. 258 − 27 **26.** 476 − 52 **27.** 398 − 23 **28.** 479 − 53

29. 5,684 − 341 **30.** 9,879 − 637

31. 7,687 − 235 **32.** 47,683 − 5,151

33. 85,298 − 32,073 **34.** 658,976 − 235,432

35. 798,000 − 522,000 **36.** 869,621 − 327,200

In 37–42, subtract:

37. 23 from 47 =24 **38.** 52 from 95=43 **39.** 24 from 87=63

40. 47 from 279 =232 **41.** 262 from 797=535 **42.** 364 from 2,987= 2623

In 43–48, find the difference between:

43. 47 and 89=42 **44.** 23 and 75=52 **45.** 42 and 98=56

46. 287 and 232 =55 **47.** 768 and 523=245 **48.** 896,785 and 71,352 825433

APPLICATION PROBLEMS

SAMPLE SOLUTION

In a presidential election, Kansas voters gave Ronald Reagan 649,423 votes, while they gave the losing candidate 321,010 votes. By how many votes did Reagan win in Kansas?

649,423 votes Reagan
− 321,010 votes loser
328,413 winning votes

1. Last year, Mr. Gonzales earned $18,324 and this year he will earn $19,765. How much more will he earn this year than last year? 1441

2. The DeLucca family bought a house for $46,253. Five years later they sold it for $68,576. How much profit did they make on the sale of the house? ________

3. Joseph bought a car for $8,875. The price of the car was $552 more this year than it was last year. What was the price of the car last year? ________

4. Angela rented a car that had an odometer (mileage indicator) reading of 42,357. When she returned the car, the reading was 77,898. What was the mileage added by Angela? ________

5. Last year's school enrollment was 2,325. This year, the enrollment is 2,879. How many more students are enrolled this year? ________

6. At the beginning of the school year, there were 5,378 textbooks stored in the Social Studies bookroom. After books were issued to students, 1,357 books remained in the bookroom. How many books were issued to students? ________

7. The average speed of the Silver Streak express train is 168 kilometers per hour, while the average speed of the Central Commuter local is 56 kilometers per hour. How much faster does the express go than the local? __________

8. At an amateur rodeo held one weekend in Boulder County, there were 2,526 paid admissions on Saturday and 3,798 on Sunday. How many more people bought tickets on Sunday than on Saturday? __________

9. The Star City airport has a runway 3,565 meters long for large jets, and a runway 2,322 meters long for smaller planes. Find the difference in the lengths of the runways. __________

10. One Tuesday, Dan's Diner had 286 orders for hamburgers and 121 orders for fish cakes. How many more people ordered hamburgers? __________

UNIT 8. Subtraction Problems With Borrowing

WORDS TO KNOW

In subtraction, you *must* subtract a smaller number from a larger number. However, the *digits* in the smaller number may be greater than those in the larger number. For example, let us try to subtract 99 from 321. You know that 99 is smaller than 321, but how do you subtract 9 from 1 in the ones column?

$$\begin{array}{r} 321 \\ -99 \\ \hline ? \end{array}$$

In this unit, you will learn the technique known as **borrowing** that enables you to perform such subtractions.

In doing subtraction problems, you may find that you have a larger digit to take from a smaller digit. In such a case, you will have to *borrow* from the digit to the left of the one you are working with.

EXAMPLE 1. Subtract 79 from 536.

Solution: Line up the digits correctly. Then, as with all adding and subtracting, begin with the ones column.

You cannot subtract 9 from 6, so you borrow from the 3 in the tens column. Now, instead of 3 tens, you have 2 tens. Cross out the 3 and write 2 above it. You have moved 1 ten, or 10 ones, to the ones column, making 16 ones. Write 1 in front of the 6.

$$\begin{array}{r} 5\overset{2}{\not{3}}{}^{1}6 \\ -79 \\ \hline \end{array}$$

Look at the tens column. You cannot subtract 7 from 2, so you borrow from the 5 in the hundreds column. Now, instead of 5 hundreds, you have 4 hundreds. Cross out the 5 and write 4 above it. The extra hundred, or 10 tens, is moved to the tens column, making 12 tens. Write 1 in front of the 2.

$$\begin{array}{r} \overset{4}{\not{5}}\overset{{}^{1}2}{\not{3}}{}^{1}6 \\ -79 \\ \hline \end{array}$$

Now it is possible to subtract 9 ones from 16 ones, 7 tens from 12 tens, and 0 hundreds from 4 hundreds.

$$\begin{array}{r} \overset{4}{\not{5}}\overset{{}^{1}2}{\not{3}}{}^{1}6 \\ -79 \\ \hline 457 \end{array}$$

Check: Does the difference plus the subtrahend equal the minuend?

457 + 79 = 536 ✓

Answer: 536 − 79 = 457

EXAMPLE 2. Subtract: 321 − 99

Solution:

In the ones column: 11 − 9 = 2
In the tens column: 11 − 9 = 2
In the hundreds column: 2 − 0 = 2

$$\begin{array}{r} \overset{2\,11}{\cancel{3}\cancel{2}}1 \\ -99 \\ \hline 222 \end{array}$$

Answer: 321 − 99 = 222

Borrow only when you need to. In the next problem, you don't need any more ones, but you will have to borrow to get more tens.

EXAMPLE 3. Subtract: 425 − 81

Solution:

$$\begin{array}{r} \overset{3\,1}{\cancel{4}}25 \\ -81 \\ \hline 344 \end{array}$$

Answer: 425 − 81 = 344

EXERCISES

In 1–40, subtract.

1. 73 54	**2.** 85 38	**3.** 93 65	**4.** 72 67	**5.** 47 29
6. 56 37	**7.** 254 38	**8.** 457 63	**9.** 539 54	**10.** 626 72
11. 138 63	**12.** 247 82	**13.** 352 47	**14.** 139 57	**15.** 265 28
16. 432 265	**17.** 621 284	**18.** 426 189	**19.** 754 376	**20.** 833 356
21. 3,232 758	**22.** 1,257 389	**23.** 1,452 785	**24.** 3,625 746	**25.** 4,351 983

26. 5,354 2,698	**27.** 4,236 3,548	**28.** 6,426 5,659	**29.** 8,435 3,878
30. 7,263 6,796	**31.** 25,431 8,273	**32.** 34,522 6,784	**33.** 27,534 6,847
34. 17,235 6,568	**35.** 23,425 13,837	**36.** 45,357 38,748	**37.** 37,432 28,855

38. 86,436 76,859	**39.** 832,738 625,465	**40.** 758,374 373,548

In 41–49, subtract and check.

41. 472,648 337,582 135066	**42.** 372,634 245,378 127256	**43.** 342,637 153,849 188 788
44. 453,455 374,457 78998	**45.** 23,453,235 18,765,859 4687376	**46.** 17,346,324 16,589,536 756788
47. 27,453,634 18,674,857 8 778 777	**48.** $24,335,463 5,747,584 18 587 879	**49.** $37,556,321 28,779,543 8 776 778

APPLICATION PROBLEMS

SAMPLE SOLUTION

In one year, the average net yearly income of farms in Florida was $31,954. In the following year, the net income was $36,322. What was the amount of the increase?

$ 36,322 *high income*
−31,954 *low income*
$ 4,368 *increase*

1. James invested $1,867 in stocks. Three years later he sold the stocks for $2,846. How much profit did he make? 979
2. A retail store sold $24,479 worth of merchandise last month. This month the sales were $25,238. How much more merchandise did the store sell this month than last month? 759
3. In the last local election, 75,678 voters went to the polls out of 123,452 registered voters. How many registered voters did not vote? 47774
4. In the last 5 years, the population of a city increased from 876,589 to 1,342,467. Find the increase in population. 465878
5. The Greens bought a boat for $24,325. If they made a down payment of $4,638, how much more remains to be paid? 19687
6. Bowling champion Betty Morris set a record for earnings in one year, when she won $35,375 for the year. This broke her own previous record of winning $26,547. What was the amount of increase in her winnings? 8828
7. A touring rock group gave a concert in San Francisco one night, and a concert in Seattle the next night. For the San Francisco concert, they sold 8,415 tickets, and for the Seattle concert, they sold 6,962 tickets. How many more tickets did they sell in San Francisco? 1453
8. Lake Winnipeg covers 9,417 square miles, and Lake Ontario covers 7,550 square miles. How much larger is Lake Winnipeg? 1867

9. One year, the Philadelphia mint coined a quantity of half-dollars worth $17,069,500, and the Denver mint coined half-dollars totaling $16,236,122 in value. How much greater was the value of the Philadelphia coinage?

833378

10. According to a population census, the population of Austin, Texas, was 345,496, and the population of Fort Worth, Texas, was 385,164. How many more people were living in Fort Worth?

39668

UNIT 9. Borrowing From Zero

Sometimes the digit you want to borrow from will be a 0. In such a case, you must move to the left until you reach a digit that is not 0.

EXAMPLE 1. Subtract 258 from 705.

Solution: Set up the problem correctly. You want to borrow a 1 for the 5 so that you can change the 5 to 15. Since you cannot borrow from the 0, the digit to the left of the 5, you must move over one more digit to the left. You then borrow 1 from the 70. Draw a box around the 70. Since 70 − 1 = 69, change the 70 to 69. Place the 1 you borrowed in front of the 5. Now subtract:

```
  705
 -258
```

```
  69,
 [70]5
 -258
  447
```

Ones column: 15 − 8 = 7
Tens column: 9 − 5 = 4
Hundreds column: 6 − 2 = 4

Check: Difference + Subtrahend = Minuend
447 + 258 = 705 ✓

Answer: 705 − 258 = 447

When there are two or more zeros to be borrowed from, follow the same procedure.

EXAMPLE 2. Subtract: 6,005 − 258

Solution: Set up the problem correctly. As before, you must borrow a 1 for the 5. But this time you must borrow from the 600. Since 600 − 1 = 599, change the 600 to 599 and place the 1 in front of the 5, changing it to 15. Subtract as before.

```
  5 99,
 [6,00]5
   -258
```

Ones column: 15 − 8 = 7
Tens column: 9 − 5 = 4
Hundreds column: 9 − 2 = 7
Thousands column: 5 − 0 = 5

```
  599,
 [6,00]5
   -258
  5,747
```

Answer: 6,005 − 258 = 5,747

EXAMPLE 3. Subtract: 80,000 − 9,873

Solution:

```
   79 99,
 [80,00]0
  -9,873
  70,127
```

Answer: 80,000 − 9,873 = 70,127

EXERCISES

In 1–38, subtract. When you must borrow from zero, draw a box as shown in the sample solutions.

SAMPLE SOLUTIONS

a.
```
  29,
 [30]7
 -169
  138
```

b.
```
  5 99,
 [6,00]3
 -3,517
  2,486
```

No.	Number	Subtract	Handwritten answer
1.	50	37	13
2.	60	43	17
3.	460	26	434
4.	703	56	647
5.	800	534	266
6.	604	278	326
7.	903	565	338
8.	600	347	253
9.	803	638	165
10.	605	329	276
11.	5,037	743	4294
12.	7,048	963	6085
13.	3,056	362	2694
14.	5,038	2,983	2055
15.	3,004	738	2266
16.	2,006	1,737	269
17.	5,000	2,267	2733
18.	1,007	479	528
19.	3,005	1,837	1168
20.	8,003	7,565	438
21.	10,005	7,328	2677
22.	20,000	14,273	5727
23.	60,005	59,376	629
24.	80,006	8,628	71 378
25.	40,001	16,383	23618
26.	50,000	9,378	40622
27.	10,004	9,248	756
28.	20,003	19,256	747
29.	30,000	14,735	15265
30.	50,503	8,736	41767
31.	80,305	12,658	67647
32.	10,500	9,753	747
33.	30,206	7,479	22727
34.	204,043	75,652	128391
35.	410,400	83,764	326636
36.	806,004	528,737	277267
37.	800,306	245,729	554577
38.	704,000	376,248	327752

APPLICATION PROBLEMS

Sample Solution

In Hawaii, there were 18,807 births in one year and 19,005 births the following year. How many more babies were born the second year?

```
  8 99
 1[9,00]5  high
-1 8,807   low
    198    difference
```

1. A real estate salesperson sold a house for $90,500, which included a commission of $6,335. What was the cost of the house without the commission? 84165
2. Elizabeth won $5,000 in a state lottery and paid $1,250 in taxes. How much money did she win, after taxes? 3750
3. Peter earns $14,000 a year and pays a total of $3,560 in taxes. How much does he earn in a year, after taxes? 10440
4. Wanda bought a color television set for $500. If she made a down payment of $125, how much does she still owe the store? 375
5. A stadium seats 40,500 people. If the attendance at a concert is 33,763, how many seats are unoccupied? 6737
6. Maria bought a radio for $42. If she paid the clerk with a hundred-dollar bill, how much change did she get? 958
7. The Atkins bought a house for $60,500. They made a down payment of $11,575, and took a mortgage for the balance. What was the amount of the mortgage? 48925
8. To pay for college tuition, Janice borrowed $10,500 from a bank. The amount of interest was $1,522. If the bank deducts the interest in advance from the amount of the loan, what was the net amount of the loan? 8978
9. Pedro bought a car for $8,508 and was given a trade-in value of $1,479 on his old car. How much does he still owe the dealer for the car? 7029
10. A color TV regularly selling for $505 was reduced by $168. What was the sale price of the TV? 337

Review of Part III (Units 7–9)

In 1–12, subtract.

1. 8,645 6,531 2114	**2.** 6,594 3,261 3333	**3.** 7,879 5,245 2634	**4.** 6,968 5,345 1623
5. 5,769 5,237 532	**6.** 35,874 13,543 22331	**7.** 87,968 25,745 62223	**8.** 69,786 35,332 34454

9. 98,679
53,255

10. 69,587
34,243

11. $15,658
14,655

12. $63,576
42,304

In 13–18, subtract.

13. 375 − 253

14. 286 − 52

15. 5,658 − 435

16. 15,875 − 4,522

17. 28,698 − 3,265

18. 19,869 − 7,425

In 19–24, find the difference between:

19. 8,658 and 3,435

20. 2,435 and 9,687

21. 3,234 and 7,867

22. 86,798 and 5,263

23. 65,659 and 23,537

24. 34,243 and 79,878

In 25–44, subtract.

25. 542
364

26. 735
387

27. 341
263

28. 457
369

29. 635
468

30. 8,345
4,268

31. 7,264
3,457

32. 5,465
3,367

33. 9,535
6,828

34. 7,524
6,846

35. 704
426

36. 800
624

37. 503
437

38. 5,032
2,745

39. 6,003
4,768

40. 25,600
12,365

41. 30,403
17,578

42. 60,000
27,235

43. 20,500
8,743

44. 30,050
23,763

For 45–55, use a calculator to check your guesses.

In 45–48, fill in the missing digits.

45. 5 8 7
− 2 □ □
□ 2 1

46. 2, 5 6 8
− 3 □ □
□, □ 4 2

47. 5 2 3
− □ □ □
2 4 8

48. □ □ □
− 3 2 1
1 4 5

In 49–51, use only the numbers 9, 28, and 73 to fill in the blanks.

49. □ − □ = 45

50. □ − □ = 64

51. □ − □ = 19

In 52–55, one digit has been replaced by zero. Find what number has been subtracted.

52. 2,345 − □ = 2,045

53. 56,789 − □ = 50,789

54. 98,765 − □ = 8,765

55. 123,456 − □ = 103,456

PART IV. Multiplying Whole Numbers

UNIT 10. Multiplication of Numbers

WORDS TO KNOW

When the number 6 is repeated five times, the five 6's add up to 30. We can also say that 6 is **multiplied** by 5, or $6 \times 5 = 30$. The symbol $\times$ indicates **multiplication.**

In multiplication, the number being multiplied is called the **multiplicand.** The number doing the multiplying is called the **multiplier.**

The answer to a multiplication problem is called the **product.**

Multiplication is a shortcut method of doing addition. When you multiply numbers, you really add groups of numbers. Multiplying 5×5, which is read "5 times 5," is the same as adding five 5's:

$$5 \times 5 = 25 \qquad 5 + 5 + 5 + 5 + 5 = 25$$

Multiplication problems are set up like this:

$$\begin{array}{rl} 534 & \leftarrow \text{multiplicand} \\ \underline{\times 3} & \leftarrow \text{multiplier} \end{array}$$

Notice that the ones digits are aligned, as in addition and subtraction. Also, the bottom number is the **multiplier** and the top number is the **multiplicand.** In solving a problem in multiplication, always multiply from right to left, starting with the ones digit in the multiplicand. The answer to a multiplication problem is called the **product.**

For speed and accuracy in multiplying, you should know the number facts for multiplying whole numbers up to 12×12. You can check yourself with a calculator.

EXAMPLE. Multiply: 534×3

Solution: Set up the problem correctly; then multiply from right to left.

Step 1: 3 times 4 equals 12. Write the 2 under the ones column and carry the 1 to the next digit, the 3.

$$\begin{array}{r} 1 \\ 534 \\ \underline{\times 3} \\ 2 \end{array}$$

Step 2: 3 times 3 equals 9, plus the 1 you carried, equals 10. Write the 0 under the tens column and carry the 1 to the 5, the next digit.

$$\begin{array}{r} 11 \\ 534 \\ \underline{\times 3} \\ 02 \end{array}$$

Step 3: 3 times 5 equals 15, plus the 1 you carried, equals 16. Write the 6 under the hundreds column and write the 1 to the left.

$$\begin{array}{r} 11 \\ 534 \\ \underline{\times 3} \\ 1,602 \end{array}$$

Answer: $534 \times 3 = 1,602$

EXERCISES

In 1–54, multiply.

1. $\begin{array}{r} 43 \\ \underline{2} \end{array}$	**2.** $\begin{array}{r} 24 \\ \underline{3} \end{array}$	**3.** $\begin{array}{r} 31 \\ \underline{4} \end{array}$	**4.** $\begin{array}{r} 53 \\ \underline{5} \end{array}$	**5.** $\begin{array}{r} 34 \\ \underline{4} \end{array}$	**6.** $\begin{array}{r} 25 \\ \underline{3} \end{array}$
7. $\begin{array}{r} 42 \\ \underline{2} \end{array}$	**8.** $\begin{array}{r} 24 \\ \underline{5} \end{array}$	**9.** $\begin{array}{r} 33 \\ \underline{4} \end{array}$	**10.** $\begin{array}{r} 26 \\ \underline{3} \end{array}$	**11.** $\begin{array}{r} 35 \\ \underline{6} \end{array}$	**12.** $\begin{array}{r} 46 \\ \underline{5} \end{array}$

13.	14.	15.	16.	17.	18.
27	38	46	52	63	84
4	3	7	8	9	6
108	114	322	416	567	504

19.	20.	21.	22.	23.	24.
95	76	46	27	46	78
5	7	8	9	6	8
475	532	368	243	276	624

25.	26.	27.	28.	29.	30.
97	69	56	48	57	79
5	7	8	9	6	9
485	483	448	432	342	711

31.	32.	33.	34.	35.	36.
69	59	367	496	786	988
7	9	5	4	3	2
483	531	1835	1984	2358	1976

37.	38.	39.	40.	41.	42.
568	784	987	756	875	987
8	6	9	7	6	8
4544	4704	8883	5292	5250	7896

43.	44.	45.	46.
3,428	2,768	5,697	6,798
5	4	3	2
17140	11072	17091	13596

47.	48.	49.	50.
9,874	5,987	23,542	35,645
9	7	6	7
88866	41909	141252	249515

51.	52.	53.	54.
65,678	76,568	87,986	67,879
8	9	7	9
525424	689112	615902	610911

APPLICATION PROBLEMS

SAMPLE SOLUTION

Henry flew from New York to Miami in exactly 3 hours. If his plane averaged 364 miles per hour, how many miles did the flight cover?

```
 11
364 miles per hour
 x 3 hours
1,092 miles
```

1. A farmer planted 37 rows of peach trees, with 8 trees in each row. How many trees did he plant? 296
2. A truck holds 678 crates of oranges. How many crates will 7 trucks hold? 4746
3. According to last year's records, a mail order house sends out an average of 968 packages each day. Calculate how many packages will be sent out during a typical 6-day week. (Since your answer will be approximate, round off your product to the nearest *ten.*) 5808
4. A retailer bought 586 blouses at $9 each. How much did she pay for all the blouses? 5274
5. A theater sold 759 tickets at $5 each. How much money was received from the sale of these tickets? 3735
6. If there are 8 frozen hamburgers in a package, how many are there in 72 packages? 576
7. Donna is earning $4 an hour. What does she earn in a 35-hour week? 140
8. A manufacturer uses 3 yards of fabric to make a set of auto seat covers. Find the number of yards of fabric that will be needed to fill an order for 235 sets. 705
9. The crates in a shipment weigh 6 pounds each. What is the total weight of a shipment of 288 crates? 1728
10. During the first 6 months of the year, a knitting shop sold an average of $21,426 worth of merchandise each month. What were the total sales for the 6-month period? Round off your product to the nearest *hundred.* 128556
129000

UNIT 11. Multiplication With Two or More Digits in the Multiplier

WORDS TO KNOW

When the multiplier has two or more digits, a problem in multiplication will have two or more **partial answers.** These partial answers must be added to obtain the product.

In a multiplication problem, the multiplier may have two or more digits. When this happens, you must multiply the multiplicand (the top number) by each digit of the multiplier separately. Each different multiplication gives a **partial answer.** If the multiplier has two digits, there will be *two* partial answers; if the multiplier has three digits, there will be *three* partial answers, and so on. When all the partial answers are added, the product is obtained.

Here are the steps for multiplying by a number that has two or more digits:

Step 1: Set up the problem correctly by lining up the ones digit of the multiplier with the ones digit of the multiplicand.

Step 2: Using the ones digit of the multiplier, multiply from right to left (exactly as you did in the preceding unit).

Step 3: Using the tens digit of the multiplier, multiply from right to left. Place the right-hand digit of this partial answer directly under the digit you are multiplying with. Repeat this step as often as necessary, depending on the number of digits in the multiplier.

Step 4: Add the partial answers to obtain the product.

EXAMPLE 1. Multiply: 235 × 23

Solution: Since the multiplier, 23, has two digits, multiply the top number two times: first by the ones digit, 3, then by the tens digit, 2.

Step 1: Set up the problem correctly.

```
 235
×23
```

Step 2: Using the digit 3 of the multiplier, multiply from right to left:

3 × 5 = 15. Write the 5 *directly under the 3,* the digit you are multiplying with. Carry the 1 to the next digit of the multiplicand, the 3.

```
  1 1
  235
 ×23
  705  ← partial answer
```

3 × 3 = 9, plus the 1 you carried, equals 10. Write the 0 and carry the 1 to the next digit, the 2.

3 × 2 = 6, plus the 1 you carried, equals 7. Write the 7. The number 705 is the first partial answer.

Step 3: Multiply with the next digit, the 2. Write the first digit of this partial answer *directly under the 2,* the digit you are multiplying with:

```
   1
  235
 ×23
  705  ← partial
 470   ← answers
```

2 × 5 = 10. Write the 0 directly under the 2 and carry the 1 to the next digit, the 3.

2 × 3 = 6, plus the 1 you carried, equals 7. Write the 7.

2 × 2 = 4. Write the 4. The number 470 is the second partial answer.

Step 4: Draw a line under the partial answers and add the two numbers to get the final answer, the product. Be sure to keep the partial answers correctly aligned.

```
   235
  ×23
   705  ← partial
  4 70  ← answers
 5,405  ← product
```

Answer: 235 × 23 = 5,405

In the following example, note how the above procedure is applied to a 3-digit multiplier.

EXAMPLE 2. Multiply: 576 × 498

Solution:

```
      32
      65  (crossed out)
      64  (crossed out)
      576
    ×498
    4 608
   51 84
  230 4
  286,848
```

Answer: 576 × 498 = 286,848

Remember

Always multiply with every digit in the multiplier. Place the first digit of each partial answer (the right-hand digit) directly under the digit you are multiplying with.

EXERCISES

In 1–34, multiply.

1. 43 × 24	2. 35 × 32	3. 63 × 53	4. 56 × 47	5. 85 × 68
6. 95 × 49	7. 87 × 68	8. 76 × 58	9. 75 × 64	10. 87 × 95
11. 59 × 34	12. 89 × 38	13. 84 × 25	14. 43 × 98	15. 81 × 94
16. 68 × 79	17. 423 × 86	18. 674 × 78	19. 937 × 46	20. 634 × 95

21. 847 × 69	22. 486 × 58	23. 2,468 × 325	24. 8,657 × 456
25. 7,564 × 675	26. 9,687 × 763	27. 7,478 × 397	28. 23,435 × 422

29. 34,675 × 568	30. 57,189 × 817	31. 86,497 × 589
32. 32,478 × 347	33. 47,684 × 568	34. 53,976 × 658

APPLICATION PROBLEMS

Sample Solution

If a year has 365 days and a day has 24 hours, how many hours are there in a year?

```
  365 days per year
x  24 hours per day
 1460
 730
8,760 hours per year
```

1. June can type an average of 49 words per minute. How many words can she type in 45 minutes? 245,196 = ______
2. If sound travels at a speed of 1,088 feet per second, how far will sound travel in 65 seconds? 5440,6528 = ______
3. If light travels at a speed of 186,282 miles per second, how far does light travel in 96 seconds? ______
4. How many envelopes are contained in 689 boxes if each box holds 144 envelopes? ______
5. A crate contains 58 apples. How many apples will there be in 267 crates? ______
6. Yolanda earns $1,568 per month. What is her yearly salary? ______
7. The members of a theater group are selling tickets for $14 each. If they fill their 235-seat theater, how much money will they take in? ______
8. A shoe store was billed for 68 pairs of running shoes, at a wholesale price of $18 a pair. What was the amount of the bill? ______
9. An auditorium has 26 rows of seats, with 42 seats in each row. What is the total number of seats? ______
10. A box factory uses 432 square inches of cardboard to manufacture its best-selling box. How many square inches of cardboard will be needed to fill an order for 20,736 boxes? ______

UNIT 12. Multiplication With 0's in the Multiplicand

Many people make mistakes with 0's when multiplying. To avoid mistakes, remember that *any number multiplied by 0 equals 0.*

If you earned no money on Monday (0 dollars), no money on Tuesday, and no money on Wednesday, how much money did you earn?

By addition: $0 + 0 + 0 = 0$.

By multiplication: $0 \times 3 = 0$.

Example 1. Multiply: 403×3

Solution: Set up the problem correctly and multiply from right to left.

$3 \times 3 = 9$. Write the 9.

$3 \times 0 = 0$. Write the 0.

$3 \times 4 = 12$. Write the 12.

```
  403
   ×3
1,209
```

Answer: $403 \times 3 = 1,209$

When the multiplier has two digits, follow the procedure in the preceding unit.

Example 2. Multiply: 403×24

Solution:

$4 \times 3 = 12$. Write the 2 directly under the 4 and carry the 1.

$4 \times 0 = 0$, plus the 1 you carried, equals 1. Write the 1.

$4 \times 4 = 16$. Write the 16.

$$\begin{array}{r} \scriptstyle 1 \\ 403 \\ \times 24 \\ \hline 1612 \end{array}$$

The first partial answer is 1,612.

$2 \times 3 = 6$. Write the 6 directly under the 2.

$2 \times 0 = 0$. Write the 0.

$2 \times 4 = 8$. Write the 8.

The second partial answer is 806. Add the two partial answers to get the product.

$$\begin{array}{r} 403 \\ \times 24 \\ \hline 1\ 612 \\ 8\ 06 \\ \hline 9{,}672 \end{array}$$

Answer: $403 \times 24 = 9{,}672$

Remember

Any number multiplied by zero equals zero.

EXERCISES

In 1–34, multiply.

1. 30 4	**2.** 50 7	**3.** 80 5	**4.** 60 8
5. 70 6	**6.** 40 9	**7.** 90 7	**8.** 80 8
9. 307 24	**10.** 509 36	**11.** 807 48	**12.** 906 75
13. 706 98	**14.** 609 76	**15.** 805 69	**16.** 5,040 48
17. 7,500 64	**18.** 8,004 98	**19.** 6,050 77	**20.** 8,008 65
21. 27,057 54	**22.** 30,205 68	**23.** 50,074 76	**24.** 76,005 48

25. 205,300 × 346

26. 460,206 × 478

27. 760,089 × 738

28. 803,702 × 915

29. $40,050 × 87

30. $83,007 × 64

31. $605,500 × 48

32. $760,020 × 76

33. $870,070 × 687

34. $600,850 × 579

APPLICATION PROBLEMS

SAMPLE SOLUTION

Dave's weekly take-home pay is $208. How much does he earn in a year? (Dave is paid for 52 weeks during the year.)

$208 per week
× 52 weeks per year
416
1040
$10,816 per year

1. If a car costs $9,080, how much will 15 cars cost? ________

2. A retailer bought 65 radios at $105 each. What is the total cost of the radios? ________

3. A construction firm is building 325 homes that will sell for $68,500 each. What will be the amount realized from the sale of the homes? ________

4. A department store bought 1,050 display cases at $87 each. Find the cost of the display cases. ________

5. A motel is installing 1,500 yards of carpeting at a cost of $8 a yard. Find the cost of the carpeting. ________

6. How much will 205 lamps cost at $27 each? ________

7. A car dealer ordered 18 cars at $6,500 each. What is the total cost of the cars? ________

8. Janice earns $1,700 each month. How much does she earn in a year? ________

9. For one performance, a rock band sold 20,050 tickets at an average price of $17. What was the total amount realized from the sale of the tickets? ________

10. The merchant fleet of a nation consists of 462 ships, carrying an average of 9,098 tons of cargo per ship. What is the carrying capacity of the whole fleet? ________

UNIT 13. Multiplication With 0's in the Multiplier

When there are 0's in the multiplier, you can simplify the multiplication by using this rule:

RULE

When you come to a 0 in the multiplier, *do not multiply by 0.* Instead, bring down the 0 *in a straight line* as a placeholder.

EXAMPLE 1. Multiply: 354 × 304

Solution: Set up the problem correctly and multiply with the right-hand digit, the 4.

```
  21
  354
 ×304
 1416
```

4 × 4 = 16. Write the 6 directly under the 4 and carry the 1.

4 × 5 = 20, plus the 1 you carried, equals 21. Write the 1 and carry the 2.

4 × 3 = 12, plus the 2 you carried, equals 14. Write the 14.

The first partial answer is 1,416.

Since the next digit is 0, *do not multiply with it.* Instead, bring down the 0 *in a straight line* as a placeholder.

```
  21
  354
 ×304
 1416
    0
```

Now, multiply with the next digit, the 3.

3 × 4 = 12. Write the 2 directly under the 3 and carry the 1.

3 × 5 = 15, plus the 1 you carried, equals 16. Write the 6 and carry the 1.

3 × 3 = 9, plus the 1 you carried, equals 10. Write the 10.

```
      11
      2̸1̸
     354
    ×304
   1 416
  106 20
 107,616
```

The second partial answer is 10,620.

Add the two partial answers to get the product.

Answer: 354 × 304 = 107,616

Here are other examples of how the rule is used when there are 0's in the multiplier:

```
    5                        361
   37         60           1,492
  ×80        ×70            ×700
2,960      4,200       1,044,400
```

EXERCISES

In 1–34, multiply.

1. 34 × 30 **2.** 47 × 50 **3.** 60 × 40 **4.** 76 × 80 **5.** 68 × 40 **6.** 90 × 90

7. 50 × 70 **8.** 49 × 80 **9.** 70 × 60 **10.** 234 × 307 **11.** 453 × 608 **12.** 765 × 460

13. 876 × 680 **14.** 605 × 306 **15.** 937 × 700 **16.** 8,075 × 350

17. 7,650
× 800

18. 4,075
× 708

19. 6,300
× 900

20. 5,000
× 407

21. 28,000
× 570

22. 46,364
× 806

23. 53,647
× 508

24. 74,030
× 450

25. 43,762
× 5,002

26. 76,628
× 6,070

27. 59,070
× 5,600

28. 78,365
× 7,006

29. $83,406 × 608

30. $42,478 × 540

31. $56,050 × 300

32. $37,000 × 507

33. $40,509 × 602

34. $70,008 × 270

APPLICATION PROBLEMS

Sample Solution

In one year, Mrs. Jacobs works 1,850 hours. How many hours will she work in 20 years?

work record
1,850 hr. per yr.
× 20 number of yr.
37,000 hr. in 20 yr.

1. A taxi company ordered 20 taxis at $8,090 each. What is the total cost of the taxis?

2. Mr. Bronowski has a *30-payment life* insurance policy. This means that he must pay yearly premiums for 30 years. If his premiums are $263 a year, how much will he pay altogether in premiums over 30 years?

3. The Wilsons own a home that has a 30-year mortgage. The yearly payments to the bank are $4,620. How much will they pay to the bank in the 30 years?

4. If a package contains 500 sheets of writing paper, how many sheets will 144 packages contain? ______

5. Envelopes are packed 300 to a box. How many envelopes are there in 175 boxes? ______

6. During a special sale, a shoe store sold 205 pairs of shoes at $20 each. What was the total amount of the sale? ______

7. An appliance store ordered 20 radios at $125 each. What was the total cost of the radios? ______

8. A bolt of velvet contains 130 yards. How many yards of velvet are there in 20 bolts? ______

9. A carton holds 200 boxes of carbon paper. If each box contains 30 sheets of carbon paper, how many sheets of carbon paper does the carton contain? ______

10. Napkins are packed 50 to a package. How many napkins are contained in 250 packages? ______

Review of Part IV (Units 10–13)

In 1–30, multiply.

1. 5,685 × 7
2. 6,765 × 8
3. 5,658 × 6
4. 8,735 × 5
5. 9,756 × 3
6. 5,798 × 9
7. 5,453 × 35
8. 7,465 × 67
9. 9,547 × 76
10. 7,658 × 565
11. 9,674 × 684
12. 4,367 × 586
13. 7,608 × 358
14. 8,500 × 564
15. 6,080 × 647
16. 23,500 × 549
17. 25,006 × 476
18. 40,500 × 645
19. 5,650 × 270
20. 7,057 × 500
21. 6,050 × 807
22. 4,657 × 570
23. 47,659 × 8,700
24. 53,968 × 8,006
25. 24,428 × 7,060
26. 82,063 × 5,207
27. 390,050 × 62,004
28. $40,050 × 310
29. $35,500 × 470
30. $70,670 × 600

For 31–43, use a calculator to check your guesses.

In 31–34, fill in the missing digits. (*Hint:* In exercise 31, by what can you multiply the 6 in the ones place of the multiplicand in order to get 8 in the ones place of the product?)

31. $46 \times \square = 138$

32. $125 \times \square = 500$

33. $132 \times \square\square = 3{,}564$

34. $1{,}230 \times \square\square = 30{,}750$

In 35–37, use only the numbers 12, 15, 18, and 25 to fill in the blanks.

35. $\square \times \square = 180$

36. $\square \times \square = 216$

37. $\square \times \square = 375$

In 38–39, find the sum in part *a.* In part *b,* fill in the missing multiplier, to get the same answer as in part *a.*

38. *a.* $43 + 43 + 43 = \square$

b. $\square \times 43 = \square$

39. *a.* $137 + 137 + 137 + 137 + 137 = \square$

b. $\square \times 137 = \square$

In 40–41, fill in the missing numbers.

40. $243 \times 7 = \square$; $200 \times 7 = \square$; $40 \times 7 = \square$; $3 \times 7 = \square$

$\square = \square + \square + \square$

41. $52 \times 36 = \square$; $2 \times 6 = \square$; $50 \times \square = \square$; $2 \times 30 = \square$; $\square\square \times 30 = \square$

$\square = \square + \square + \square + \square$

42. Complete the multiplication table. A sample answer is shown.

×	25	37	89
16			
42			
51		1,887	

43. Multiply to get two important dates in American history.

32,159 × 464

__________ ; __________

PART V. Dividing Whole Numbers

UNIT 14. Division of Numbers

WORDS TO KNOW

When a pack of 52 playing cards is dealt to 4 players, each player receives 13 cards. We say that the pack has been **divided** among the 4 players.

When any number is broken down into smaller groups of numbers, the process is called **division.**

In division, the number being divided is called the **dividend.** The number doing the dividing is called the **divisor.**

The answer to a division problem is called the **quotient.**

$$\begin{array}{r} 13 \leftarrow \text{quotient} \\ \text{divisor} \rightarrow 4\overline{)52} \leftarrow \text{dividend} \end{array}$$

The symbol $\overline{)\quad}$ is called the **division box.**

Division is the opposite of multiplication. When you perform division, you find out how many times a smaller number (the **divisor**) is contained in a larger number (the **dividend**). For example, "How many 5's are there in 30?" Recalling that $5 \times 6 = 30$ and that division is the opposite of multiplication, you answer, "There are six 5's in 30." In other words, "30 divided by 5 equals 6." The answer, 6, is the **quotient.**

In symbols, you can indicate division problems in several ways. Using the symbol ÷, which means "divided by," you can write the division problem like this:

$$30 \div 5$$

As you will learn (Unit 22), you can also write the division problem as a *fraction:*

$$\frac{30}{5}$$

However, the most common way of indicating a division problem is to use the **division box:**

$$5\overline{)30}$$

Let us solve a division problem by means of the division box.

EXAMPLE 1. Divide: $2\overline{)256}$

Solution: This problem asks, "How many 2's are there in 256?" A calculator can give you the answer in a fraction of a second, but the human mind cannot work so fast. In order to break down the dividend, 256, you must perform a series of *separate division problems.* To get started, it is useful to draw a vertical line according to the following rule:

RULE

In the dividend, moving from left to right, move as many digits as is necessary to make a *new dividend* that is as large as, or larger than, the divisor. Then draw a vertical line.

Step 1: In the example, draw a vertical line after the 2, the first digit in the dividend, because it is as large as the divisor. The division problem now looks like this:

$$2\overline{)2|56}$$

Step 2: Consider the digit to the left of the line to be a *new dividend.* Instead of asking, "How many 2's are there in 256?" you ask, "How many 2's are there in 2?" The answer is 1, so you place the 1 *directly above the digit you are using,* the 2.

$$\begin{array}{r} 1 \\ 2\overline{)2|56} \end{array}$$

new dividend → 2

Step 3: Multiply this answer, 1, by the divisor, 2. The product is 2, and you place it under the 2 of the dividend.

```
    1
 2)256
   2
   -
```

Step 4: Subtract the bottom number from the top number, which results in 0. (It is not necessary to write the 0.) The next step is to bring down the next digit of the dividend, the 5, in a straight line. We call the 5 a *new dividend,* because you will now divide the 2 into the 5.

```
    1
 2)256
   2↓
   -
    5 ←—— new dividend
```

Step 5: Now you have a new problem: "How many times does 2 go into 5?" The answer is 2. You place the 2 *directly above the 5 in the division box.*

```
    12
 2)256
   2
   -
    5
```

Step 6: Multiply this answer, 2, by the divisor, which is also 2. The product is 4, and you place it under the 5. Subtract the 4 from the 5, and the result is 1.

```
    12
 2)256
   2
   -
    5
    4
    -
    1
```

Step 7: Bring down the next digit, the 6, and place it next to the 1. This forms a *new dividend* of 16.

```
    12
 2)256
   2
   -
    5
    4
    -
    16 ←—— new dividend
```

Step 8: You again have a new problem: "How many times does 2 go into 16?" The answer is 8, and you place the 8 *directly above the 6 in the division box.*

```
    128
 2)256
   2
   -
    5
    4
    -
    16
    16
    --
```

Step 9: Multiply the 8 by the divisor, 2, giving you 16. Write the 16 under the new dividend, the 16, and subtract. The result is 0. Since there are no more digits to bring down, there is no new dividend.

Answer: 2)256 = 128

We say that, "There are 128 2's in 256." In other words, when 2 is divided into 256, the quotient is 128. According to the calculations, the number 256 contains 128 2's.

Since division is the opposite of multiplication, you can check a division problem by multiplying. Multiply the answer to the problem (the quotient) by the divisor. The product should equal the dividend.

```
  128
   ×2
  ---
  256 ✓
```

Remember

Quotient × Divisor = Dividend

When the divisor contains two or more digits, you solve division problems by following the same steps as when the divisor contains one digit:

EXAMPLE 2.

```
      16
 16)256
    16
    --
     96
     96
     --
```

EXAMPLE 3.

```
           21
 137)2,877
     2 74
     ----
       137
       137
       ---
```

When drawing the vertical line to form the first new dividend, move at least as many places in the dividend as there are digits in the divisor. In

Example 3, for instance, the divisor has three digits and the new dividend also has three digits. However, there are times when you will have to move one place more in the dividend in order to make the new dividend larger than the divisor.

Example 4. Divide: $17\overline{)1{,}496}$

Solution: Draw the vertical line after the *third* digit in the dividend in order to make the new dividend larger than the divisor.

$$17\overline{)1{,}49|6}$$

Now, complete the division as before.

$$\begin{array}{r} 88 \\ 17\overline{)1{,}49|6} \\ \underline{1\ 36} \\ 136 \\ \underline{136} \end{array}$$

Answer: $17\overline{)1{,}496} = 88$

EXERCISES

In 1–40, divide.

1. $2\overline{)682}$ 2. $3\overline{)693}$ 3. $7\overline{)504}$ 4. $4\overline{)1{,}284}$

5. $5\overline{)3{,}155}$ 6. $7\overline{)5{,}243}$ 7. $6\overline{)4{,}584}$ 8. $5\overline{)7{,}490}$

9. $6\overline{)84{,}726}$ 10. $8\overline{)60{,}968}$ 11. $8\overline{)72{,}976}$ 12. $7\overline{)96{,}544}$

13. $15\overline{)45}$ 14. $16\overline{)80}$ 15. $24\overline{)96}$ 16. $27\overline{)567}$

17. $26\overline{)8{,}164}$ 18. $37\overline{)4{,}551}$ 19. $48\overline{)74{,}976}$ 20. $213\overline{)639}$

21. $128\overline{)640}$

22. $640\overline{)4,480}$

23. $173\overline{)2,422}$

24. $321\overline{)7,383}$

25. $213\overline{)7,455}$

26. $478\overline{)21,988}$

27. $706\overline{)37,418}$

28. $2,324\overline{)27,888}$

29. $3,542\overline{)85,008}$

30. $4,306\overline{)150,710}$

31. $8,704 \div 34$

32. $16,758 \div 49$

33. $53,751 \div 123$

34. $\$83,660 \div 235$

35. $\$168,532 \div 463$

36. $\$161,142 \div 753$

37. $116,580 divided by 435

38. $244,224 divided by 636

39. $213,160 divided by 584

40. $159,766 divided by 629

APPLICATION PROBLEMS

SAMPLE SOLUTION

If a supermarket sells 7,524 eggs in a week, how many dozens of eggs are sold?

$$\begin{array}{r} 627 \\ 12\overline{)7{,}524} \\ \underline{72} \\ 32 \\ \underline{24} \\ 84 \\ \underline{84} \end{array}$$

627 dozen

1. Mrs. Smith bought a refrigerator for $636. She paid for it with 12 equal monthly installments. How much was each payment? ______
2. A sales representative traveled 525 miles, using 35 gallons of gasoline. How many miles did she travel on one gallon of gasoline? (Assume that each gallon gave the same mileage.) ______
3. A retail store bought 72 lamps for $1,296. What is the cost of each lamp? ______
4. How much is one yard of carpeting if 45 yards cost $405? ______
5. If a box holds 24 cans of juice, how many boxes are needed to pack 552 cans? ______
6. After making a down payment on a car, Terence owed a balance of $9,288. If the balance is paid off in 36 equal payments, how much will each payment be? ______
7. The yearly mortgage payment on a home is $5,436. How much are the monthly payments? ______
8. Rita earns $25,220 a year. What is her weekly salary? ______
9. A jet airliner traveled 11,250 miles. If it averaged 750 miles per hour, how many hours did it travel? ______
10. Paul and Liz bought a bedroom set at a sale price of $2,328. How much will each payment be if it is paid off in 24 equal payments? ______

UNIT 15. Division With a Remainder

WORDS TO KNOW

You know that the answer to a subtraction problem is called the **difference.** In the preceding unit, when you performed the final subtraction in each division problem, the difference was 0. When you perform division and the final difference is *not* 0, we call this **division with a remainder.**

When a division problem does not "come out even," we say the problem has a **remainder.** When you are asked, "What is half of 5?" you reply, "Half of 5 is $2\frac{1}{2}$." What you have done is to divide 2 into 5 and then express the remainder in a fraction.

```
    2
 2)5
   4
   1 ←—— remainder
```

To express the remainder in a fraction, draw a line under the remainder and write the divisor under the line:

```
    2
┌─2)5
│   4
│   1
└──→2
```

Then bring the fraction, $\frac{1}{2}$, up to the answer:

```
   2 1/2 ←┐
 2)5      │
   4      │
   1      │
   2 ─────┘
```

Thus, $5 \div 2 = 2\frac{1}{2}$.

Now let us do a more difficult division: 5) 1,183

Divide as you have learned to do:

```
      236
 5)1,183
   1 0
     18
     15
      33
      30
remainder ——→ 3
```

However, there is a remainder of 3.

To write the remainder in a fraction, draw a line under the remainder and write the divisor under the line:

```
      236
 5)1,183
 │ 1 0
 │   18
 │   15
 │    33
 │    30
 │     3
 └────→5
```

Finally, you bring the $\frac{3}{5}$ up to your answer, making the fraction a part of the quotient:

```
      236 3/5 ←┐
 5)1,183       │
   1 0         │
     18        │
     15        │
      33       │
      30       │
       3       │
       5 ──────┘
```

Thus, 5) 1,183 $= 236\frac{3}{5}$.

Following are two more examples of division with a remainder.

EXAMPLE 1.

```
      36 13/24
 24)877
 │   72
 │   157
 │   144
 │    13
 └───→24
```

EXAMPLE 2.

```
          12 1/210
 210)2,521
 │   2 10
 │     421
 │     420
 │       1
 └─────→210
```

EXERCISES

In 1–38, divide. Write each remainder in a fraction.

1. $4\overline{)929}$ **2.** $6\overline{)829}$ **3.** $5\overline{)748}$ **4.** $3\overline{)275}$ **5.** $5\overline{)317}$

6. $4\overline{)945}$ **7.** $6\overline{)967}$ **8.** $8\overline{)971}$ **9.** $7\overline{)799}$ **10.** $6\overline{)851}$

11. $3\overline{)821}$ **12.** $4\overline{)939}$ **13.** $5\overline{)747}$ **14.** $7\overline{)886}$ **15.** $6\overline{)923}$

16. $9\overline{)733}$ **17.** $6\overline{)247}$ **18.** $8\overline{)257}$ **19.** $14\overline{)327}$ **20.** $23\overline{)590}$

21. $18\overline{)659}$ **22.** $34\overline{)865}$ **23.** $28\overline{)3,481}$

24. $16\overline{)3,941}$ **25.** $46\overline{)9,915}$ **26.** $38\overline{)8,517}$

27. $234\overline{)74,931}$ **28.** $423\overline{)91,483}$ **29.** $186\overline{)79,125}$

30. $416\overline{)142,597}$ **31.** $358\overline{)223,619}$ **32.** $197\overline{)126,761}$

33. $99,933 \div 234$ **34.** $241,547 \div 429$ **35.** $171,090 \div 536$

36. $164,070 \div 386$ **37.** $198,090 \div 528$ **38.** $216,280 \div 625$

APPLICATION PROBLEMS

In 1–3, divide, and show each remainder in a fraction.

1. A truck carries a load of 35 cases. If the load weighs 11,385 pounds, how much does each case weigh? __________
2. A car averages 18 miles per gallon of gasoline. How many gallons of gasoline will it need to travel 1,139 miles? __________
3. An oil well produces 6,751 barrels of oil in a year. How many barrels are produced in a month? (Assume that the production is the same each month.) __________

4. A school has an allotment of $5,800 to buy typewriters. If the price of a typewriter is $245, how many typewriters can the school buy? ____________

   How much money will be left over? (*Hint:* The remainder will be the amount of money left over.) ____________

5. The Curtis family bought carpeting at $9 a yard. The total bill of $412 included a delivery charge of under $10. How many yards did the Curtis family buy? ____________

   How much was the delivery charge? (*Hint:* The remainder will be the dollar amount of the delivery charge.) ____________

6. A dress manufacturer uses 3 yards of fabric to make a dress. How many dresses can be made with 218 yards of fabric? ____________

   How many yards will remain? ____________

7. How many yards are there in 9,135 inches? ____________

   How many inches are left? (*Hint:* 1 yard = 36 inches) ____________

In everyday life, it is not always convenient to express a remainder in a fraction. In problems 8–10, your answers will be whole numbers.

SAMPLE SOLUTION

There are 4 quarts in a gallon. How many 1-gallon containers will you need to hold 182 quarts of kerosene? (*Hint:* You will need one 1-gallon container to hold a fraction of a gallon.)

$$\begin{array}{r} 45\frac{2}{4} \\ 4\overline{)182} \\ \underline{16} \\ 22 \\ \underline{20} \\ 2 \end{array}$$

need 1 container to hold $\frac{2}{4}$ gal.

$45 + 1 = 46$

46 1-gallon containers

8. A poultry farmer produces 3,833 eggs. How many boxes that hold a dozen eggs will he need to box all the eggs? (*Hint:* The farmer will need 1 box to hold a fraction of a dozen eggs.) ____________

9. A school has an average class size of 35 students. How many classrooms does the school need if the total number of students is 2,634? (*Hint:* The school will need 1 classroom for a group of students, even if the group is smaller than 35.) ____________

10. Pencils are packed 24 in a box. How many boxes will be needed to pack 3,447 pencils? ____________

UNIT 16. Division With 0's in the Quotient

WORDS TO KNOW

We sometimes think of 0 as meaning "nothing" or as having no value. But when 0 is a digit in a number, we must be careful not to overlook it.

Consider the number 202. Although the 0 has no value in itself, its presence affects the value of the number: 202 = 2 hundreds + 0 tens + 2 ones. Suppose we accidentally leave out the 0. The number 202 becomes 22, and has a value of only 2 tens + 2 ones. The 0 makes quite a difference.

Because the digit 0 affects the value of a number by "holding the other digits in place," 0 is called a **placeholder.**

A very common mistake in division problems is to forget to place a 0 in the answer when you should do so.

EXAMPLE 1. 2,912 ÷ 14

Solution: Divide as you have learned:

```
        2
14 )2,912
    2 8↓
new dividend ——→ 11
```

Note that the new dividend, 11, is *smaller* than the divisor.

Since 14 cannot go into 11 (or goes into 11 zero times), *place a 0 in the quotient as a placeholder, directly above the 1 you just brought down:*

```
        20
14 )2,912
    2 8
     11
```

Now, to form a new dividend, bring down the next digit, 2, and place it next to the 11:

```
        20
14 )2,912
    2 8 ↓
     112
```

Since the new dividend, 112, is now larger than the divisor, 14, you can complete the division:

```
       208
14 )2,912
    2 8
     112
     112
```

Answer: 208

Part of the way through this problem, you realized that the divisor, 14, was too large to go into the new dividend, 11. You could have said, "14 goes into 11 zero times." Thinking like this will remind you to place a 0 in the quotient whenever the divisor is too large for the new dividend.

Here is another situation that requires you to place a 0 in the quotient:

EXAMPLE 2. 4,530 ÷ 15

Solution:

```
        3
15 )4,530
    4 5↓
new dividend ——→ 3
```

This time, when you subtract 45 from 45, the result is 0. And, when you bring down the 3 to form the new dividend, you see that 15 will not go into 3.

You say, "15 goes into 3 zero times," and *place a 0 in the quotient as a placeholder,* directly above the 3 you just brought down:

```
        30
15 )4,530
    4 5
      3
```

To form a new dividend, bring down the next digit, 0, and place it next to the 3:

```
       302
15 )4,530
    4 5 ↓
      30
```

Since the new dividend, 30, is now larger than 15, you can complete the division:

```
      302
15 )4,530
    4 5
      30
      30
```

Answer: 302

Do you see the similarity between this example and Example 1?

EXAMPLE 1.

```
                 208
             14 )2,912
                 2 8
first new
dividend  ----> (1)12
                 112
```

EXAMPLE 2.

```
                 302
             15 )4,530
                 4 5
first new
dividend  ----> (3)0
                   30
```

In Example 1, the first new dividend had two digits; in Example 2, the first new dividend had only *one* digit, the digit that was brought down from the dividend. But, in both cases, the first new dividend was smaller than the divisor.

Sometimes a 0 in the dividend will require you to place a 0 in the quotient.

EXAMPLE 3. 1,208 ÷ 4

Solution: Divide as you have learned:

```
    3
4 )1,208
   1 2
```

Since the next digit to be brought down is a 0, *you don't even bother bringing it down* because 4 into 0 goes 0 times. *Place a 0 in the quotient as a placeholder directly above the 0 in the dividend.*

```
    30
4 )1,208
   1 2
```

To form a new dividend, bring down the next digit, 8. Then complete the division.

```
                  302
              4 )1,208
                 1 2
new dividend ------> 8
                     8
```

Answer: 302

EXAMPLE 4. 3,200 ÷ 8

Solution:

```
    400
8 )3,200
   3 2
```

Answer: 400

You have learned how to express the *remainder* of a division problem in a *fraction.* Sometimes you must place a 0 in the quotient before writing the remainder fraction.

EXAMPLE 5. 2,485 ÷ 8

Solution:

Divide as you have learned. The divisor, 8, cannot go into the new dividend, 5. Also, *there are no more digits to bring down.* Therefore, you say, "8 goes into 5 zero times."

```
                   31
               8 )2,485
                  2 4
                    8
                    8
new dividend -------> 5
```

You place a 0 in the quotient as a placeholder, directly above the 5 you just brought down. Keep in mind that the new dividend, 5, does not become a remainder *until you place a 0 in the quotient.*

```
    310
8 )2,485
   2 4
     8
     8
      5
```

Complete the division by writing the remainder in a fraction, $\frac{5}{8}$. Place the $\frac{5}{8}$ next to the answer, making the fraction a part of the quotient.

```
    310 5/8
8 )2,485
   2 4
     8
     8
     5/8
```

Answer: $310\frac{5}{8}$

Remember

In division, do not leave a blank space above any digit you bring down. When necessary, use zero as a placeholder.

EXERCISES

In 1–36, divide, and place zeros in the quotient as necessary. Show any remainder in a fraction.

1. 16)6,528
2. 23)7,015
3. 34)7,004
4. 42)8,736
5. 23)9,407
6. 19)5,757
7. 36)7,524
8. 21)8,526
9. 28)8,512
10. 26)8,034
11. 31)9,517
12. 27)8,316
13. 18)3,672
14. 15)1,545
15. 23)4,669
16. 19)9,557
17. 27)10,854
18. 24)14,472
19. 32)12,896
20. 35)10,570
21. 22)15,466
22. 38)11,476
23. 42)25,242
24. 46)13,892
25. 49)245,147
26. 37)222,148
27. 52)208,312

28. $53\overline{)3,180,106}$ **29.** $44\overline{)1,760,220}$ **30.** $63\overline{)1,260,252}$

31. $45\overline{)271,358}$ **32.** $56\overline{)281,687}$ **33.** $223\overline{)1,121,693}$

34. $337,449 \div 48$ **35.** $426,128 \div 53$ **36.** $260,118 \div 37$

APPLICATION PROBLEMS

SAMPLE SOLUTION

How many 15-man squads can be formed from 1,620 men?

$$\begin{array}{r} 108 \\ 15\overline{)1,620} \\ 15 \\ \hline 120 \\ 120 \\ \hline \end{array}$$

108 15-man squads

1. A machine, working at a constant rate, produces 2,592 articles in 24 hours. How many articles are produced in one hour? ____________

2. A bookstore sold 3,708 dictionaries in a year. How many dictionaries did the store sell in one month? (Assume that the sales were the same each month.) ____________

3. Five schools have a total of 4,535 students. If each school has the same number of students, how many students does each school have? ____________

4. Last year, Julio's salary was $10,452. How much did he earn per week? ____________

5. Four members of a family won a prize of $1,220. If they share the prize equally, how much will each member get? ____________

6. Clarissa's car averages 18 miles to a gallon of gasoline. How many gallons did the car use to travel 1,085 miles? ____________

7. Melons are packed 12 to a box. How many boxes will be needed to pack 72,012 melons? ____________

8. Five water tanks of equal capacity hold 12,554 gallons of water. How many gallons does one tank hold? ____________

9. A dealer received a shipment of cloth that cost $8,123. How many yards of cloth were received if the cost was $4 per yard? ____________

10. Mrs. Swenson earns $12,036 a year, on a steady salary. How much does she earn in a month? ____________

UNIT 17. Division of Numbers That Contain End Zeros

WORDS TO KNOW

When the last digits of a number are zeros, these digits are called **end zeros.**

When you divide one number that has **end zeros** by another number that has end zeros, you can simplify the division by getting rid of the end zeros.

To simplify a division problem involving end zeros, follow this rule:

RULE

Cross out the end zeros, one for one, both in the divisor and in the dividend.

EXAMPLE 1. Divide: 600)2,400

Solution: To simplify the division, get rid of the two zeros in the divisor and the two zeros in the dividend.

```
6ØØ)2,4ØØ
```

Complete the division, ignoring all the zeros you crossed out.

```
           4
6ØØ)2,4ØØ
         2 4
```

Answer: 4

What you have done by getting rid of the end zeros is to divide both the divisor and the dividend by 100.

```
       6              24
100)600        100)2,400
      600             2 00
                        400
                        400
```

Since 600 ÷ 100 = 6 and 2,400 ÷ 100 = 24, then 2,400 ÷ 600 is the same as 24 ÷ 6.

```
              4
600)2,400
        2 400

        4
    6)24
       24
```

Remember

You must cross out an equal number of end zeros in both the divisor and the dividend.

Example 2. Divide: $50\overline{)2{,}500}$

Solution: You get rid of one 0 in the divisor and one 0 in the dividend.

$$5\cancel{0}\overline{)2{,}50\cancel{0}}$$

Complete the division, ignoring the zeros you got rid of.

$$\begin{array}{r} 50 \\ 5\cancel{0}\overline{)2{,}50\cancel{0}} \\ \underline{2\ 5} \end{array}$$

Answer: 50

Of course, it is possible to have a remainder.

Example 3. Divide: $70\overline{)4{,}730}$

Solution: Get rid of one 0 in the divisor and one 0 in the dividend.

$$7\cancel{0}\overline{)4{,}73\cancel{0}}$$

Complete the division.

$$\begin{array}{r} 67\frac{4}{7} \\ 7\cancel{0}\overline{)4{,}73\cancel{0}} \\ \underline{4\ 2} \\ 53 \\ \underline{49} \\ 4 \\ \rightarrow \overline{7} \end{array}$$

Answer: $67\frac{4}{7}$

EXERCISES

In 1–18, divide.

1. $30\overline{)650}$
2. $50\overline{)230}$
3. $80\overline{)4{,}760}$
4. $200\overline{)2{,}450}$
5. $300\overline{)5{,}400}$
6. $240\overline{)3{,}600}$
7. $300\overline{)156{,}000}$
8. $370\overline{)5{,}690}$
9. $5{,}700\overline{)67{,}000}$
10. $900\overline{)87{,}000}$
11. $240\overline{)37{,}650}$
12. $470\overline{)63{,}390}$

13. 260,000 ÷ 400 14. 463,500 ÷ 7,000 15. 253,500 ÷ 600

16. 47,050 ÷ 400 17. $205,000 ÷ 5,000 18. $250,000 ÷ 10,000

APPLICATION PROBLEMS

SAMPLE SOLUTION

The distance from the earth to the moon is sometimes expressed in terms of the radius of the earth. If the earth's radius is 4,000 miles and the distance from the earth to the moon is 240,000 miles, how many "earth-radii" will be needed to express the distance from the earth to the moon?

60
4,000)240,000
24

60 earth-radii

1. Tom's car averages 20 miles to a gallon of gasoline. How many gallons will Tom need to travel 780 miles? ________

2. A machine can produce 300 articles in one day. How many days will it take to produce 16,500 articles? ________

3. A jet airliner averages 600 miles per hour. How many hours will it take to travel 7,000 miles? ________

4. On a certain flight, astronauts averaged 2,000 miles per hour traveling to and from the moon. At this speed, how many hours will it take them to travel 480,000 miles? ________

5. A ton is equal to 2,000 pounds. How many tons are there in 127,000 pounds? ________

6. The Greenwood family has a mortgage on their home for which the yearly payments to the bank are $5,700. For the total mortgage, they expect to pay $171,000. For how many years is the mortgage? ________

7. A concert promoter sold 3,800 tickets for a total amount of $57,000. What was the average price of admission? ________

8. A hospital bought 50 nurses' uniforms for $750. What was the cost of each uniform? ________

9. Sheila earns $480 for a 40-hour week. What is her hourly rate of pay? ________

10. A church group sold 1,500 raffle books for a total of $22,500. What was the price of each raffle book? ________

Review of Part V (Units 14–17)

In 1–24, divide. Show any remainder in a fraction.

1. $7\overline{)252}$ **2.** $8\overline{)2{,}056}$ **3.** $23\overline{)966}$

4. $37\overline{)8{,}214}$ **5.** $683\overline{)842{,}822}$ **6.** $31\overline{)64{,}325}$

7. $253\overline{)513{,}590}$ **8.** $42\overline{)42{,}042}$ **9.** $520\overline{)119{,}600}$

10. $7{,}200\overline{)4{,}680{,}000}$ **11.** $25\overline{)4{,}056}$ **12.** $822\overline{)20{,}555}$

13. $53\overline{)32{,}120}$ **14.** $28\overline{)28{,}565}$ **15.** $88\overline{)177{,}765}$

16. $39{,}483 \div 321$ **17.** $5{,}630 \div 75$ **18.** $474{,}474 \div 237$

19. $15,650 divided by 25 **20.** $59,475 divided by 61 **21.** $18,600 divided by 300

22. $37,100 divided by 70 **23.** $542\overline{)54{,}199{,}458}$ **24.** $999\overline{)20{,}181{,}798}$

For 25–34, use a calculator.

In 25–27, divide. Be careful to put the dividend into the calculator first, then the divisor.

25. $2{,}345 \div 67$ **26.** $36\overline{)9{,}288}$ **27.** $145\overline{)388{,}600}$

In 28–30, fill in the missing digits.

28. $\begin{array}{r} 7\square \\ 8\overline{)5\square 4} \end{array}$ **29.** $\begin{array}{r} \square 2 \\ 27\overline{)86\square} \end{array}$ **30.** $\begin{array}{r} 14 \\ 1\square 5\overline{)1{,}47\square} \end{array}$

In 31–33, use only the numbers 16, 144, and 720 to fill in the blanks.

31. $\square \div \square = 5$ **32.** $\square \div \square = 9$ **33.** $\square \div \square = 45$

34. Complete the division table. A sample answer is shown.

÷	360	432	1,080
18		24	
24			
72			

PART VI. Word Problems

UNIT 18. Solving Word Problems

WORDS TO KNOW

In everyday life, it is often necessary to figure out problems such as the total cost of a purchase or how long it will take to make a trip. In such situations, you are not told whether to *add, subtract, multiply,* or *divide.* Rather, you must decide what arithmetic operation to use. This type of problem, in which you must decide *how* to solve it, is called a **word problem.**

In order to solve a **word problem,** you must first find the facts of the problem and understand what the problem is asking you to find, and then decide on the arithmetic operation you will use in solving the problem.

To solve a word problem step by step, ask yourself the following questions:

Question 1: *Do you understand what the problem is telling you?* It may be necessary to read the problem several times to understand it.

Question 2: *What are the facts given in the problem?* Write down the known facts in the problem. This will help you to understand how they are related.

Question 3: *What is the problem asking you to find?* Have a clear understanding of what you must find.

Question 4: *What arithmetic operation will you have to use to solve the problem?* Study the given information and decide whether you should *add, subtract, multiply,* or *divide* to solve the problem. Then perform the calculation.

Question 5: *Is your solution correct? Does it seem reasonable?* Check your answer, the unknown fact, against the known facts in the problem.

Most students find Question 4 to be the most difficult to answer. To help you decide which arithmetic operation to use, keep in mind that every problem has *key words* that will tell you whether to add, subtract, multiply, or divide.

Use *addition* if the problem uses key words like these:

1. Find the *sum* . . . .
2. What is the *total* . . . .
3. Determine the cost of *all* . . . .
4. How much *altogether* . . . .

Use *subtraction* if the problem uses key words like these:

1. What is the *difference* . . . .
2. How much *more* (*larger, greater*) . . . .
3. How much *less* . . . .
4. What was the *increase* . . . .

Use *multiplication* if the problem uses key words like these:

1. Find the cost of *ten items* if the cost of *one item* is . . . .
2. If *each item* weighs . . . , how much will *all* of them weigh?
3. If the *hourly* wage is . . . , find the *weekly* wage.
4. Find the *total* . . . . (Note that the word "total" may tell you to add *or* to multiply. Your choice will depend on the facts in the particular problem.)

Use *division* if the problem uses key words like these:

1. How much is *each* . . . .
2. Find the cost *per* . . . .
3. How much does *one* . . . .

No matter how difficult a problem may seem, there are only *four* arithmetic operations to choose from.

EXAMPLE 1. John worked 40 hours last week and his rate of pay was $4 per hour. Find his wages for the entire week.

Solution: Solve this word problem step by step by answering the preceding five questions:

1. The problem tells you John's hourly rate of pay and the number of hours he has worked.

2. The given facts are: (1) John worked 40 hours and (2) his rate of pay is $4 per hour.

3. The unknown fact is John's total wages for last week.

4. You must use *multiplication* because the problem gives you John's hourly wage and asks you to find his weekly wage. The number of hours worked multiplied by the rate per hour will equal the total wages for the week. Multiplying:

40	hours worked
× $4	rate per hour
$160	total wages

John's total wages last week were $160.

5. Check your answer by dividing the total wages for the week by 40 hours. You get an hourly rate of $4, which is the same as the given fact.

$$40\overline{)\$160} = \$4$$

16

In each of the following examples, tell whether you would solve the problem by using addition, subtraction, multiplication, or division. Then solve the problem.

Example 2. Sally bought a toaster for $13, an electric can opener for $12, and a coffee maker for $29. How much did she spend altogether?

Solution: Add the three costs:

$$\$13 + \$12 + \$29 = \$54$$

Answer: $54

Example 3. Sally paid for her purchases in Example 2 with a $100 bill. How much change did she receive?

Solution: Subtract the total cost of $54 from $100:

$$\$100 - 54 = \$46$$

Answer: $46

Example 4. Aaron earns $250 a week. How much does he earn in a year?

Solution: Since there are 52 weeks in a year, multiply $250 by 52:

$$\begin{array}{r} \$250 \\ \times 52 \\ \hline 500 \\ 12\,50 \\ \hline \$13{,}000 \end{array}$$

Answer: $13,000

Example 5. A dozen pencils cost 96¢. How much does one pencil cost?

Solution: Divide the cost of 96¢ by 12:

$$12\overline{)96¢} = 8¢$$

96

Answer: 8¢

Exercises

In 1–10, complete each sentence, telling whether you must use *addition, subtraction, multiplication,* or *division* to solve the problem. **Do not actually compute the answer.**

Sample Solution

A radio originally priced at $65 was marked down to $47. To find the amount saved, *subtract the marked-down price from the original price*.

1. To find the price of one shirt if 3 shirts cost $42, ______________________________.

2. A television set sells for $232 and the sales tax is $18. To find the total cost, ______________________________.

3. To find the cost of 8 pairs of socks at a price of $2 per pair, ______________________________.

4. A coat selling for $85 was reduced to $65. To find the amount of the reduction, ______________________________.

5. Susan earned $17, Maria earned $11, and Alice earned $14. To find how much all the girls earned, ______________________________.

6. To find the cost of 4 tires if one tire costs $35, ______________________________.

7. Tom earns $10,650 a year and his brother earns $8,750 a year. To find how much more Tom earns than his brother, ______________________________.

8. To find the monthly rent of a family that pays $4,536 in rent each year, ______________________________.

9. To find the number of inches in 52 feet, ______________________________.

10. To find the number of yards in 72 feet, ______________________________.

APPLICATION PROBLEMS

In 1–10, compute the required answer.

1. How much will 8 typewriters cost if they are priced at $315 each? ____________

2. Last year, a salesperson earned a salary of $12,682, and received commissions of $14,915. What were the total earnings for the year? ____________

3. A winter coat originally priced at $275 was reduced to $169. Find the amount of the reduction. ____________

4. Coretta paid $37 for a weekly commutation ticket. If she did not buy tickets during her three-week vacation, how much did her commutation tickets cost her for the year? ____________

5. In 1927, Charles Lindbergh flew the first solo transatlantic flight, from Mineola, N.Y., to Paris, France. He flew the distance of about 3,604 miles in about 34 hours. How many miles did he average per hour? ____________

6. Before school started in the fall, Pat went shopping for clothes. She bought a skirt for $35, a sweater for $22, a blouse for $19, and shoes for $44. What was the total amount of her purchases? ____________

7. A high school held its graduation exercises in a local theater, and contracted for a rental fee of $1,200. A deposit of $225 was paid. What balance still had to be paid? ____________

8. In 1933, Wiley Post made a record-breaking flight around the world via the Arctic Circle. He traveled about 116 hours at an average speed of 136 miles per hour. What distance did he fly? ____________

9. Will Barr drove to three different towns on a recent business trip. On Monday he drove 167 miles, Tuesday he drove 215 miles, Wednesday 93 miles, and on Thursday he drove 118 miles to get back home. What was his total mileage? ____________

10. Tim had one year in which to pay off the $1,476 balance he owed on his boat. If he paid in equal monthly installments, how much did he pay each month? ____________

UNIT 19. Problem-Solving Strategy: Breaking Down a Problem

Numerical problems in real life often cannot be solved by just one arithmetic operation. A helpful strategy is to break down the problem into simpler, single-operation problems.

The step-by-step approach is the same as before, except that you will have to read the problem even more carefully, and will have to choose more than one operation to work it out. A lot of common sense goes into deciding what operations to use.

Example 1. Joe Carter bought some farm property in Minnesota for $35,750, and made a down payment of $12,902. If he must pay the balance in 96 equal monthly payments, how much will he have to pay each month?

Solution:

1. Read the problem over until you understand it. You are told the full purchase price, the amount of the down payment, and the number of equal monthly payments.
2. The full price was $35,750, and $12,902 was paid. The rest will be split into 96 equal payments.
3. You need to find the amount of each payment.
4. Break down the problem to see what operations to use.

   a. Of the total price of $35,750, Joe already paid $12,902.

   Subtract to find how much he still owes.

   $$\begin{array}{r} \$35{,}750 \\ -12{,}902 \\ \hline \$22{,}848 \end{array}$$

   b. The amount still owed will be paid in 96 equal payments.

   Divide the result of part *a* by 96 to find the amount of each payment.

   $$\begin{array}{r} \$238 \\ 96\overline{)\$22{,}848} \\ 19\ 2 \\ \hline 3\ 64 \\ 2\ 88 \\ \hline 768 \\ 768 \\ \hline \end{array}$$

5. Check the division by multiplying.

   $$\begin{array}{r} \$238 \\ \times 96 \\ \hline 1\ 428 \\ 21\ 42 \\ \hline \$22{,}848 \end{array}$$

   Then check the subtraction by adding.

   $$\begin{array}{r} \$22{,}848 \\ +12{,}902 \\ \hline \$35{,}750 \end{array}$$

Answer: $238 a month

In each of the following examples, tell how you would solve the problem by using addition, subtraction, multiplication, or division. Then answer the question.

Example 2. Tickets to the school play were priced at $4 and $3. If 137 tickets were sold at $4, and 210 tickets at $3, how much money was taken in?

Solution: Multiply the $4 price by the number of $4 tickets sold, and multiply the $3 price by the number of $3 tickets sold. Add the products.

$$\begin{array}{r} 137 \\ \times \$4 \\ \hline \$548 \end{array} \qquad \begin{array}{r} 210 \\ \times \$3 \\ \hline \$630 \end{array} \qquad \begin{array}{r} \$548 \\ +\$630 \\ \hline \$1,178 \end{array}$$

Answer: $1,178

Example 3. A dry-goods store sold 4 bolts of a cotton fabric at $3 a yard. If each bolt held 100 yards, what was the total amount of the sale?

Solution: Multiply 100 yards by the price of $3 a yard to get the selling price of each bolt. Then multiply by 4 bolts.

$$\begin{array}{r} 100 \\ \times \$3 \\ \hline \$300 \end{array} \qquad \begin{array}{r} \$300 \\ \times 4 \\ \hline \$1,200 \end{array}$$

Answer: $1,200

EXERCISES

In 1–10, you must use more than one arithmetic operation to solve each problem. Tell what mathematical steps you must use. **Do not actually compute the answer.**

Sample Solution

Bess bought 14 packages of paper plates at $2 per package. She paid $18 down and promised to pay the rest next week. How much does she owe? *multiply the number of packages by the cost per package, 14 × $2. Then subtract $18.00 from this product.*

1. Harry bought a sports jacket for $40, a hat for $10, and three ties for $2 each. He agreed to pay for the clothing in four equal payments. How much is each payment? ______________________________

2. The Jensens bought a freezer for $550. They paid $70 down and agreed to pay the balance in 12 equal monthly payments. What is the amount of each monthly payment? ______________________________

3. Each month, Sam pays $450 in rent, $36 for his telephone, and $24 for gas and electricity. What is the annual cost of his rent, telephone, and gas and electricity? ______________________________

4. Mr. Lipsky is entitled to half of the profits of a grocery store. Last year, the store took in $55,000 and had operating expenses of $40,000. How much is Mr. Lipsky's share of the profits? ______________________________

5. Three girls agree to share equally the rent of an apartment. If the total monthly rent is $360, how much will each girl pay in rent over one entire year? ______________________________

6. Sally used 112 inches of ribbon to make 8 class-night decorations. She wants to make 3 more. How much more ribbon will she need? ______________________________

7. To enclose a backyard, Mr. Collins needs eight lengths of fencing 8 feet long, two 10-foot lengths, and one piece 4 feet long. What is the total length of the fencing he needs? ______________________________

8. Mrs. Brown bought 45 yards of carpeting at $9 per yard. If she put down a deposit of $175, how much must she pay when the carpeting is delivered? (There is no delivery charge.) ______________________________

9. Mr. Gold bought a refrigerator for $310 and made a down payment of $130. If the balance will be paid in 12 equal monthly installments, how much will each payment be? ______________________________

10. Mrs. Stanford bought a living room set for $1,265. Her down payment was $350 and she made two payments of $45 each. How much does she still owe for the set? ______________________________

APPLICATION PROBLEMS

In 1–10, compute the required answer.

1. A retailer ordered the following items: 36 blouses at $18 each, 48 dresses at $29 each, and 96 belts at $7 each. What was the total cost of the order? ______________

2. The Cruz family bought a living room set for $1,050. They made a down payment of $275 and then 8 payments of $87 each. How much did they still owe for the living room set? ______________

3. A retailer ordered 48 boxes of shirts. If each shirt costs $7 and each box holds 3 shirts, what was the total cost of the order? ______________

4. A salesperson spends an average of $58 a day when traveling. How much is spent in a 7-day week? ________

What are the expenses on an 8-week trip? ________

5. Marilyn bought a color television set with a down payment of $78 and 6 payments of $59 each. What was the total cost of the set? ________

6. Anthony bought the following items: a shirt for $10, a pair of shoes for $32, and a pair of slacks for $35. If he paid the clerk with a one-hundred-dollar bill, how much change did he get? ________

7. A theater has 56 orchestra rows with 37 seats in each row, 23 balcony rows with 29 seats in each row, and 15 box seat areas with 7 seats in each box. What is the total seating capacity of the theater? ________

8. Victor bought a new car with a down payment of $575, the balance to be paid off in 36 monthly installments of $249 each. What is the total cost of the car? ________

9. An insurance company ordered 7 electric typewriters at $689 each and 3 photocopiers at $1,467 each. What is the total cost of the office equipment? ________

10. The average monthly rent in a 56-unit apartment building is $397. What is the *yearly* rent collected on the 56 apartments? ________

UNIT 20. Problem-Solving Strategy: Estimating

However you do your calculating—mentally, using pencil and paper, or using a calculator—it is always helpful to have a rough idea of how large your answer should be. Anyone can make an error in mental computation or push a wrong button on a calculator. If you have a "ballpark estimate" of what your result should be, it is possible to catch such errors.

ROUNDING OFF

One way to get an **estimate** is by **rounding off.** You learned how to round off numbers earlier in this book. Review Unit 2 if you need to.

When you round off, it may help to picture the location of your number on a number line between two numbers. Is it halfway between them, less than halfway, or more than halfway?

Suppose you are rounding off 77 to the nearest ten. Which two tens is 77 between?

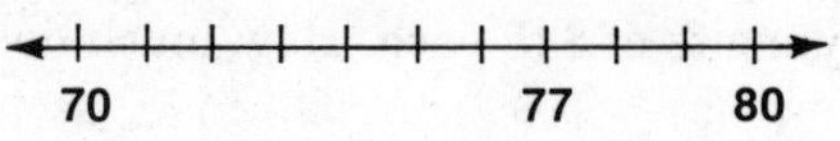

77 is between 70 and 80.
The halfway number is 75.

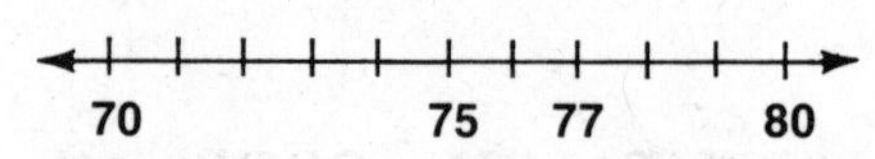

Since 77 is more than halfway from 70 to 80, 77 is rounded up to 80.

EXAMPLE 1. Find an estimate for the sum of 581 and 409 by rounding off each number to the nearest *hundred.*

Solution:

$$\begin{array}{rcr} 581 & \longrightarrow & 600 \\ 409 & \longrightarrow & 400 \\ \hline & & 1{,}000 \end{array}$$

Answer: $581 + 409 \approx 1{,}000$
("$\approx$" stands for "is approximately equal to.")

EXAMPLE 2. Of the choices below, which gives a better estimate of the product 6×48?

(a) 6×40 (b) 6×50

Solution: Since 48 rounded off to the nearest ten is 50, the second estimate is better.

Answer: (b) 6×50

Sometimes an estimate is enough to answer a word problem.

EXAMPLE 3. Sally owes 11 payments of $356 each on her car. She has about $2,000 in the bank. Does she have enough money to cover all of her car payments?

Solution: About how much does she owe altogether? Estimate by rounding off the numbers to the nearest ten, and multiply.

$$\begin{array}{rcr} \$356 & \longrightarrow & \$360 \\ 11 & \longrightarrow & \times 10 \\ \hline & & \$3,600 \end{array}$$

Answer: Since the estimate is more than the amount she has in the bank, she does not have enough money to cover her car payments.

EXAMPLE 4. On the way home from work, Ted stopped at the supermarket. He got a half gallon of milk for $1.19, a dozen eggs for $.96, hamburger meat for $3.12, ketchup for $.99, a loaf of bread for $1.19, and margarine for $.79. On line at the checkout counter, he found he had eight dollars and change with him. Was that enough to pay for his purchases?

Solution: Ted took an estimate by rounding off the amount of each purchase to the nearest dollar and adding mentally.

$$\begin{array}{rcr} \$1.19 & \rightarrow & \$1 \\ .96 & \rightarrow & 1 \\ 3.12 & \rightarrow & 3 \\ .99 & \rightarrow & 1 \\ 1.19 & \rightarrow & 1 \\ .79 & \rightarrow & 1 \\ \hline & & \$8 \end{array}$$

Answer: Ted thought he probably had enough money. It turned out that the bill was $8.24, which he was able to pay.

REFERENCE NUMBERS

A method other than rounding off that can be used in estimating, is working with **reference numbers** that are easier to compute.

EXAMPLE 5. About how much is 61 divided by 7?

Solution: 61 is not divisible by 7. But 61 is close to 63, which is divisible by 7.

$$61 \div 7 \approx 63 \div 7 = 9$$

Answer: 61 divided by 7 is about equal to 9.

EXAMPLE 6. The Sun Shu Restaurant has a dinner special that costs $49.95 for six people. If six friends share the cost, about how much will each one pay?

Solution: $49.95 is a little more than $48, which is divisible by 6.

$$\$49.95 \div 6 \approx \$48 \div 6 = \$8$$

Answer: Each person must pay a little more than $8.

EXERCISES

In 1–6, circle the rounded-off expression, (A), (B), or (C), that gives the best estimate, compute mentally, and write the estimated answer.

SAMPLE SOLUTION

$43 + 58$

(A) $40 + 50$

(B) $40 + 60$ (circled)

(C) $50 + 60$

$43 + 58 \approx 100$

1. 87 − 31
(A) 90 − 30
(B) 80 − 30
(C) 90 − 40

____

2. 67 × 92
(A) 60 × 90
(B) 70 × 100
(C) 70 × 90

____

3. 117 ÷ 21
(A) 120 ÷ 20
(B) 110 ÷ 20
(C) 120 ÷ 30

____

4. 372 + 594 + 826
(A) 300 + 500 + 800
(B) 400 + 600 + 900
(C) 400 + 600 + 800

____

5. 2,893 − 1,821
(A) 2,800 − 1,800
(B) 2,900 − 1,800
(C) 2,900 − 1,900

____

6. 491 × 217
(A) 500 × 200
(B) 400 × 200
(C) 500 × 300

____

In 7–9, circle the expression, (A), (B), or (C), having the reference numbers that are easiest to work with, compute mentally, and write the estimated answer.

7. 146 ÷ 12
(A) 150 ÷ 12
(B) 140 ÷ 12
(C) 144 ÷ 12

____

8. 73 ÷ 8
(A) 72 ÷ 8
(B) 74 ÷ 8
(C) 75 ÷ 8

____

9. 6,514 ÷ 82
(A) 6,500 ÷ 80
(B) 6,400 ÷ 80
(C) 7,000 ÷ 80

____

10. Round off each amount in the table to the nearest *ten dollars*, and write the rounded-off number in the table of Estimated Earnings shown below.

	DAILY EARNINGS INCLUDING OVERTIME				
Employee	Monday	Tuesday	Wednesday	Thursday	Friday
Al	$51.45	55.13	58.80	51.45	51.45
Betty	51.45	51.45	55.13	51.45	51.45
Charles	53.76	47.04	47.04	48.72	47.04
Debbie	44.55	41.58	47.52	41.58	41.58

Add horizontally to estimate the week's earnings for each employee, and add vertically to estimate the total earned by all employees each day. Get the grand totals horizontally and vertically. (The grand totals should be the same.)

	ESTIMATED EARNINGS					
Employee	Monday	Tuesday	Wednesday	Thursday	Friday	Total
Al						
Betty						
Charles						
Debbie						
Total						

APPLICATION PROBLEMS

In 1–10, give estimated answers without doing the exact computation. Use a mental estimate.

1. Jennifer wants to buy a bracelet for $3.25 and a scarf for $2.69. To the nearest *dollar*, how much money does she need? ______

2. Electric light bulbs that cost 59¢ each are selling in a package of 4 for $1.99. How much could be saved, to the nearest *ten cents*, by buying the package? ______

3. Ann is driving to Mexico City from her home in San Francisco, a distance of 1,188 miles. She plans to cover about 300 miles a day. How many days should the trip take? ______

4. There are 362 calories in a chocolate milk shake, compared with 257 calories in a fruit-flavored yogurt. About how many more calories are there in the milk shake? ______

5. Stuart's monthly rent is $396. His bill for utilities averages $42 each month, and his phone bill averages $29. Estimate, to the nearest *ten dollars*, the total of these items per month. ______

6. Tess and Tom Tolliver are planning a two-week charter vacation in Scandinavia. The base price is $699 each, and a room with a private bath in Oslo will cost $29 extra for two. Estimate the total amount the trip will cost Tess and Tom. ______

7. Mrs. Chase is buying a collection of 4 framed photographs for a total cost of $87.50. What is the cost of one photograph? (*Hint:* Use a reference number for this estimate.) ______

8. Charles Sawyer earns $23,575 a year. About how much does he make a month? ______

9. Steve gives guitar lessons for a fee of $150 for eight 1-hour lessons. About how much is he charging for one lesson? ______

10. Susie went shopping for souvenirs in New Orleans. She bought 6 placemats at $1.95 each, a Cajun cookbook for $5.25, a set of postcards for $1.98, and a pen for $2.69. What was the cost of her purchases, to the nearest *dollar*? ______

UNIT 21. Other Problem-Solving Strategies

GUESS AND CHECK

Often a good way to solve a problem is a method we call **guess and check.** In this method, you *guess* at what seems a reasonable answer to the problem, and then solve the problem using your guess. Even if it does not work out the first time, the result will probably give you a clue to a better second guess. You just keep refining your guesses until you find the answer.

EXAMPLE 1. Sean has grades of 71 in English, 81 in Science, and 68 in Social Studies. What grade does he need in Math to get an average of 75 for the four subjects?

Solution: You get an average by adding all the numbers and then dividing by how many numbers there are.

$$\frac{71 + 81 + 68 + ?}{4} = 75$$

What seems like a good guess? You might try 72.

$$\frac{71 + 81 + 68 + 72}{4} = \frac{292}{4} = 73$$

This gives an average of 73. Not a bad guess, but you need something higher. You might try 75, but that gives an average of $73\frac{3}{4}$, still not enough, so you need something even higher. Using a guess of 80 results in the average of 75.

$$\frac{71 + 81 + 68 + 80}{4} = \frac{300}{4} = 75$$

Answer: 80

Working Backwards

Another useful strategy for problem solving is **working backwards.** In this method, you retrace the steps of the problem, starting from the end, and do the opposite of every arithmetic operation along the way.

Example 2. Joe is thinking of a number. If you multiply the number by 2, then add 14, then divide by 3, the answer is 12. What is Joe's number?

Solution: Work from the end. Since the answer of 12 was obtained by dividing by 3, *multiply* by 3.

$$\begin{array}{r} 12 \\ \times 3 \\ \hline 36 \end{array}$$

Since, in the step before that, 14 was added, *subtract* 14.

$$\begin{array}{r} 36 \\ -14 \\ \hline 22 \end{array}$$

Since, at the beginning, the number was multiplied by 2, *divide* by 2.

$$22 \div 2 = 11$$

Answer: Joe's number is 11.

Exercises

Use a calculator for these exercises.

In 1–10, find the missing number by the guess-and-check method or by working backwards.

1. $15 + 25 + \square = 60$

2. $340 - \square = 290$

3. $300 \div \square = 12$

4. $\square \times 5 = 210$

5. $122 + \square + 47 = 200$

6. $\dfrac{8 + \square}{2} = 10$

7. $\dfrac{75 + 85 + \square}{3} = 81$

8. $\square \times 12 \div 8 = 3$

9. $468 - 18 + \square = 510$

10. In the product $402 \times 3\square = \square5{,}\square7\square$, each of the digits 0 through 9 appears once and only once. Fill in the blanks to make the problem correct.

In 11–13, guess and insert the signs (+, −, ×, or ÷) that will make each problem correct. Check, using a calculator.

11. $1{,}258 \square 524 \square 234 = 500$

12. $69{,}104 \square 56 = 1{,}234$

13. $98 \square 76 = 7{,}448$

APPLICATION PROBLEMS

In 1–10, guess and check, or work backwards to find the answer.

1. Peggy bought a book bag for $6. She spent half the money she had left for a video cassette. After that purchase, she had $8 left. How much money did she start with? ______

2. Sam and Susie are twins. Sam has as many brothers as he has sisters. Susie has twice as many brothers as sisters. How many boys and how many girls are in the family? ______

3. In a 42-minute lunch period, Wayne was asleep for 10 more minutes than he was awake. How long was he asleep? ______

4. It is a 4-block walk from Amy's house to the J&M Supermarket where she works. One route is shown in the diagram. How many other 4-block routes can she take? ______

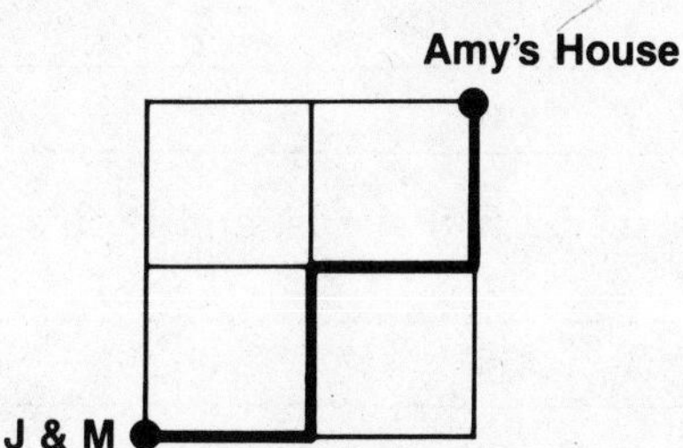

5. The teacher gave a number and told the class to multiply the number by 3, then divide by 2, then add 18. Don did the problem in his head and got the correct answer of 30, but he forgot what the original number was. What was the number? ______

6. The following table shows Dale's weight loss per month. How much did she lose in June so that her average weight loss was 3 pounds a month for the 6 months? ______

Jan	Feb	Mar	Apr	May	Jun
6	3	2	1	2	?

7. Jack has 22 more comic books than Jill does. Together, they have 60 comic books. How many does Jack have? ______

8. Grandpa gave Kenny a penny on Sunday, and promised to double the amount each day for a week, so that Kenny got 2 cents on Monday, 4 cents on Tuesday, and so on. How much money altogether will Kenny receive by the end of the 7 days? ______

9. Linda is saving up for a $200 ski trip during the February school vacation. At the beginning of the year, she had $42 in her account. The first week in January, she deposited $16, the second week $23, the third week $17, and the fourth week $28. How much more money does she need? ______

10. Artie lent half his money to Bob, and Bob used half the money he borrowed from Artie to pay a debt he owed Corey. Corey spent $4 of what Bob paid him, and had $3 left of what Bob paid. How much money did Artie start with? ______

Review of Part VI (Units 18–21)

In 1–5, tell how you would solve the problem. **Do not actually compute the answer.**

1. Al's new car had a base price of $6,449. He took the optional AM/FM radio at $165 and air conditioning at $432. What was the total price? ______________________________

2. A shipment of 3 dozen shirts cost the retailer $270. What was the cost per shirt? ______________________________

3. Sandy planted 3 rows of crocus bulbs, with 12 in a row, and 2 rows of tulip bulbs, with 8 in a row. How many bulbs were planted altogether? ______________________________

4. Anna's English composition contained 1,248 words, and Beth's had 972 words. How many words longer was Anna's composition? ______________________________

5. Michael was taking a bike trip, a distance of 385 miles, to visit a friend at college. The first day he went 145 miles, and the second day he went 156 miles. How much further did he have to go?
____

In 6–10, circle the best estimate, (A), (B), or (C).

6. Martha's commutation ticket costs $98.90 a month. At that rate, a year's commuting costs about:

(A) $1,200 (B) $1,000 (C) $500

7. Barbara bought some furniture at a neighborhood thrift shop. She got an oak table for $129, a rocking chair for $41, and a chest of drawers for $88. The three items cost about:

(A) $220 (B) $240 (C) $260

8. On one shift, Steve fried 158 chicken parts to fill 39 orders. The average number of chicken parts per order was about:

(A) 4 (B) 5 (C) 10

9. Carter's annual salary is $33,773, and Downey's is $33,624. About how much more does Carter make a year?

(A) $100 (B) $150 (C) $200

10. The Playtime Nursery School has an enrollment of 28 three-year-olds, 33 four-year-olds, and 37 five-year-olds. The total number of children is about:

(A) 80 (B) 90 (C) 100

In 11–13, circle the rounded expression, (A), (B), or (C), that gives the best estimate. Compute mentally, and write the estimated answer.

11. $132 - 27$

(A) $130 - 20$
(B) $140 - 30$
(C) $130 - 30$

12. 159×21

(A) 150×20
(B) 160×20
(C) 160×30

13. $47 + 53 + 29$

(A) $50 + 50 + 30$
(B) $40 + 50 + 20$
(C) $40 + 50 + 30$

In 14–15, circle the expression, (A), (B), or (C), having the reference numbers that are easiest to work with, compute mentally, and write the estimated answer.

14. $61 \div 9$
(A) $65 \div 9$
(B) $63 \div 9$
(C) $60 \div 9$

15. $493 \div 69$
(A) $500 \div 60$
(B) $500 \div 70$
(C) $490 \div 70$

In 16–18, guess and insert the signs (+, −, ×, or ÷) that will make each problem correct. Check, using a calculator.

16. 42 ☐ 27 ☐ 16 = 53

17. 14 ☐ 6 ☐ 7 = 12

18. 357 ☐ 443 = 800

In 19–21, use a calculator to compute the answer.

19. In a promotion by a car rental agency, Carole Gomez won 6,000 kilometers of free travel distance. She planned to travel cross country as far as possible without exceeding the 6,000-kilometer limit.

This was the route she took: New York to Atlanta, 1,218 km; Atlanta to New Orleans, 683 km; New Orleans to Dallas, 681 km; Dallas to Denver, 1,069 km; Denver to Phoenix, 948 km; Phoenix to San Diego, 489 km; San Diego to Los Angeles, 163 km; and Los Angeles to San Francisco, 571 km. How much less than the full 6,000 kilometers did she travel? ____________

20. Take a sheet of paper, and fold it in half. You now have 2 layers of paper. If you fold it in half a second time, you will have 4 layers. If you fold it in half a third time, you will have 8 layers. Each fold doubles the number of layers.

Use a calculator to compute the number of layers of paper you get with 20 folds. ____________

21. On a trip to Tokyo, tourist Tim Tyler exchanged $200 for 52,000 Japanese yen. He spent 1,456 yen for lunch; 6,890 yen for souvenirs; 5,980 yen for dinner; 14,820 yen for a hotel room; and 975 yen for breakfast. How much Japanese money did he have left? ____________

Cumulative Review of Parts I–VI

In 1–2, write each number as a word phrase.

1. 41,701 ______________________________

2. 6,020,108 ______________________________

In 3–4, write the value of each digit.

3. 34,500 = ____ten-thousands, ____thousands, ____hundreds, ____tens, ____ones

4. 3,007,005 = ____millions, ____hundred-thousands, ____ten-thousands, ____thousands, ____hundreds, ____tens, ____ones

In 5–6, write each word phrase as a number.

5. Seven thousand, twenty-eight ____________

6. Ten million, forty thousand, forty ____________

7. Round off 50,642 to the nearest *thousand.* ____________

8. Round off 14,990 to the nearest *ten-thousand.* ____________

9. Round off 19,462,325 to the nearest *million.* ____________

In 10–11, add and check.

10. 8,735 + 976 + 43 + 7

11. 563,296 + 798,978 + 58,787 + 9,438 + 496

12. Arrange the numbers in descending order and add. 863 + 48 + 24,876 + 4 + 363,586

In 13–14, add horizontally and vertically. Verify the accuracy by checking that the vertical and horizontal grand totals are the same.

13.
87 + 43 + 65 + 94 =
45 + 72 + 86 + 58 =
93 + 64 + 78 + 57 =
86 + 53 + 79 + 97 = ____
\+ \+ \+ =

14.
493 + 675 + 820 + 507 =
900 + 708 + 630 + 815 =
706 + 530 + 780 + 600 =
480 + 609 + 800 + 608 = ____
\+ \+ \+ =

In 15–17, subtract.

15. \$28,675 − 6,342

16. \$573,456 − 30,032

17. \$46,324 − 7,586

In 18–20, multiply.

18. 6,472 × 48

19. 23,579 × 685

20. 305,070 × 658

In 21–23, divide, and write any remainder in a *fraction.*

21. 13,851 ÷ 57

22. 122,275 ÷ 335

23. 168,900 ÷ 475

In 24–32, perform the indicated operations.

24. $\begin{array}{r}\$50,070\\ -6,575\\ \hline\end{array}$

25. $98\overline{)13,328}$

26. $\begin{array}{r}42,437\\ \times 674\\ \hline\end{array}$

27. $702,346 \div 234$

28. $\begin{array}{r}407,687\\ \times 4,090\\ \hline\end{array}$

29. $\begin{array}{r}\$556,865\\ +528,598\\ \hline\end{array}$

30. $230\overline{)691,000}$

31. $\begin{array}{r}\$20,403,060\\ -6,538,978\\ \hline\end{array}$

32. $\begin{array}{r}87,967\\ \times 8,007\\ \hline\end{array}$

33. Arrange the numbers in descending order and add. 18 + 29,634 + 7 + 3,586 + 378,596 + 563

34. Find the sum of 10,972 and 54,923.

35. Divide 9,283 by 421.

36. From 72,913, subtract 4,281.

37. Multiply 2,378 by 65.

38. Subtract 92 from 487.

39. Add 853 to 7,986.

40. Find the difference between 853 and 795.

41. How much larger is 12,980 than 9,892?

42. From the sum of $267,542 and $5,283, subtract $198,725.

43. Add 85 to the product of 285 and 23.

44. From the quotient of 14,910 divided by 42, subtract 250.

For 45–50, use a calculator to check whether the answer is correct. If the answer is wrong, correct it, and find which of the buttons, +, −, ×, or ÷, was pushed by mistake to get the wrong answer.

SAMPLE SOLUTIONS

a. 1,573 − 842 = 731

Correct

b. 347 + 24 = 8,328 *wrong*

The answer is 371.
The X button was pushed by mistake.

45. 573 × 215 = 123,195

46. 5,405 − 235 = 23

47. 52 + 9,085 = 9,137

48. 12,243 + 1,802 = 10,441

49. 1,188 ÷ 22 = 1,166

50. 2,357 × 51 = 2,408

For 51–54, guess which sign of operation, +, −, ×, or ÷, must be placed in each box, and use a calculator to check.

51. 240 ☐ 120 ☐ 82 = 442

52. 240 ☐ 120 ☐ 82 = 2,361,600

53. 240 ☐ 20 ☐ 2 = 6

54. 240 ☐ 120 ☐ 82 = 202

Use a calculator for 55 and 56. Write the numerical answer and the word that it makes upside down. The word should fit the clue.

55. 5,787 ÷ 3 × 4; fish part

56. 12,345 + 22,664; honk, honk

57. Perform the calculations without a calculator, and complete the cross-number puzzle. If any of your ACROSS and DOWN entries do not fit together, use the calculator to find errors.

ACROSS

1. 153 − 115
3. 17 × 4
5. 2 × 31 × 35
9. 1,855 ÷ 53
11. 14 + 14 + 14
12. 3 × 29
13. 10,388 ÷ 371
15. 15 × 82 ÷ 123
16. 5,029 − 4,984
17. 27 × 3
18. 1,043 − 987
19. 4,554 ÷ 99
21. 5 × 13
23. 3,771 + 1,234
25. 8,649 ÷ 93
26. 33 + 55

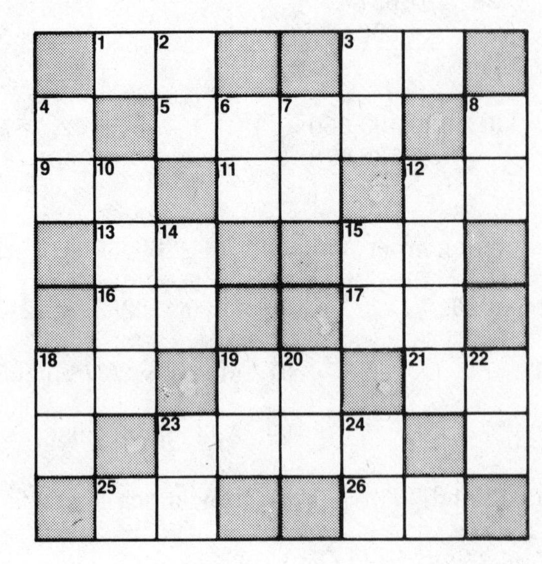

DOWN

2. 29 + 53
3. 7,680 ÷ 128
4. 648 − 555
6. 13,986 ÷ 999
7. 18 × 4
8. 19 × 3
10. 2,999 + 1,234 + 1,013
12. 167 × 48
14. 984 − 899
15. 3,996 ÷ 222
18. 17 + 17 + 17
19. 722 − 682
20. 4 × 15
22. 3,900 ÷ 75
23. 5,194 ÷ 7 ÷ 14
24. 29 + 29

In 58–63, tell how you would use the arithmetic operations to solve each problem. **Do not actually compute the answer.**

58. For a village fair, the village of Highland ordered 8 cases of soda at $12 a case, 14 dozen frankfurters at $5 a dozen, and 14 dozen frankfurter buns at $1 a dozen. What was the total cost of the order?

____

____

59. Gina bought a $98 coat on the layaway plan. She paid $14 down, and the rest in 12 equal weekly installments. How much did she pay each week?

____

____

60. The Stompin' Sounds hired a 1,200-seat hall for their concert. They sold $15 tickets for half the seats and $12 tickets for the rest of the seats. If all the tickets were sold, what were the total receipts?

____

____

61. Four students shared an off-campus apartment that rented for $700 a month. If they shared the expense equally, what was the *annual* cost to each student?

____

____

62. A mountain climbers' club climbed a 14,336-foot mountain in Colorado in three days. The first day they climbed 8,820 feet, the second day 3,571 feet, and the third day they reached the peak.

How many feet did they climb the third day? ____

____

63. A lecture hall had 18 rows, with 16 seats in each row. For a popular lecture, an additional 48 folding chairs were brought in. What was the total number of seats?

In 64–67, solve each problem.

64. Last year, a sales representative earned a salary of $16,567 together with commissions of $17,829. What were her total earnings?

65. A taxi company bought four taxis for $38,000. How much did each taxi cost?

66. A washing machine selling for $435 was reduced to $348.

a. Find the amount of the reduction.

b. If the washing machine will be paid for in 12 equal installments, how much will each payment be?

67. The sales office of a condominium development sold 5 apartments priced at $50,500 each. The price included a commission of $2,525 on each apartment. Find:

a. the total purchase price for the 5 apartments.

b. the total commissions on the 5 apartments.

c. the net amount after deducting the commissions from the purchase price.

In 68–72, the questions refer to the table, which shows the results of a four-month contest for best salesperson. Use a calculator to compute.

NAME	AMOUNT OF SALES				
	April	May	June	July	Total
Carla	$22,054	30,297	24,383	18,660	
Fran	28,352	19,820	25,047	20,928	
Jack	18,622	21,340	40,548	15,223	
Ed	20,268	39,223	16,820	19,985	
Marlene	27,456	26,580	31,383	24,221	
Total					

For 68–70, write your answers in the table.

68. Find the total of sales for each month.

69. Find the total of the sales for all four months.

70. Find the total of sales for each person.

71. Who is the winning salesperson?

72. Find the vertical grand total. (It should be the same as the horizontal grand total found in exercise 69.)

PART VII. Fractions

UNIT 22. Understanding Fractions

Words to Know

In everyday language, a *fraction* is a "part of something." We talk of "a fraction of an inch" or of a new track record that beat the old record by "a fraction of a second."

In mathematics, we call the inch or the second a ***unit.*** Units may be inches, seconds, pounds, cartons of eggs, or anything at all that we can count or measure.

We define a **fraction** as a "part of a unit."

Recall how we express the remainder of a division problem in a fraction:

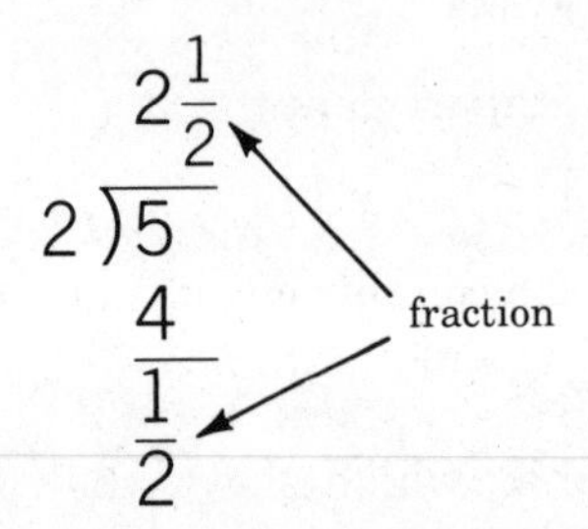

A fraction consists of two separate numbers written with the aid of a **fraction bar.** The top number is called the **numerator.** The bottom number is called the **denominator.**

numerator ⟶ 1
denominator ⟶ 2
⟵ fraction bar

Thus far, you have worked mostly with whole numbers. In your application problems, you have often used whole ***units*** such as miles, gallons, and television sets. Many of your answers "came out even." If you found that a dealer bought "$12\frac{1}{2}$ television sets," you realized that something was wrong.

However, parts of units are very common. You can buy $\frac{1}{2}$ of a gallon of milk, jog for $\frac{3}{4}$ of a mile, or pitch for $\frac{2}{3}$ of an inning. Parts of units are called **fractions.**

We write fractions as two numbers separated by a **fraction bar,** like this: $\frac{3}{4}$

The top number, the **numerator,** tells you *how many parts are represented.* In the fraction $\frac{3}{4}$, the numerator tells you that *three* quarters are represented.

The bottom number, the **denominator,** tells you into *how many equal parts* the original unit has been divided. In the fraction $\frac{3}{4}$, the denominator tells you that the unit has been divided into *four* equal parts, or into *fourths.*

If you jog for a distance of $\frac{3}{4}$ mile, the fraction $\frac{3}{4}$ tells you that the mile has been divided into *four* equal quarter-miles and that you jogged for *three* of these quarter-miles.

Suppose you have a pie and you cut it into two equal parts. You now have a whole pie that is made up of two half-pies. If you take one piece (one half-pie) from the two half-pies, you have one half-pie left:

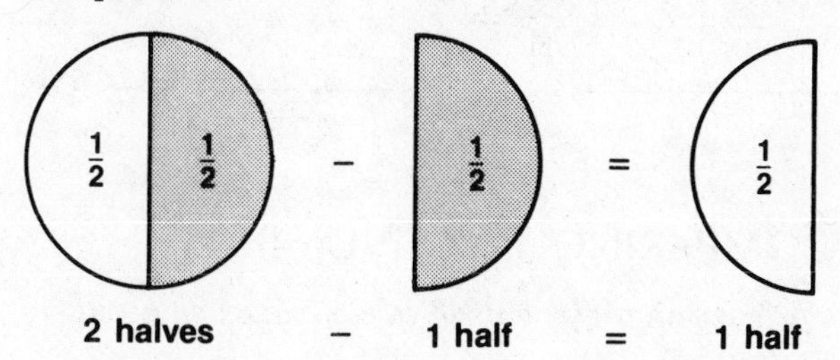

In fractions:

$$\frac{2}{2} - \frac{1}{2} = \frac{1}{2}$$

When you add the two half-pies together, you again have a whole pie made of two half-pies:

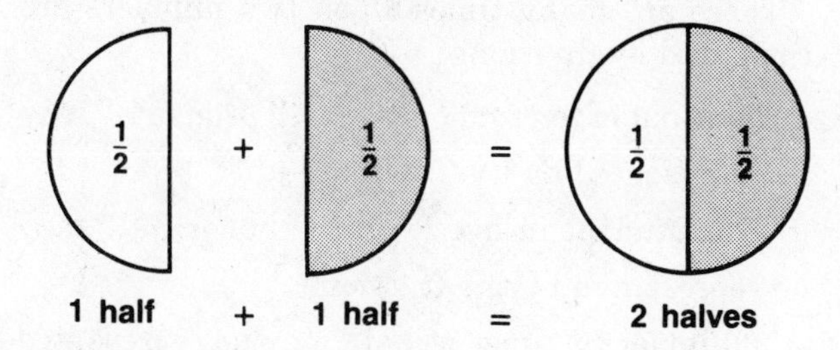

In fractions:

$$\frac{1}{2} + \frac{1}{2} = \frac{2}{2}$$

(Note that 2 half-pies equal a whole pie, or $\frac{2}{2} = 1$.)

Take another pie and cut it into three equal parts. If you take away one piece $\left(\frac{1}{3}\right.$ of the pie$\left.\right)$, you have $\frac{2}{3}$ of the pie left:

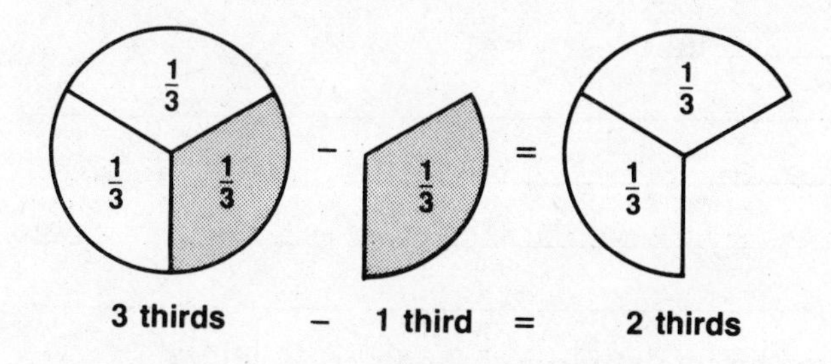

In fractions:

$$\frac{3}{3} - \frac{1}{3} = \frac{2}{3}$$

And, when you add the $\frac{1}{3}$ back to the $\frac{2}{3}$, you have $\frac{3}{3}$, or the whole pie:

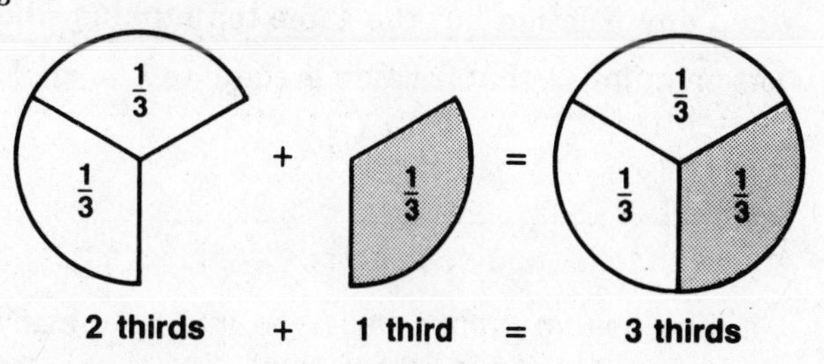

In fractions:

$$\frac{2}{3} + \frac{1}{3} = \frac{3}{3}$$

(Note that 3 thirds equal a whole pie, or $\frac{3}{3} = 1$.)

Cut another pie into 8 equal parts and take away 3 pieces. When you take $\frac{3}{8}$ from $\frac{8}{8}$, you have $\frac{5}{8}$ left:

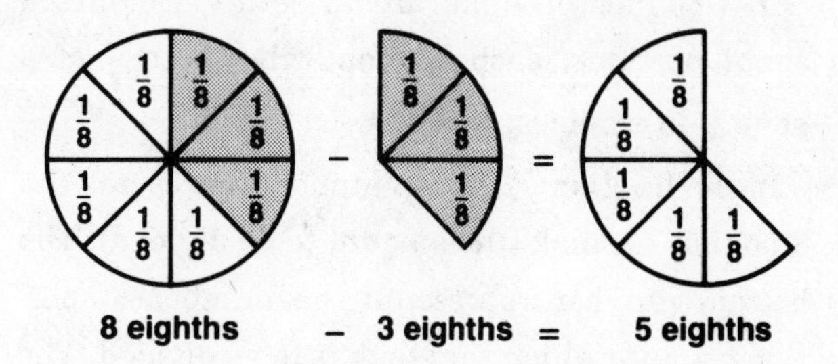

In fractions:

$$\frac{8}{8} - \frac{3}{8} = \frac{5}{8}$$

Finally, when you add the $\frac{3}{8}$ back to the $\frac{5}{8}$, you get $\frac{8}{8}$, the whole pie:

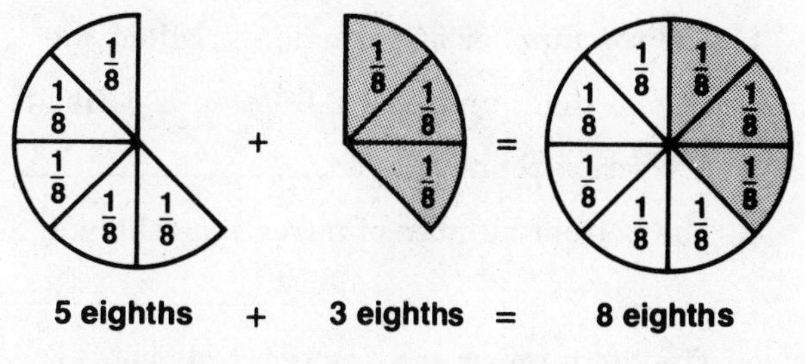

In fractions:

$$\frac{5}{8} + \frac{3}{8} = \frac{8}{8}$$

(Note that 8 eighths equal a whole pie, or $\frac{8}{8} = 1$.)

From the preceding examples, you can see that $\frac{2}{2} = 1$, that $\frac{3}{3} = 1$, and that $\frac{8}{8} = 1$. In general, when any fraction has the same top number and bottom number, that fraction is equal to 1 (a whole unit).

Remember

A fraction tells you two things:

1. The *bottom number* tells you into how many parts the unit has been divided.
2. The *top number* tells you how many of those parts are represented by the fraction.

Representing Division

When you studied how to divide, you used the symbols ÷ and $\overline{)\quad}$ to indicate division. In a fraction, *the fraction bar also represents the division of one number by another number.*

If 3 pounds of steak are to be divided into 4 equal portions, each portion will weigh $\frac{3}{4}$ of a pound. In symbols, $3 \div 4 = \frac{3}{4}$.

In the fraction $\frac{3}{4}$, the top number represents the 3 pounds of steak (the amount to be divided). The bottom number represents the number of portions, 4, into which the steak is to be divided. The fraction bar means "divided by," and the fraction $\frac{3}{4}$ is read, "3 divided by 4" or "three pounds divided into four portions."

Remember

The fraction bar and the division symbol ÷ both mean "divided by."

$$\frac{3}{4} = 3 \div 4$$

Comparing Two Numbers

A fraction may be used to *compare two numbers by division.*

If, on a test, you score 17 right answers out of 20 problems, you have answered correctly $\frac{17}{20}$ of the test. The fraction $\frac{17}{20}$ compares *the number of questions answered correctly with the total number of questions.*

There are many times when two numbers are compared by division:

"Nine out of ten sophomores will graduate" may be represented by the fraction $\frac{9}{10}$.

"One student in five is on the honor roll" may be represented by the fraction $\frac{1}{5}$.

"9 hits for 28 times at bat" may be represented by the fraction $\frac{9}{28}$.

Notice that the larger number is generally the bottom number. When comparing two numbers by division, it is usual to put the larger number (the "all" or "total" number) in the denominator.

Exercises

In 1–14, fill in the correct word or phrase that will best complete the statement or answer the question.

1. The top number of a fraction is called the ____________.
2. The bottom number of a fraction is called the ____________.
3. A fraction represents a ____________ unit.
4. The bottom number of a fraction tells you ____________.
5. The top number of a fraction tells you ____________.
6. If a whole unit is divided into five equal parts, the bottom number of the fraction will be ______.
7. If you took two of the parts in Exercise 6, you would write this fraction as ____________.
8. If the top number of a fraction is the same as the bottom number, that fraction is equal to ______.
9. If you divide a unit into eight equal parts, what will the bottom number of the fraction be? ______

10. If you divide a unit into *sevenths* and take $\frac{3}{7}$ away, how many sevenths will remain? ________

11. In Exercise 10, if you took $\frac{4}{7}$ more away, how many *sevenths* have you taken all together? ______

12. Write a fraction that represents a whole unit. (A whole unit has a value of 1.) ____________

13. In the fraction $\frac{5}{16}$, into how many parts has the unit been divided? ________________

How many parts were taken (from the whole unit)? ________________

How many parts of the whole unit remain? ________________

14. How many *sixteenths* have a value of 1? $\left(\frac{?}{16} = 1\right)$ ________________

In 15–20, represent the given facts as fractions.

Sample Solution

Seven out of ten doctors recommend "Paladin, the wonder drug." $\frac{7}{10}$

15. Tarkington completed 19 passes in 29 attempts. ____________

16. The National Safety Council predicted that 50,000,000 motorists would be on the road, and that there would be 400 traffic fatalities. ____________

17. The Lakers have won only 3 of their last 10 games. ____________

18. You have one chance in a million of winning the prize. ____________

19. Because of the influenza epidemic, only 20 pupils were present, while 15 pupils were absent. Compare the number of pupils present to the total class. (*Hint:* The total class consists of the sum of the number of pupils present and the number of pupils absent.) ____________

20. The betting was "5 to 4" in favor of the Chiefs. Express the Chiefs' winning chances as a fraction. (*Hint:* "5 to 4" means that, out of 9 chances, there are 5 chances of winning and 4 chances of losing. The Chiefs have 5 winning chances out of 9 chances.) ____________

APPLICATION PROBLEMS

1. Ten students are absent from a class of 35. Write as a fraction the number of students absent, compared to the total number of students. ____________

   Write as a fraction the number of students present, compared to the total number of students. ____________

2. Fred bought a dozen rolls and ate 3 of them. What fraction of the rolls did he eat? ____________

3. A bushel basket contains 150 pieces of fruit consisting of apples and pears. If there are 85 apples, what fraction of the bushel is apples? ____________

   What fraction of the bushel is pears? ____________

4. John bought a pound (1 pound = 16 ounces) of ham and used 4 ounces to make a sandwich. What fraction of the pound of ham *remained*? ____________

5. The Wilsons bought a house for \$60,000 and made a down payment of \$20,000. The down payment was what fraction of the price of the house? ________

6. Janice worked 50 hours last week. If 10 hours were overtime hours, what fraction of the total was overtime? ________

7. Mr. Wilson earns \$150 a week. He spends \$45 on food, deposits \$35 in his savings account, and deposits the rest of his money in his checking account. Write the above amounts as fractions by comparing each item with Mr. Wilson's weekly salary.
   Food: ________
   Savings: ________
   Checking: ________

8. The Bowlers won 6 games out of 10 games played. What fraction of the games did they win? ________

   What fraction of the games did they lose? ________

9. Crowley College does not hold classes during the months of June, July, August, and January. For what fraction of the year does the college not hold classes? ________

10. Trudy was not prepared for her science test, but guessed right on 22 questions. If there were 50 questions on the test, what fraction was right? ________

UNIT 23. Reducing Fractions to Lowest Terms

WORDS TO KNOW

The numerator and the denominator of a fraction are called the **terms** of the fraction. The fractions $\frac{2}{4}$ and $\frac{1}{2}$ have the same *value,* but they are written with different *terms.*

Expressing $\frac{2}{4}$ of a pound as $\frac{1}{2}$ of a pound is *reducing* the $\frac{2}{4}$ to its *lowest terms* of $\frac{1}{2}$.

In general, when you change fractions to simplest forms, you say that you **reduce fractions to lowest terms.**

When you cut a pizza into 8 equal pieces and eat 4 of them, you have eaten $\frac{4}{8}$ of the pizza. However, you have also eaten $\frac{1}{2}$ of the pizza because $\frac{4}{8} = \frac{1}{2}$. But it is easier to say you ate "half a pizza" than "four eighths of a pizza." When you change $\frac{4}{8}$ to $\frac{1}{2}$, you **reduce the fraction to its lowest terms.** Whenever you simplify a fraction (reduce it), you do not change the value of the fraction.

The following rule will enable you to reduce fractions:

RULE

You can divide the numerator and the denominator of a fraction by the *same number* without changing the value of that fraction.

See how using this rule enables you to reduce $\frac{4}{16}$ by dividing the numerator and the denominator by 4.

$$\frac{4}{16} = \frac{4 \div 4}{16 \div 4} = \frac{1}{4}$$

We say that when $\frac{4}{16}$ is written as $\frac{1}{4}$, it is *reduced to lowest terms.*

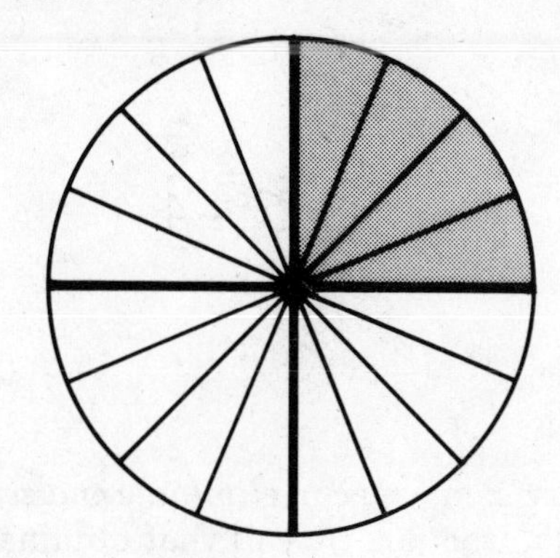

You can see from the diagram that $\frac{1}{4}$ is the same amount of pizza as $\frac{4}{16}$.

Example 1. Reduce $\frac{5}{15}$ to lowest terms.

Solution: You must find a number that will divide into both the top and bottom numbers *evenly,* without any remainder. Find this number by trial and error:

Will 2 go into 5 and 15? No.
Will 3 go into 5 and 15? No; into 15 only.
Will 4 go into 5 and 15? No.
Will 5 go into 5 and 15? Yes; $5 \div 5 = 1$ and $15 \div 5 = 3$.

Using the preceding rule: $\frac{5}{15} = \frac{5 \div 5}{15 \div 5} = \frac{1}{3}$

Answer: $\frac{5}{15} = \frac{1}{3}$

Alternate Solution: A popular method of reducing a fraction to lowest terms is to draw a vertical line to the left of the fraction to indicate division. To reduce $\frac{5}{15}$ to lowest terms, you would write $\left|\frac{5}{15}\right.$. The notation "$\left|\frac{5}{15}\right.$" means, "What number divides evenly into 5 and 15?"

Some people prefer to use this more detailed notation:

$$? \left|\frac{5}{15}\right.$$

When the divisor is found, by trial and error, write it to the left of the vertical line, as shown:

$$5\left|\frac{5}{15}\right.$$

Then indicate the two divisions with the following notation:

"5 into 5 goes 1" is indicated as $5\left|\frac{\overset{1}{\cancel{5}}}{15}\right.$

"5 into 15 goes 3" is indicated as $5\left|\frac{5}{\underset{3}{\cancel{15}}}\right.$

This notation permits you to write the solution very compactly. Here is how the entire solution of "Reduce $\frac{5}{15}$ to lowest terms" would appear:

Solution:

$$5\left|\frac{\overset{1}{\cancel{5}}}{\underset{3}{\cancel{15}}}\right. = \frac{1}{3}$$

Answer: $\frac{5}{15} = \frac{1}{3}$

Example 2. Reduce $\frac{9}{24}$ to lowest terms.

Solution:

$$3\left|\frac{\overset{3}{\cancel{9}}}{\underset{8}{\cancel{24}}}\right. = \frac{3}{8}$$

Answer: $\frac{9}{24} = \frac{3}{8}$

Whenever the top and bottom numbers of a fraction are even numbers (2, 4, 6, 8, 10, 12 . . .), you can always divide by 2 to reduce the fraction. However, you must keep dividing by 2 until 2 no longer divides into the top and bottom numbers.

Example 3. Reduce $\frac{16}{24}$ by dividing by 2.

Solution:

$$2\left|\frac{\overset{8}{\cancel{16}}}{\underset{12}{\cancel{24}}}\right. = \frac{8}{12} \quad 2\left|\frac{\overset{4}{\cancel{8}}}{\underset{6}{\cancel{12}}}\right. = \frac{4}{6} \quad 2\left|\frac{\overset{2}{\cancel{4}}}{\underset{3}{\cancel{6}}}\right. = \frac{2}{3}$$

When reducing a fraction by this method, your solution could be written in one step, as shown:

$$\begin{array}{r|l} & 2 \\ & \cancel{4} \\ & \cancel{8} \\ 2 & \dfrac{\cancel{16}}{\cancel{24}} = \dfrac{2}{3} \\ & \cancel{12} \\ & \cancel{6} \\ & 3 \end{array}$$

Answer: $\frac{16}{24} = \frac{2}{3}$

Note: If, in Example 3, you see that 8 divides evenly into both 16 and 24, you can reduce $\frac{16}{24}$ much more easily:

$$\begin{array}{r|l} & 2 \\ 8 & \dfrac{\cancel{16}}{\cancel{24}} = \dfrac{2}{3} \\ & 3 \end{array}$$

Be sure to look at each answer carefully to make certain your new fraction cannot be reduced still more.

EXAMPLE 4. Reduce $\frac{36}{48}$ to lowest terms.

Solution:

$$\begin{array}{r|l} & 9 \\ & \cancel{18} \\ 2 & \dfrac{\cancel{36}}{\cancel{48}} = \dfrac{9}{12} \\ & \cancel{24} \\ & 12 \end{array}$$

The fraction $\frac{9}{12}$ is not the answer, since $\frac{9}{12}$ can be reduced further.

$$\begin{array}{r|l} & 3 \\ 3 & \dfrac{\cancel{9}}{\cancel{12}} = \dfrac{3}{4} \\ & 4 \end{array}$$

Answer: $\frac{36}{48} = \frac{3}{4}$

As in division, you can *eliminate end zeros* when reducing a fraction. (Recall that eliminating end zeros is the same as dividing by powers of 10.)

EXAMPLE 5. What fraction of the distance from the earth to the moon is 20,000 miles? (Use 240,000 miles as the distance.)

Solution: Get rid of end zeros, one for one, in the numerator and in the denominator.

$$\frac{2\cancel{0},\cancel{000}}{24\cancel{0},\cancel{000}} = \frac{2}{24} \qquad \begin{array}{r|l} & 1 \\ 2 & \dfrac{\cancel{2}}{\cancel{24}} = \dfrac{1}{12} \\ & 12 \end{array}$$

Answer: $\frac{1}{12}$

Remember

To reduce a fraction to lowest terms, you must divide out every number that divides evenly into both the numerator and the denominator.

EXERCISES

In 1–36, reduce each fraction to lowest terms.

SAMPLE SOLUTIONS

a. $\begin{array}{r|l} & 1 \\ 3 & \dfrac{\cancel{3}}{\cancel{9}} = \dfrac{1}{3} \\ & 3 \end{array}$ **b.** $\begin{array}{r|l} & 2 \\ 4 & \dfrac{\cancel{8}}{\cancel{12}} = \dfrac{2}{3} \\ & 3 \end{array}$ **c.** $\begin{array}{r|l} & 1 \\ 6 & \dfrac{\cancel{6}\cancel{00}}{\cancel{1{,}8}\cancel{00}} = \dfrac{1}{3} \\ & 3 \end{array}$ **d.** $\begin{array}{r|l} & 18 \\ 2 & \dfrac{\cancel{36}}{\cancel{84}} \\ & 42 \end{array} = \begin{array}{r|l} & 3 \\ 6 & \dfrac{\cancel{18}}{\cancel{42}} = \dfrac{3}{7} \\ & 7 \end{array}$

1. $\frac{4}{8}$

2. $\frac{7}{14}$

3. $\frac{16}{32}$

4. $\frac{10}{30}$

5. $\frac{15}{20}$

6. $\frac{9}{36}$

7. $\frac{45}{60}$

8. $\frac{16}{24}$

9. $\frac{12}{16}$

10. $\frac{15}{60}$

11. $\frac{13}{52}$

12. $\frac{24}{32}$

13. $\frac{24}{64}$

14. $\frac{8}{18}$

15. $\frac{4}{14}$

16. $\frac{24}{38}$

17. $\frac{102}{106}$

18. $\frac{212}{214}$

19. $\frac{35}{45}$

20. $\frac{16}{60}$

21. $\frac{8}{48}$

22. $\frac{72}{84}$

23. $\frac{45}{54}$

24. $\frac{42}{70}$

25. $\frac{42}{63}$

26. $\frac{22}{24}$

27. $\frac{36}{54}$

28. $\frac{120}{360}$

29. $\frac{36}{144}$

30. $\frac{25}{150}$

31. $\frac{35}{75}$

32. $\frac{48}{110}$

33. $\frac{72}{90}$

34. $\frac{135}{645}$

35. $\frac{15{,}000}{60{,}000}$

36. $\frac{18{,}000}{24{,}600}$

APPLICATION PROBLEMS

SAMPLE SOLUTION

What fraction of a 24-hour day is 20 minutes?

1 hr. = 60 min.

$$\begin{array}{r} 24 \\ \times\ 60 \\ \hline 1{,}440 \end{array}$$ min. per day

$2\,|\,\frac{20}{1{,}440} = \frac{1}{72}$

$\frac{1}{72}$

In 1–10, reduce all answers to lowest terms.

1. Jane walked for 15 minutes. What fraction of an hour did she walk? (60 minutes = 1 hour) ________
2. Mary opened a box of 24 dishes and found that four of them were broken. What fraction was broken? ________
3. Debra needed 24 inches of material to add sleeves to a blouse. What fraction of a yard does she need? (36 inches = 1 yard) ________
4. John spent 6 hours at the beach. What fraction of a day did he spend at the beach? (24 hours = 1 day) ________
5. Sandy had $35 and spent $15. What fraction of her money did she spend? ________
6. A television set originally selling for $380 was reduced by $95. What fraction of the original price is the reduction? ________
7. Out of 864 graduating seniors, 576 were accepted to state colleges. What fraction of the senior class was accepted to state colleges? ________
8. The Ramirez family bought a home for $68,000 and made a down payment of $13,600. What fraction of the price is the down payment? ________
9. A dress shop had 240 dresses in stock. During a week-long special sale, it sold 168 dresses. What fraction of the dresses *remained*? ________
10. Ted bought a $325 camera at a sale price of $195. What fraction of the original price was the *amount that he saved*? ________

UNIT 24. Raising Fractions to Higher Equivalents

Words to Know

When fractions have the same value, we call them **equivalent fractions** or simply **equivalents.** Thus, $\frac{1}{4}$ and $\frac{4}{16}$ are equivalent fractions. The fraction with the larger numerator and denominator is called the **higher equivalent.**

When fractions are added or subtracted, it is often necessary to raise a given fraction to a **higher equivalent fraction.**

To raise a fraction to a higher equivalent fraction, use the following rule:

Rule

You can multiply the numerator and the denominator of a fraction by the same number without changing the value of that fraction.

The fraction $\frac{2}{4}$ is a higher equivalent of the fraction $\frac{1}{2}$ because $\frac{2}{4}$ has a larger numerator and denominator and the same value as $\frac{1}{2}$. You change $\frac{1}{2}$ to $\frac{2}{4}$ by multiplying the numerator and denominator by 2:

$$\frac{1}{2} = \frac{1 \times 2}{2 \times 2} = \frac{2}{4}$$

What you really do when you multiply the numerator and denominator each by 2, is multiply the fraction by $\frac{2}{2}$, which is equal to 1. Multiplying by a fraction equal to 1 changes the original fraction to an equivalent fraction without changing its value.

Now you will use a procedure, based on the rule, that will enable you to raise a fraction to a higher equivalent that has a given denominator.

Example 1. Raise $\frac{3}{4}$ to a higher equivalent that has a denominator of 16.

Solution: To raise $\frac{3}{4}$ to 16ths, you must find how many 16ths are contained in $\frac{3}{4}$. Set up the problem like this:

$$\frac{3}{4} = \frac{?}{16}$$

Step 1: To find how the 4 (the denominator of $\frac{3}{4}$) is changed to 16, divide the 4 into 16. The quotient is 4.

$$\frac{3}{4 \xrightarrow{\text{into}}} = \frac{?}{16} \xrightarrow{\text{is}} 4$$

Step 2: Multiply the numerator, 3, by the quotient you just found, 4. The product is 12. Write this product over the 16.

$$\frac{3 \times 4 \xrightarrow{\text{is}}}{4 \xrightarrow{\text{into}}} = \frac{12}{16} \xrightarrow{\text{is}} 4$$

Answer: $\frac{3}{4} = \frac{12}{16}$

Example 2. Raise $\frac{4}{5}$ to 20ths.

Solution: Set up the problem as before, and divide the denominator of $\frac{4}{5}$ into 20. The quotient is 4. Multiply the numerator, 4, by the quotient you just found, 4. The product is 16. Write this product over the 20.

$$\frac{4 \times 4 \xrightarrow{\text{is}}}{5 \xrightarrow{\text{into}}} = \frac{16}{20} \xrightarrow{\text{is}} 4$$

Answer: $\frac{4}{5} = \frac{16}{20}$

Remember

To raise a fraction to a higher equivalent that has a given denominator:

1. Divide the denominator of the fraction into the given denominator of the higher equivalent.
2. Multiply the numerator of the fraction by the quotient you found in step 1, and write this product over the given denominator.

Raising Fractions Mentally

Because the procedure for raising fractions to higher equivalents is used over and over again in the addition and subtraction of fractions, it is useful to be able to perform the operation **mentally.**

Example 3. Raise $\frac{2}{3}$, $\frac{3}{4}$, and $\frac{5}{6}$ to 12ths.

Solution: Set up each problem as before, writing 12 as the given denominator.

Think: "3 into 12 = 4, and 4 × 2 = 8." Write the 8 over the 12.

$$\frac{2}{3} \rightarrow \frac{}{12} \qquad \frac{2}{3} = \frac{8}{12}$$

Think: "4 into 12 = 3, and 3 × 3 = 9." Write the 9 over the 12.

$$\frac{3}{4} \rightarrow \frac{}{12} \qquad \frac{3}{4} = \frac{9}{12}$$

Think: "6 into 12 = 2, and 2 × 5 = 10." Write the 10 over the 12.

$$\frac{5}{6} \rightarrow \frac{}{12} \qquad \frac{5}{6} = \frac{10}{12}$$

Answer: $\frac{2}{3} = \frac{8}{12}, \frac{3}{4} = \frac{9}{12}, \frac{5}{6} = \frac{10}{12}$

EXERCISES

In 1–12, change each fraction to a higher equivalent that has the given denominator.

1. $\frac{2}{3} = \frac{}{12}$ **2.** $\frac{1}{4} = \frac{}{12}$ **3.** $\frac{2}{5} = \frac{}{20}$ **4.** $\frac{4}{7} = \frac{}{14}$

5. $\frac{3}{5} = \frac{}{20}$ **6.** $\frac{7}{16} = \frac{}{32}$ **7.** $\frac{1}{3} = \frac{}{9}$ **8.** $\frac{4}{5} = \frac{}{20}$

9. $\frac{3}{4} = \frac{}{12}$ **10.** $\frac{5}{24} = \frac{}{48}$ **11.** $\frac{7}{12} = \frac{}{36}$ **12.** $\frac{5}{8} = \frac{}{16}$

In 13–24, change:

13. $\frac{5}{8}$ to 32nds **14.** $\frac{7}{15}$ to 30ths **15.** $\frac{5}{24}$ to 48ths

16. $\frac{8}{17}$ to 34ths **17.** $\frac{23}{32}$ to 64ths **18.** $\frac{17}{28}$ to 56ths

19. $\frac{23}{33}$ to 66ths

20. $\frac{7}{10}$ to 50ths

21. $\frac{9}{15}$ to 45ths

22. $\frac{5}{9}$ to 27ths

23. $\frac{8}{11}$ to 44ths

24. $\frac{19}{24}$ to 72nds

APPLICATION PROBLEMS

SAMPLE SOLUTION

Sal's pizza was cut into 8 slices. If he ate $\frac{3}{4}$ of the pizza, how many slices did he eat?

$\frac{3}{4} = \frac{?}{8}$ $\frac{3}{4} = \frac{6}{8}$

6 slices

1. How many 32nds of an inch are there in $\frac{3}{4}$ of an inch? ________
2. Change $\frac{1}{4}$ of a pound to ounces (16ths). ________
3. Mrs. Smith bought $\frac{3}{4}$ of a yard of velvet. How many inches (36ths) did she buy? ________
4. How many inches (12ths) are there in half of a foot? ________
5. How many sheets of paper are there in $\frac{3}{4}$ of a ream? (A ream contains 500 sheets of paper.) ________
6. In a class of 36 students, $\frac{2}{3}$ are girls. How many of the students are girls? ________
7. A store is having a "$\frac{1}{4}$ off" sale. What is the sale price of a jacket originally selling for $48? (*Hint:* If the store takes $\frac{1}{4}$ off the original price, then the buyer pays $\frac{3}{4}$ of the price.) ________
8. If 4 out of 5 people approve of the President's foreign policy, how many of the 250,000 surveyed do *not* approve? ________
9. A fast-food restaurant uses 4 ounces of ground beef to make a hamburger. How many pounds of ground beef will be needed to make 500 hamburgers? (*Hint:* 4 ounces is what fraction of a pound?) ________
10. A retailer sells a dishwasher for $450. If the dishwasher cost him $\frac{3}{5}$ of the selling price, how much did the retailer pay for the dishwasher? ________

UNIT 25. Comparing Fractions

Words to Know

When you are working with fractions, it is often necessary for all the fractions in a problem to have the same denominator. When fractions have the same denominator, it is called a **common denominator.**

You can compare two or more fractions to determine which fraction has the largest value.

To compare the values of two or more fractions:

Step 1: Change the given fractions to equivalent fractions that have a *common denominator.*

Step 2: Compare the *numerators.* The fraction that has the largest numerator has the largest value.

Example 1. Which fraction is larger, $\frac{1}{2}$ or $\frac{3}{8}$?

Solution:

Step 1: $\frac{1}{2} = \frac{4}{8}$ and $\frac{3}{8} = \frac{3}{8}$.

Step 2: Since 4 is larger than 3, then $\frac{4}{8}$ is larger than $\frac{3}{8}$.

Answer: $\frac{1}{2}$ is larger than $\frac{3}{8}$.

Example 2. Which fraction is smaller, $\frac{3}{4}$ or $\frac{4}{5}$?

Solution: Find a number that can serve as a common denominator for the two given fractions.

To find a common denominator, multiply the largest given denominator by 2, 3, 4, 5, 6, etc., until you get a number that the given denominators divide into evenly without any remainder.

$5 \times 2 = 10$. 10 cannot be used.

$5 \times 3 = 15$. 15 cannot be used.

$5 \times 4 = 20$. 20 *can* be used because the given denominators, 4 and 5, divide evenly into 20.

Step 1: $\frac{3}{4} = \frac{15}{20}$ and $\frac{4}{5} = \frac{16}{20}$.

Step 2: $\frac{15}{20}$ is smaller than $\frac{16}{20}$.

Answer: $\frac{3}{4}$ is smaller than $\frac{4}{5}$.

Example 3. Arrange the following fractions in descending order of value (largest first):

$$\frac{1}{6} \quad \frac{2}{3} \quad \frac{3}{5}$$

Solution: Find a common denominator for the three given fractions.

Using the same procedure as in Example 2, you find that $6 \times 5 = 30$ can be used as a common denominator because the three given denominators, 3, 5, and 6, all divide evenly into 30.

Step 1: $\frac{1}{6} = \frac{5}{30} \quad \frac{2}{3} = \frac{20}{30} \quad \frac{3}{5} = \frac{18}{30}$

Step 2: Arrange the equivalent fractions in descending order of value according to their numerators.

$$\frac{2}{3} = \frac{20}{30} \quad \frac{3}{5} = \frac{18}{30} \quad \frac{1}{6} = \frac{5}{30}$$

Answer: $\frac{2}{3} \quad \frac{3}{5} \quad \frac{1}{6}$

EXERCISES

In 1–8, circle the *larger* fraction.

Sample Solutions

a. $\frac{1}{2}$ ($\frac{5}{8}$) $\frac{1}{2} = \frac{4}{8}$, smaller than $\frac{5}{8}$

b. $\frac{9}{16}$ ($\frac{5}{8}$) $\frac{5}{8} = \frac{10}{16}$, larger than $\frac{9}{16}$

1. $\frac{4}{5}$ $\frac{13}{15}$

2. $\frac{5}{7}$ $\frac{9}{14}$

3. $\frac{3}{5}$ $\frac{5}{10}$

4. $\frac{5}{9}$ $\frac{13}{18}$

5. $\frac{8}{10}$ $\frac{15}{20}$

6. $\frac{9}{12}$ $\frac{15}{24}$

7. $\frac{2}{3}$ $\frac{11}{15}$

8. $\frac{3}{5}$ $\frac{11}{20}$

In 9–16, circle the *smaller* fraction.

Sample Solutions

c. $\frac{3}{4}$ (circled) $\frac{13}{16}$ $\frac{3}{4} = \frac{12}{16}$, smaller than $\frac{13}{16}$

d. $\frac{5}{8}$ (circled) $\frac{3}{4}$ $\frac{3}{4} = \frac{6}{8}$, larger than $\frac{5}{8}$

9. $\frac{3}{5}$ $\frac{7}{15}$

10. $\frac{7}{9}$ $\frac{15}{18}$

11. $\frac{9}{12}$ $\frac{17}{24}$

12. $\frac{2}{6}$ $\frac{2}{12}$

13. $\frac{3}{7}$ $\frac{8}{21}$

14. $\frac{7}{13}$ $\frac{15}{26}$

15. $\frac{3}{8}$ $\frac{8}{24}$

16. $\frac{5}{6}$ $\frac{13}{18}$

In 17–24, arrange each set of fractions in descending order of values (largest first).

Sample Solution

$\frac{3}{4}$ $\frac{5}{6}$ $\frac{7}{12}$ $\frac{3}{4} = \frac{9}{12}$ $\frac{5}{6} = \frac{10}{12}$ $\frac{7}{12} = \frac{7}{12}$

The descending order is $\frac{10}{12}$ $\frac{9}{12}$ $\frac{7}{12}$, or $\frac{5}{6}$ $\frac{3}{4}$ $\frac{7}{12}$.

17. $\frac{7}{9}$ $\frac{17}{18}$ $\frac{5}{6}$

18. $\frac{5}{8}$ $\frac{3}{4}$ $\frac{11}{16}$

19. $\frac{2}{3}$ $\frac{3}{4}$ $\frac{5}{6}$

20. $\frac{3}{5}$ $\frac{8}{15}$ $\frac{7}{10}$

21. $\frac{13}{14}$ $\frac{5}{7}$ $\frac{23}{28}$

22. $\frac{5}{8}$ $\frac{2}{3}$ $\frac{5}{6}$

23. $\frac{7}{10}$ $\frac{19}{25}$ $\frac{4}{5}$

24. $\frac{2}{3}$ $\frac{3}{5}$ $\frac{7}{9}$

APPLICATION PROBLEMS

SAMPLE SOLUTIONS

a. In the final week of the baseball season, catcher Michael Mahoney had 9 hits for 28 times at bat and pitcher Larry Schultz had 2 hits for 7 times at bat. Who had the better batting average?

Mahoney: 9 for 28 = $\frac{9}{28}$

Schultz: 2 for 7 = $\frac{2}{7} = \frac{8}{28}$

$\frac{9}{28}$ is larger than $\frac{8}{28}$.

Mahoney

b. Baby Jane weighed $6\frac{13}{16}$ pounds at birth, and Baby Joan weighed $6\frac{7}{8}$ pounds. Which baby weighed more?

Since the whole number is 6 in each case, just compare the fractions.

$\frac{7}{8} = \frac{14}{16}$ is larger than $\frac{13}{16}$.

Baby Joan

1. John lives $\frac{7}{8}$ of a mile from school and James lives $\frac{13}{16}$ of a mile from school. Who lives closer to the school? __________

2. Two brands of coffee are the same price. If brand A contains $\frac{6}{7}$ of a pound and brand B contains $\frac{11}{14}$ of a pound, which brand is the better buy? __________

3. One piece of pipe measures $8\frac{3}{8}$ inches in diameter and another piece of pipe measures $8\frac{15}{32}$ inches in diameter. Which pipe has the larger diameter? __________

4. One savings bank offers $5\frac{3}{4}\%$ interest and another bank offers $5\frac{5}{8}\%$ interest. Which rate of interest is higher? __________

5. Maria ran a race in $4\frac{7}{10}$ minutes and Ann ran the same race in $4\frac{3}{5}$ minutes. Who won the race? __________

6. Which is a better buy, $\frac{2}{3}$ of a yard of ribbon or 25 inches of the same ribbon, if both sell for the same price? (*Hint:* What fraction of a yard is 25 inches?) __________

7. Which is more, $\frac{3}{4}$ of a pound or 11 ounces? __________

8. Which is a shorter period of time, $\frac{3}{4}$ of a year or 10 months? __________

9. Three pistons have the following measurements: $2\frac{5}{16}$ inches, $2\frac{11}{32}$ inches, and $2\frac{9}{64}$ inches. Which piston is the largest? ______________

10. A share of stock sold for $\$23\frac{7}{8}$. The next day it sold for $\$23\frac{3}{4}$. Did the price of the stock go up or down? ______________

UNIT 26. Improper Fractions and Mixed Numbers

Words to Know

We have defined a fraction as a part of a unit. However, there are times when a fraction means *more* than a unit. When you buy 6 quarts of milk, for example, you have $\frac{6}{4}$ of a gallon. But $\frac{6}{4}$ of a gallon is the same as $1\frac{1}{2}$ gallons (1 gallon = 4 quarts). We call $\frac{6}{4}$ an **improper fraction.** An improper fraction is a fraction whose numerator is *as large as or larger than* its denominator.

We call $1\frac{1}{2}$ a **mixed number,** and define a mixed number as a number made up of whole units plus a fraction of a unit.

A fraction that is a part of a unit (whose numerator is smaller than its denominator) is called a **proper fraction.**

Since fractions represent parts of whole units, it is said to be "improper" to write a fraction that contains more than one whole unit. Suppose you want to watch a television drama that lasts for 90 minutes, and you want to know what time the show will be over. Before figuring out the problem, you change the time from $\frac{90}{60}$ of an hour to $1\frac{1}{2}$ hours. What you have done is to change an *improper fraction* to a *mixed number.*

A fraction whose numerator is as large as or larger than its denominator is called an **improper fraction.** Examples of improper fractions are $\frac{6}{4}$, $\frac{5}{4}$, and $\frac{4}{4}$.

A number that contains both whole units and a fraction of a unit is called a **mixed number.** Examples of mixed numbers are $3\frac{2}{3}$, $2\frac{1}{4}$, and $1\frac{1}{8}$.

When writing the answer to any calculation, we usually change any improper fractions to mixed numbers or to whole numbers. To change an improper fraction to a mixed number or to a whole number, carry out the indicated division by *dividing the denominator into the numerator.*

A good way to remember which part of the fraction goes inside the division box is to draw the division box next to the bottom number of the fraction and let the top number "fall into" the division box.

Example 1. Change $\frac{12}{3}$ to a whole number.

Solution:

Step 1: Draw the division box next to the denominator and let the numerator "fall in."

$$\frac{12}{3}\overline{)12}$$

Step 2: Do the division.

$$\begin{array}{r} 4 \\ 3\overline{)12} \\ \underline{12} \end{array}$$

Answer: $\frac{12}{3} = 4$

EXAMPLE 2. Change $\frac{19}{4}$ to a mixed number.

Solution:

$$\frac{19}{4} \rightarrow 4\overline{)19} \qquad 4\overline{)19}^{\,4\frac{3}{4}}$$

$$\begin{array}{r} 16 \\ \hline 3 \end{array} \rightarrow \frac{3}{4}$$

Answer: $\frac{19}{4} = 4\frac{3}{4}$

There are times when it is necessary to change a mixed number to an improper fraction.

EXAMPLE 3. Change $5\frac{3}{8}$ to an improper fraction.

Solution: To do this problem, you are really finding how many eighths are contained in $5\frac{3}{8}$. Set up the problem by rewriting the same denominator:

$$5\frac{3}{8} = \frac{}{8}$$

Step 1: To find the new numerator, *multiply the denominator in the fraction part of the mixed number by the whole number.* The denominator of $\frac{3}{8}$ is 8. Therefore, $8 \times 5 = 40$.

$$8 \times 5 = 40 \quad 5\frac{3}{8} = \frac{}{8}$$

Step 2: Now, *add the numerator of the fraction part of the mixed number* to the product you just got: $40 + 3 = 43$. Write this new numerator over the same denominator.

$$8 \times 5 = 40 \quad 5\frac{3}{8} = \frac{43}{8} \quad 40 + 3 = 43$$

Answer: $5\frac{3}{8} = \frac{43}{8}$

With practice, you can find the new numerator mentally. In the above problem, you would think: "$8 \times 5 = 40$, and $40 + 3 = 43$."

$$5\frac{3}{8} = \frac{43}{8}$$

EXAMPLE 4. Change $6\frac{2}{3}$ to an improper fraction.

Solution: Set up the problem:

$$6\frac{2}{3} = \frac{}{3}$$

Think: "$3 \times 6 = 18$, and $18 + 2 = 20$." Write the 20 over the same denominator.

$$6\frac{2}{3} = \frac{}{3} \qquad 6\frac{2}{3} = \frac{20}{3}$$

Answer: $6\frac{2}{3} = \frac{20}{3}$

Remember

1. To change an improper fraction to a mixed number: Divide the denominator into the numerator.
2. To change a mixed number to an improper fraction: Multiply the denominator by the whole number and add the numerator. This gives you the new numerator, which you write over the same denominator.

EXERCISES

In 1–10, change each improper fraction to a mixed number or to a whole number.

SAMPLE SOLUTIONS

a. $\frac{17}{5}$ $\quad 5\overline{)17}^{\,3\frac{2}{5}}$, 15, $\frac{2}{5}$

b. $\frac{12}{4}$ $\quad 4\overline{)12}^{\,3}$, 12

1. $\frac{11}{8}$ 2. $\frac{15}{3}$ 3. $\frac{16}{15}$ 4. $\frac{17}{8}$ 5. $\frac{18}{3}$

6. $\frac{23}{6}$ 7. $\frac{17}{7}$ 8. $\frac{23}{8}$ 9. $\frac{32}{6}$ 10. $\frac{15}{4}$

In 11–20, change each mixed number to an improper fraction.

Sample Solutions

a. $3\frac{2}{3} = \frac{11}{3}$
$3 \times 3 = 9$, and $9 + 2 = 11$.

b. $1\frac{5}{8} = \frac{13}{8}$
$8 \times 1 = 8$, and $8 + 5 = 13$.

11. $5\frac{5}{6}$ 12. $12\frac{2}{3}$ 13. $6\frac{3}{4}$ 14. $4\frac{3}{7}$ 15. $15\frac{1}{2}$

16. $7\frac{4}{5}$ 17. $6\frac{5}{16}$ 18. $3\frac{9}{12}$ 19. $8\frac{1}{3}$ 20. $10\frac{3}{4}$

APPLICATION PROBLEMS

Sample Solution

A film version of *War and Peace* runs for 390 minutes. Using the fact that 1 minute $= \frac{1}{60}$ of an hour, change the running time to hours.

$390 \text{ min.} = \frac{390}{60} = \frac{39}{6}$ hr.

$\frac{39}{6} = 6\overline{)39}$; $6\frac{1}{2}$; 36; 3; $\frac{3}{6}$

$6\frac{1}{2}$ hr.

1. Mr. Wilson gave out 12 quarters to his grandchildren. How much money in dollars did he give out?
(*Hint:* One quarter $= \frac{1}{4}$ of a dollar.) ________

2. Cathy wants 50 slices of pizza for a club meeting. If a slice is $\frac{1}{8}$ of a pizza, how many pizzas are needed? ________

3. A pail holds 12 quarts of water. What is the capacity in gallons? (*Hint:* 1 quart $= \frac{1}{4}$ gallon.) ________

4. Julio lived in Trenton 42 months. How many years was that? ________

5. A living room rug is 14 feet long by 10 feet wide. How many yards long is the rug? (*Hint:* 1 foot $= \frac{1}{3}$ yard.) ________

   How many yards wide is the rug? ________

SAMPLE SOLUTION

How many $\frac{1}{4}$ cups of flour can you get from a bag that holds $6\frac{3}{4}$ cups?

$6\frac{3}{4} = \frac{27}{4}$

27 $\frac{1}{4}$ cups

6. How many $\frac{1}{8}$-inch-thick slices of cheese can be cut from a piece of cheese that measures $4\frac{3}{8}$ inches? ________

7. How many $\frac{1}{2}$ teaspoons of honey can be poured from a jar that holds $12\frac{1}{2}$ teaspoons? ________

8. Small nails are packed in $\frac{1}{4}$-pound boxes. How many $\frac{1}{4}$-pound boxes can be packed from $15\frac{3}{4}$ pounds of nails? ________

9. How many $\frac{1}{2}$ dozen eggs are there in $6\frac{1}{2}$ dozen eggs? ________

10. A box contains 15 packages of batteries. If each package holds 3 batteries, how many dozen batteries are there in the box? (*Hint:* What fraction of a dozen is 3?) ________

Review of Part VII (Units 22–26)

In 1–12, change each fraction to a higher equivalent that has the given denominator.

1. $\frac{4}{5} = \frac{\ \ }{20}$
2. $\frac{3}{7} = \frac{\ \ }{28}$
3. $\frac{3}{5} = \frac{\ \ }{30}$
4. $\frac{5}{8} = \frac{\ \ }{40}$
5. $\frac{5}{6} = \frac{\ \ }{42}$
6. $\frac{9}{15} = \frac{\ \ }{45}$
7. $\frac{7}{12} = \frac{\ \ }{48}$
8. $\frac{11}{16} = \frac{\ \ }{48}$

9. $\frac{5}{24} = \frac{\quad}{72}$ 10. $\frac{7}{8} = \frac{\quad}{64}$ 11. $\frac{4}{7} = \frac{\quad}{35}$ 12. $\frac{5}{9} = \frac{\quad}{54}$

In 13–24, reduce each fraction to lowest terms.

13. $\frac{5}{15}$ 14. $\frac{14}{28}$ 15. $\frac{12}{60}$ 16. $\frac{24}{48}$

17. $\frac{30}{40}$ 18. $\frac{16}{18}$ 19. $\frac{20}{34}$ 20. $\frac{25}{75}$

21. $\frac{35}{45}$ 22. $\frac{18}{27}$ 23. $\frac{21}{49}$ 24. $\frac{33}{99}$

In 25–32, circle the larger fraction.

25. $\frac{3}{7}$ $\frac{7}{21}$ 26. $\frac{4}{5}$ $\frac{17}{20}$ 27. $\frac{3}{4}$ $\frac{10}{16}$ 28. $\frac{21}{28}$ $\frac{5}{7}$

29. $\frac{3}{4}$ $\frac{4}{5}$ (Use 20ths.) 30. $\frac{5}{6}$ $\frac{7}{8}$ (Use 48ths.) 31. $\frac{2}{3}$ $\frac{3}{5}$ (Use 15ths.) 32. $\frac{3}{8}$ $\frac{4}{7}$ (Use 56ths.)

In 33–40, change each improper fraction to a whole number or to a mixed number.

33. $\frac{23}{5}$ 34. $\frac{15}{3}$ 35. $\frac{19}{4}$ 36. $\frac{43}{8}$

37. $\frac{57}{12}$ 38. $\frac{48}{14}$ 39. $\frac{37}{3}$ 40. $\frac{65}{16}$

In 41–48, change each mixed number to an improper fraction.

41. $6\frac{3}{5}$ 42. $8\frac{2}{3}$ 43. $12\frac{1}{4}$ 44. $15\frac{3}{8}$

45. $13\frac{4}{5}$ 46. $11\frac{5}{8}$ 47. $7\frac{5}{16}$ 48. $14\frac{5}{6}$

PART VIII. Adding and Subtracting Fractions

UNIT 27. Addition of Fractions That Have the Same Denominator

When two or more fractions are added, the sum is another fraction that is larger than any of the addends. Whenever the sum is an improper fraction, change it to a mixed number or a whole number.

The easiest fractions to add are those with the same denominator. To add fractions that have the same denominator, *add the numerators and rewrite the same denominator.*

EXAMPLE 1. Add: $\frac{3}{16} + \frac{5}{16} + \frac{7}{16}$

Solution: To add these fractions, you are really finding how many 16ths there are in all the fractions to be added.

Step 1: Add the numerators: $3 + 5 + 7 = 15$

Step 2: Rewrite the same denominator with this sum over it.

Answer: $\frac{3}{16} + \frac{5}{16} + \frac{7}{16} = \frac{15}{16}$

EXAMPLE 2. Add: $\frac{5}{8} + \frac{7}{8} + \frac{3}{8} + \frac{5}{8}$

Solution:

Step 1: Add the numerators:
$5 + 7 + 3 + 5 = 20$

Step 2: Rewrite the same denominator, 8, with this sum over it.

$$\frac{20}{8}$$

Step 3: Since the answer is an improper fraction, change it to a mixed number.

$$\frac{20}{8} \qquad 8\overline{)20} = 2\frac{1}{2}, \quad 20 - 16 = 4, \quad \frac{4}{8} = \frac{1}{2}$$

Answer: $\frac{5}{8} + \frac{7}{8} + \frac{3}{8} + \frac{5}{8} = 2\frac{1}{2}$

Remember

When adding fractions that have the same denominator, add only the top numbers.

EXERCISES

In 1–16, add. Change improper fractions to mixed numbers or to whole numbers. Reduce answers to lowest terms.

SAMPLE SOLUTIONS

a. $\frac{3}{5} + \frac{2}{5} + \frac{4}{5} = \frac{9}{5} = 1\frac{4}{5}$

$\frac{9}{5} \qquad 5\overline{)9} = 1\frac{4}{5}, \quad 9 - 5 = 4, \quad \frac{4}{5}$

b. $\frac{1}{4} + \frac{2}{4} + \frac{3}{4} + \frac{2}{4} = \frac{8}{4} = \frac{2}{1} = 2$

1. $\frac{3}{7} + \frac{5}{7} + \frac{2}{7} + \frac{6}{7}$

2. $\frac{5}{8} + \frac{3}{8} + \frac{1}{8} + \frac{7}{8}$

3. $\frac{3}{6} + \frac{5}{6} + \frac{1}{6} + \frac{5}{6}$

4. $\frac{3}{9} + \frac{7}{9} + \frac{5}{9} + \frac{4}{9}$

5. $\frac{5}{12} + \frac{7}{12} + \frac{9}{12} + \frac{13}{12}$

6. $\frac{4}{11} + \frac{5}{11} + \frac{8}{11} + \frac{10}{11}$

7. $\frac{13}{24} + \frac{7}{24} + \frac{9}{24} + \frac{21}{24}$

8. $\frac{8}{15} + \frac{12}{15} + \frac{7}{15} + \frac{14}{15}$

9. $\frac{8}{18} + \frac{11}{18} + \frac{15}{18} + \frac{10}{18}$

10. $\frac{13}{20} + \frac{17}{20} + \frac{11}{20} + \frac{15}{20}$

11. $\frac{18}{25} + \frac{19}{25} + \frac{13}{25} + \frac{21}{25}$

12. $\frac{21}{30} + \frac{15}{30} + \frac{18}{30} + \frac{25}{30}$

13. $\frac{25}{32} + \frac{19}{32} + \frac{27}{32} + \frac{30}{32}$

14. $\frac{15}{36} + \frac{28}{36} + \frac{25}{36} + \frac{34}{36}$

15. $\frac{27}{35} + \frac{32}{35} + \frac{30}{35} + \frac{19}{35}$

16. $\frac{35}{40} + \frac{28}{40} + \frac{21}{40} + \frac{36}{40}$

APPLICATION PROBLEMS

SAMPLE SOLUTION

In Pete's Pizza Parlor, each whole pizza is cut into 8 equal slices. Find how many pizzas Pete needs to fill the following orders: 3 slices, 5 slices, 2 slices, 1 slice, 1 slice.

1 slice is $\frac{1}{8}$ of a pizza, so 3 slices = $\frac{3}{8}$, 5 slices = $\frac{5}{8}$, 2 slices = $\frac{2}{8}$, 1 slice = $\frac{1}{8}$, 1 slice = $\frac{1}{8}$

$$\frac{3}{8} + \frac{5}{8} + \frac{2}{8} + \frac{1}{8} + \frac{1}{8} = \frac{\overset{3}{\cancel{12}}}{\underset{2}{\cancel{8}}} = \frac{3}{2}$$

$1\frac{1}{2}$ pizzas

1. Mary baked a cake and added $\frac{1}{2}$ cup sugar, $\frac{1}{2}$ cup shortening, $\frac{1}{2}$ cup chocolate, and $\frac{1}{2}$ cup milk. How many cups did she add? __________

2. A plumber cut off the following pieces of pipe: $\frac{3}{8}$ inch, $\frac{7}{8}$ inch, $\frac{5}{8}$ inch, and $\frac{7}{8}$ inch. How many inches of pipe did he cut off? __________

3. Tom needs the following pieces of lumber to finish a bookcase: $\frac{3}{4}$ foot, $\frac{1}{4}$ foot, $\frac{2}{4}$ foot. How many feet of lumber does he need? __________

4. Jane is making a dress and needs the following pieces of material: $\frac{1}{5}$ yard for the collar, $\frac{4}{5}$ yard for the front, $\frac{4}{5}$ yard for the back, and $\frac{3}{5}$ yard for the sleeves. How many yards of material does she need? __________

5. A clerk sold the following weights of cheese: $\frac{5}{16}$ pound, $\frac{7}{16}$ pound, $\frac{13}{16}$ pound, and $\frac{9}{16}$ pound. What was the total weight sold? __________

6. Arthur worked overtime the following numbers of minutes: 45, 30, 15, 45, 30. How many overtime hours did he work? (*Hint:* 1 minute is $\frac{1}{60}$ of an hour.) __________

7. Mrs. Barnes bought the following provisions for a party: 15 ounces of ham, 13 ounces of cheese, 9 ounces of salami, 15 ounces of bologna, and 22 ounces of potato salad. How many pounds did she buy? (*Hint:* 1 ounce is what fraction of a pound?) __________

8. A clerk sold the following lengths of fabric: 28 inches, 34 inches, 19 inches, 23 inches, and 26 inches. How many yards did the clerk sell? (*Hint:* 1 inch = $\frac{1}{36}$ yard.) __________

9. A party mix contains 11 ounces of hazelnuts, 15 ounces of cashew nuts, 13 ounces of raisins, and 15 ounces of peanuts. How many pounds does the mix weigh? __________

10. A bakery used the following numbers of eggs for various cake mixes: 9, 10, 7, 8, and 8. How many dozen eggs were used? (*Hint:* 1 egg is what fraction of a dozen?) __________

UNIT 28. Addition of Fractions That Have Unlike Denominators

In order to add a group of fractions that have unlike denominators, you must make the fractions "alike." That is, you must find a **common denominator** and change each given fraction to an equivalent fraction that has this same denominator.

The largest denominator in the problem can often be used as the common denominator.

Example 1. Add: $\frac{3}{4} + \frac{5}{8} + \frac{5}{16}$

Solution:

Step 1: Arrange the given fractions, one under the other, as shown. To the right, draw the fraction bars for the equivalent fractions and for the sum.

$$\frac{3}{4} = \frac{\quad}{\quad}$$
$$\frac{5}{8} = \frac{\quad}{\quad}$$
$$\frac{5}{16} = \frac{\quad}{\quad}$$
$$\overline{\qquad\qquad}$$
$$\frac{\quad}{\quad}$$

Step 2: Notice that the denominator of $\frac{5}{16}$ is the largest denominator in the problem. Therefore, you will raise $\frac{3}{4}$ and $\frac{5}{8}$ to equivalent fractions that have a denominator of 16. You can do this because the denominator of $\frac{3}{4}$ divides evenly into 16 (without a remainder) and the denominator of $\frac{5}{8}$ divides evenly into 16. Write 16 as the common denominator for all the fractions, including the fraction that will be your answer.

$$\frac{3}{4} = \frac{\quad}{16}$$
$$\frac{5}{8} = \frac{\quad}{16}$$
$$\frac{5}{16} = \frac{\quad}{16}$$
$$\overline{\qquad\qquad}$$
$$\frac{\quad}{16}$$

Step 3: Divide each given denominator into the common denominator and then multiply the quotient by each numerator. The resulting product is the new numerator of each equivalent fraction.

$$\frac{3 \times 4 \rightarrow 12}{4 \longrightarrow 16 \rightarrow 4}$$

Think: "4 into 16 = 4, and 3 × 4 = 12." Write the 12.

$$\frac{5 \times 2 \rightarrow 10}{8 \longrightarrow 16 \rightarrow 2}$$

Think: "8 into 16 = 2, and 5 × 2 = 10." Write the 10.

$$\frac{5 \times 1 \rightarrow 5}{16 \longrightarrow 16 \rightarrow 1}$$

Think: "16 into 16 = 1, and 5 × 1 = 5." Write the 5.

$$\overline{\qquad\qquad}$$
$$\frac{\quad}{16}$$

Step 4: Add the top numbers of the equivalent fractions and put this sum over the common denominator, 16. Your answer is $\frac{27}{16}$, an improper fraction.

$$\begin{aligned} \frac{3}{4} &= \frac{12}{16} \\ \frac{5}{8} &= \frac{10}{16} \\ \frac{5}{16} &= \frac{5}{16} \\ \hline & \frac{27}{16} \end{aligned}$$

Step 5: Change $\frac{27}{16}$ to a mixed number.

$$\frac{27}{16} = 1\frac{11}{16}$$

Answer: $\frac{3}{4} + \frac{5}{8} + \frac{5}{16} = 1\frac{11}{16}$

EXAMPLE 2. Add: $\frac{2}{3} + \frac{1}{4} + \frac{5}{12}$

Solution: Raise the addends to 12ths mentally.

$$\begin{aligned} \frac{2}{3} &\rightarrow \frac{8}{12} \\ \frac{1}{4} &\rightarrow \frac{3}{12} \\ \frac{5}{12} &\rightarrow \frac{5}{12} \\ \hline & \frac{16}{12} = 1\frac{4}{12} = 1\frac{1}{3} \end{aligned}$$

Answer: $\frac{2}{3} + \frac{1}{4} + \frac{5}{12} = 1\frac{1}{3}$

At times, the largest denominator in the problem cannot be used as a common denominator because one or more of the given denominators will not divide evenly into it. When this happens, you multiply the largest denominator by 2, 3, 4, 5, 6, etc., until you get a number that can be used as a common denominator.

EXAMPLE 3. Add: $\frac{3}{4} + \frac{1}{2} + \frac{3}{5}$

Solution: Note that neither 4 nor 2 will divide evenly into 5. Thus, 5 cannot be used as the common denominator. Multiply the 5 to get trial common denominators:

$5 \times 2 = 10$, which cannot be used.
$5 \times 3 = 15$, which cannot be used.
$5 \times 4 = 20$. Use 20 as the common denominator, because all the given denominators (4, 2, and 5) divide evenly into 20.

Set up the problem as in examples 1 and 2 and solve:

$$\begin{aligned} \frac{3}{4} &= \frac{15}{20} \\ \frac{1}{2} &= \frac{10}{20} \\ \frac{3}{5} &= \frac{12}{20} \\ \hline & \frac{37}{20} = 1\frac{17}{20} \end{aligned}$$

Answer: $\frac{3}{4} + \frac{1}{2} + \frac{3}{5} = 1\frac{17}{20}$

EXERCISES

In 1–16, add. Change improper fractions to mixed numbers or to whole numbers. Reduce answers to lowest terms.

SAMPLE SOLUTIONS

a.
$$\begin{aligned} \frac{3}{4} &= \frac{6}{8} \\ \frac{5}{8} &= \frac{5}{8} \\ \hline & \frac{11}{8} = 1\frac{3}{8} \end{aligned}$$

$$8\overline{)11} = 1\frac{3}{8}; \quad 11 - 8 = 3$$

b.
$$\begin{aligned} \frac{1}{2} &= \frac{6}{12} \\ \frac{2}{3} &= \frac{8}{12} \\ \frac{3}{4} &= \frac{9}{12} \\ \hline & \frac{23}{12} = 1\frac{11}{12} \end{aligned}$$

$$12\overline{)23} = 1\frac{11}{12}; \quad 23 - 12 = 11$$

1. $\frac{5}{8}$ $\frac{9}{16}$ <u>   </u>

2. $\frac{2}{3}$ $\frac{5}{6}$ <u>   </u>

3. $\frac{3}{4}$ $\frac{11}{16}$ <u>   </u>

4. $\frac{5}{7}$ $\frac{19}{21}$ <u>   </u>

5. $\frac{5}{6}$ $\frac{15}{18}$ <u>   </u>

6. $\frac{2}{5}$ $\frac{3}{15}$ <u>   </u>

7. $\frac{2}{3}$ $\frac{3}{9}$ <u>   </u>

8. $\frac{3}{7}$ $\frac{9}{28}$ <u>   </u>

9. $\frac{4}{5}$ $\frac{7}{15}$ $\frac{13}{20}$ <u>   </u>

10. $\frac{4}{5}$ $\frac{7}{10}$ $\frac{13}{20}$ <u>   </u>

11. $\frac{5}{8}$ $\frac{7}{16}$ $\frac{25}{32}$ <u>   </u>

12. $\frac{5}{8}$ $\frac{3}{5}$ $\frac{9}{10}$ <u>   </u>

13. $\frac{2}{3}$ $\frac{5}{8}$ $\frac{7}{12}$ <u>   </u>

14. $\frac{3}{4}$ $\frac{7}{8}$ $\frac{4}{5}$ <u>   </u>

15. $\frac{5}{8}$ $\frac{3}{4}$ $\frac{9}{12}$ <u>   </u>

16. $\frac{3}{4}$ $\frac{5}{16}$ $\frac{7}{12}$ <u>   </u>

APPLICATION PROBLEMS

SAMPLE SOLUTION

Mrs. Brown needed $\frac{3}{4}$ cup of flour, $\frac{2}{3}$ cup of milk, and $\frac{1}{6}$ cup of butter for a mixture she made to bread pork chops. How many cups did the mixture contain?

$$\frac{3}{4} = \frac{9}{12}$$
$$\frac{2}{3} = \frac{8}{12}$$
$$\frac{1}{6} = \frac{2}{12}$$
$$\frac{19}{12}$$

$$12 \overline{)19} = 1\frac{7}{12}$$
$$\frac{12}{7}, \quad \frac{7}{12}$$

$1\frac{7}{12}$ cups

1. Helen bought $\frac{1}{2}$ pound of candy, $\frac{3}{4}$ pound of nuts, and $\frac{5}{8}$ pound of raisins. How many pounds did her purchases weigh altogether? __________

2. John walked $\frac{3}{4}$ mile to school, $\frac{1}{3}$ mile to town, and $\frac{1}{2}$ mile to a friend's house. How many miles did he walk? __________

3. Mrs. Brown needs the following pieces of material to make a blouse: $\frac{2}{7}$ yard for the front and back, $\frac{1}{3}$ yard for the sleeves, and $\frac{1}{6}$ yard for the collar. How much material is needed for the blouse? __________

4. The following pieces were cut from a pipe: $\frac{1}{5}$ foot, $\frac{2}{3}$ foot, $\frac{1}{6}$ foot. How many feet of pipe were cut? __________

5. Henry spent $\frac{1}{2}$ hour on his arithmetic homework, $\frac{1}{3}$ hour on history, and $\frac{1}{6}$ hour on English. How long did it take him to do all of his homework? __________

6. Cynthia bought $\frac{3}{4}$ kilogram of apples, $\frac{2}{3}$ kilogram of pears, and $\frac{5}{6}$ kilogram of cherries. How many kilograms of fruit did she buy? __________

7. The sides of a playground measure $\frac{2}{3}$ mile, $\frac{5}{8}$ mile, $\frac{1}{2}$ mile, and $\frac{3}{4}$ mile. What is the distance in miles around the playground? __________

8. A retailer ordered the following numbers of blouses: $\frac{3}{4}$ dozen blue, $\frac{5}{6}$ dozen red, and $\frac{1}{3}$ dozen beige. How many dozen blouses were ordered? ____________

9. A carpenter needed a $\frac{3}{4}$-meter board, a $\frac{5}{8}$-meter board, and a $\frac{9}{12}$-meter board. How many meters of board did he need? ____________

10. Ms. Scott used $\frac{3}{4}$ cup flour, $\frac{1}{3}$ cup oil, and $\frac{1}{5}$ cup vinegar, as called for in a recipe. How many cups of ingredients did she use? ____________

UNIT 29. Addition of Mixed Numbers

Since a mixed number consists of a whole number and a fraction, you add mixed numbers in separate steps.

To add mixed numbers:

Step 1: Add the whole numbers.

Step 2: Add the fractions.

Step 3: Combine the two answers into a final sum.

EXAMPLE 1. Add: $12\frac{3}{4} + 15\frac{7}{8} + 14\frac{1}{2}$

Solution: Add the whole numbers first. Then add the fractions.

Step 1:

$$\begin{array}{r|l} 12 & \frac{3}{4} \\ 15 & \frac{7}{8} \\ 14 & \frac{1}{2} \\ \hline 41 & \end{array}$$

Step 2:

$$\begin{array}{r|l} 12 & \frac{3}{4} = \frac{6}{8} \\ 15 & \frac{7}{8} = \frac{7}{8} \\ 14 & \frac{1}{2} = \frac{4}{8} \\ \hline 41 & \quad \frac{17}{8} = 2\frac{1}{8} \end{array}$$

Step 3: Combine the two answers:

$$41 + 2\frac{1}{8} = 43\frac{1}{8}$$

Answer: $12\frac{3}{4} + 15\frac{7}{8} + 14\frac{1}{2} = 43\frac{1}{8}$

EXAMPLE 2. Add: $2\frac{1}{3} + 5\frac{1}{2} + 7\frac{1}{4}$

Solution:

$$\begin{array}{r|l} 2 & \frac{1}{3} = \frac{4}{12} \\ 5 & \frac{1}{2} = \frac{6}{12} \\ 7 & \frac{1}{4} = \frac{3}{12} \\ \hline 14 & \quad \frac{13}{12} = 1\frac{1}{12} \\ +1\frac{1}{12} & \\ \hline 15\frac{1}{12} & \end{array}$$

Answer: $2\frac{1}{3} + 5\frac{1}{2} + 7\frac{1}{4} = 15\frac{1}{12}$

EXERCISES

In 1–16, add. Change improper fractions to mixed numbers or to whole numbers. Reduce answers to lowest terms.

Sample Solution

$$\begin{array}{r|l} 5 & \frac{2}{3} = \frac{12}{18} \\ 7 & \frac{5}{6} = \frac{15}{18} \\ 4 & \frac{7}{9} = \frac{14}{18} \\ \hline 16 & \frac{41}{18} = 2\frac{5}{18} \end{array}$$

$$\frac{41}{18} \quad 18\overline{)41} = 2\frac{5}{18}, \quad 41 - 36 = 5, \quad \frac{5}{18}$$

$$16 + 2\frac{5}{18} = 18\frac{5}{18}$$

1. $8\frac{3}{4}$, $11\frac{3}{8}$, $9\frac{1}{2}$

2. $13\frac{3}{4}$, $7\frac{4}{5}$, $15\frac{1}{2}$

3. $15\frac{2}{3}$, $12\frac{1}{4}$, $14\frac{5}{6}$

4. $23\frac{9}{16}$, $15\frac{5}{8}$, $24\frac{30}{32}$

5. $19\frac{4}{5}$, $10\frac{15}{20}$, $7\frac{3}{4}$

6. $25\frac{5}{7}$, $23\frac{10}{14}$, $47\frac{19}{28}$

7. $17\frac{1}{3}$, $15\frac{5}{6}$, $18\frac{5}{8}$

8. $12\frac{4}{5}$, $9\frac{12}{15}$, $11\frac{5}{6}$

9. $18\frac{2}{3}$, $21\frac{4}{5}$, $25\frac{7}{10}$

10. $14\frac{11}{16}$, $9\frac{3}{4}$, $15\frac{25}{64}$

11. $23\frac{3}{4}$, $15\frac{5}{8}$, $19\frac{4}{5}$

12. $24\frac{5}{7}$, $19\frac{2}{3}$, $25\frac{11}{14}$

13. $25\frac{5}{14}$, 23, $21\frac{5}{7}$, $26\frac{3}{4}$

14. $12\frac{2}{3}$, $15\frac{5}{6}$, $14\frac{7}{9}$, 16

15. 28, $24\frac{9}{10}$, $23\frac{3}{4}$, $20\frac{4}{5}$

16. $18\frac{13}{14}$, $23\frac{5}{7}$, 21, $25\frac{1}{4}$

APPLICATION PROBLEMS

1. A plumber needs 3 lengths of pipe measuring $6\frac{3}{16}$ inches, $8\frac{5}{8}$ inches, and $4\frac{1}{2}$ inches. What length of pipe does she need to cut the 3 pieces? ________

2. A salesperson sold the following pieces of velvet: $3\frac{1}{3}$ yards, $5\frac{3}{4}$ yards, $4\frac{1}{2}$ yards. How many yards did she sell? ________

3. Mary bought $2\frac{1}{2}$ pounds of apples, $1\frac{3}{4}$ pounds of cherries, and $2\frac{3}{8}$ pounds of pears. How many pounds of fruit did she buy? ________

4. A carpenter needs 3 pieces of board measuring $24\frac{3}{4}$ inches, $15\frac{7}{8}$ inches, and $13\frac{15}{16}$ inches. How many inches of board does he need? ________

5. Jane is making a 2-piece dress and needs $2\frac{1}{5}$ yards for one part and $3\frac{2}{3}$ yards for the other part. How much material does she need? ________

6. Mindy worked the following overtime hours: $2\frac{1}{4}$, $1\frac{2}{3}$, and $2\frac{5}{6}$. How many hours of overtime did she work? ________

7. Mrs. Wright bought three packages of meat weighing $3\frac{2}{3}$ kilograms, $4\frac{3}{4}$ kilograms, and $3\frac{5}{6}$ kilograms. What was the total weight of the three packages of meat? ________

8. Joseph worked the following hours last week: $8\frac{3}{4}$, 9, $9\frac{1}{2}$, 8, and $8\frac{2}{3}$. How many hours did Joseph work last week? ________

9. A baker used the following numbers of eggs for various mixes: $3\frac{3}{4}$ dozen, $4\frac{1}{3}$ dozen, 5 dozen, and $3\frac{5}{6}$ dozen. How many dozen eggs did he use? ________

10. In one day, a salesperson traveled the following distances to visit customers: $23\frac{4}{7}$ miles, 15 miles, $24\frac{2}{3}$ miles, and $22\frac{3}{4}$ miles. How many miles did he travel? ________

UNIT 30. Subtraction of Fractions

Subtracting fractions is very similar to adding fractions. Arrange the fractions, one under the other, find the common denominator, and raise to equivalent fractions. Then you *subtract the numerators.* (Recall that in addition you *added* the numerators.) Write this answer over the common denominator to get your final difference.

EXAMPLE 1. Subtract: $\frac{3}{4} - \frac{2}{3}$

Solution: Set up the problem as for the addition of fractions, find the common denominator, and raise to equivalent fractions. (Be sure to write the common denominator in the space where your answer will be.)

$$\begin{array}{r} \frac{3}{4} = \frac{9}{12} \\ -\frac{2}{3} = \frac{8}{12} \\ \hline \frac{}{12} \end{array}$$

Now *subtract* the numerators.

$$\begin{array}{r} \frac{3}{4} = \frac{9}{12} \\ -\frac{2}{3} = \frac{8}{12} \\ \hline \frac{1}{12} \end{array}$$

Answer: $\frac{3}{4} - \frac{2}{3} = \frac{1}{12}$

EXAMPLE 2. Subtract: $\frac{7}{8} - \frac{5}{16}$

Solution:

$$\begin{array}{r} \frac{7}{8} = \frac{14}{16} \\ -\frac{5}{16} = \frac{5}{16} \\ \hline \frac{9}{16} \end{array}$$

Answer: $\frac{7}{8} - \frac{5}{16} = \frac{9}{16}$

EXERCISES

In 1–24, subtract. Reduce answers to lowest terms.

SAMPLE SOLUTIONS

a.
$$\begin{array}{r} \frac{2}{3} = \frac{4}{6} \\ \frac{1}{6} = \frac{1}{6} \\ \hline \frac{3}{6} = \frac{1}{2} \end{array}$$

b.
$$\begin{array}{r} \frac{3}{4} = \frac{15}{20} \\ \frac{3}{5} = \frac{12}{20} \\ \hline \frac{3}{20} \end{array}$$

Trial denominator:
5×2=10 (no)
5×3=15 (no)
5×4=20 (yes)

1. $\begin{array}{r} \frac{7}{8} \\ \frac{3}{4} \\ \hline \end{array}$

2. $\begin{array}{r} \frac{1}{2} \\ \frac{1}{3} \\ \hline \end{array}$

3. $\begin{array}{r} \frac{2}{3} \\ \frac{3}{5} \\ \hline \end{array}$

4. $\begin{array}{r} \frac{5}{6} \\ \frac{3}{8} \\ \hline \end{array}$

5. $\begin{array}{r} \frac{5}{8} \\ \frac{7}{12} \\ \hline \end{array}$

6. $\begin{array}{r} \frac{4}{5} \\ \frac{3}{4} \\ \hline \end{array}$

7. $\begin{array}{r} \frac{2}{3} \\ \frac{5}{16} \\ \hline \end{array}$

8. $\begin{array}{r} \frac{3}{5} \\ \frac{5}{12} \\ \hline \end{array}$

9. $\begin{array}{r} \frac{2}{3} \\ \frac{3}{7} \\ \hline \end{array}$

10. $\begin{array}{r} \frac{4}{5} \\ \frac{4}{6} \\ \hline \end{array}$

11. $\begin{array}{r} \frac{1}{4} \\ \frac{1}{5} \\ \hline \end{array}$

12. $\begin{array}{r} \frac{2}{3} \\ \frac{5}{9} \\ \hline \end{array}$

13. $\begin{array}{r} \frac{5}{6} \\ \frac{5}{8} \\ \hline \end{array}$

14. $\begin{array}{r} \frac{3}{4} \\ \frac{2}{3} \\ \hline \end{array}$

15. $\begin{array}{r} \frac{5}{7} \\ \frac{1}{3} \\ \hline \end{array}$

16. $\begin{array}{r} \frac{5}{8} \\ \frac{30}{64} \\ \hline \end{array}$

17. $\begin{array}{r} \frac{1}{3} \\ \frac{1}{5} \\ \hline \end{array}$

18. $\begin{array}{r} \frac{7}{9} \\ \frac{1}{4} \\ \hline \end{array}$

19. $\begin{array}{r} \frac{3}{5} \\ \frac{2}{6} \\ \hline \end{array}$

20. $\begin{array}{r} \frac{7}{8} \\ \frac{2}{3} \\ \hline \end{array}$

21. $\frac{7}{8} - \frac{1}{3}$

22. $\frac{3}{4} - \frac{9}{16}$

23. $\frac{4}{5} - \frac{9}{15}$

24. $\frac{9}{11} - \frac{2}{3}$

APPLICATION PROBLEMS

1. The Wilsons bought $\frac{3}{4}$ ton of coal and used up $\frac{1}{2}$ ton. How much coal was left? __________

2. Jane bought $\frac{7}{8}$ of a yard of material to make a pair of shorts. If she only used $\frac{3}{4}$ of a yard, how much material was left? __________

3. Mary had $\frac{3}{4}$ pound of sugar and used $\frac{9}{16}$ pound. How much sugar was left? __________

4. John lives $\frac{9}{10}$ of a mile from school and Tom lives $\frac{4}{5}$ of a mile from school. How much closer does Tom live to the school? __________

5. Carol used $\frac{3}{4}$ cup of milk and $\frac{1}{2}$ cup of sugar to bake a cake. How much more milk than sugar did she use? __________

6. Samantha bought $\frac{3}{4}$ pound of ham and ate $\frac{5}{8}$ pound for lunch. How much of the ham was left? __________

7. William is $\frac{7}{8}$ meter tall and Lucy is $\frac{2}{3}$ meter tall. How much taller is William? __________

8. Mrs. Santos had $\frac{3}{4}$ gallon of cooking oil, and used $\frac{1}{8}$, $\frac{3}{16}$, and $\frac{5}{16}$ gallon. How much cooking oil was left? __________

9. Sara had $\frac{3}{4}$ yard of fabric. If she used $\frac{1}{6}$ yard, $\frac{5}{12}$ yard, and $\frac{1}{12}$ yard, how much fabric was left? __________

10. A bag contained $\frac{7}{8}$ pound of sugar. If $\frac{5}{16}$, $\frac{1}{8}$, and $\frac{1}{4}$ pound were used, how much of the sugar remained? __________

UNIT 31. Subtraction of Mixed Numbers

Subtracting mixed numbers is very similar to adding mixed numbers. Because subtraction often involves *borrowing,* however, you must *subtract the fractions first.* Then you subtract the whole numbers.

EXAMPLE 1. Subtract: $7\frac{5}{8} - 5\frac{7}{16}$

Solution: Set up the mixed numbers, one under the other, and find the common denominator. Subtract the equivalent fractions first. Then subtract the whole numbers and combine these two results into the final answer.

$$\begin{array}{r|l} 7 & \frac{5}{8} = \frac{10}{16} \\ -5 & \frac{7}{16} = \frac{7}{16} \\ \hline 2 & \phantom{\frac{7}{16} =} \frac{3}{16} \end{array}$$

Answer: $7\frac{5}{8} - 5\frac{7}{16} = 2\frac{3}{16}$

EXAMPLE 2. Subtract: $15\frac{1}{5} - 8\frac{3}{4}$

Solution:

$$\begin{array}{r|l} 15 & \frac{1}{5} = \frac{4}{20} \\ -8 & \frac{3}{4} = \frac{15}{20} \\ \hline & \phantom{\frac{3}{4} =} \frac{}{20} \end{array}$$

Notice that the numerator of $\frac{15}{20}$ is *larger* than the numerator of $\frac{4}{20}$. Since you cannot take away 15 from 4, *borrow* 1 from 15, and change 15 to 14.

Place the borrowed 1 in front of $\frac{4}{20}$, getting the mixed number $1\frac{4}{20}$.

$$\begin{array}{r|l} \overset{14}{\cancel{15}} & \frac{1}{5} = 1\frac{4}{20} \\ -8 & \frac{3}{4} = \frac{15}{20} \\ \hline & \phantom{\frac{3}{4} = 1} \frac{}{20} \end{array}$$

Change $1\frac{4}{20}$ to an improper fraction:

$$1\frac{4}{20} = \frac{24}{20}$$

Think: "20 × 1 = 20, and 20 + 4 = 24."

Replace $1\frac{4}{20}$ with $\frac{24}{20}$ in the problem and subtract $\frac{15}{20}$ from $\frac{24}{20}$. Then subtract the whole numbers and combine the two results into the final answer.

$$\begin{array}{r|l} \overset{14}{\cancel{15}} & \frac{1}{5} = 1\frac{4}{20} \rightarrow \frac{24}{20} \\ -8 & \frac{3}{4} = \frac{15}{20} \rightarrow \frac{15}{20} \\ \hline 6 & \phantom{\frac{3}{4} =} \frac{}{20} \rightarrow \frac{9}{20} = 6\frac{9}{20} \end{array}$$

Answer: $15\frac{1}{5} - 8\frac{3}{4} = 6\frac{9}{20}$

Alternate Solution: Another way of borrowing from the whole number is to change the borrowed 1 to an equivalent fraction that has the common denominator. Let us use this method to subtract $8\frac{3}{4}$ from $15\frac{1}{5}$.

Start the problem as in the previous method.

$$\begin{array}{r|l} 15 & \frac{1}{5} = \frac{4}{20} \\ -8 & \frac{3}{4} = \frac{15}{20} \\ \hline & \quad\ \ \frac{\ }{20} \end{array}$$

As before, borrow 1 from the 15. However, raise the borrowed 1 to an equivalent fraction that has a denominator of 20, namely, $1 = \frac{20}{20}$. (Recall that the value of any fraction that has the same number in the numerator and the denominator is 1.) Add the borrowed $\frac{20}{20}$ to $\frac{4}{20}$ and subtract as before.

$$\begin{array}{r|l} \overset{14}{\cancel{15}} & \frac{1}{5} = \frac{4}{20} + \frac{20}{20} \longrightarrow \frac{24}{20} \\ -8 & \frac{3}{4} = \frac{15}{20} \qquad\ \longrightarrow \frac{15}{20} \\ \hline 6 & \quad\ \ \frac{\ }{20} \qquad\quad\ \longrightarrow \frac{9}{20} = 6\frac{9}{20} \end{array}$$

Answer: $15\frac{1}{5} - 8\frac{3}{4} = 6\frac{9}{20}$

Remember

Any whole number divided by itself is a fraction equivalent to 1.

$\frac{2}{2} = 1 \quad \frac{5}{5} = 1 \quad \frac{20}{20} = 1 \quad \frac{100}{100} = 1$ etc.

EXERCISES

In 1–24, subtract. Reduce answers to lowest terms.

Sample Solutions

a. $$\begin{array}{r|l} 15 & \frac{3}{4} = \frac{6}{8} \\ 7 & \frac{3}{8} = \frac{3}{8} \\ \hline 8 & \quad\ \ \frac{3}{8} = 8\frac{3}{8} \end{array}$$

b. $$\begin{array}{r|l} \overset{17}{\cancel{18}} & \frac{1}{2} = \frac{4}{8} + \frac{8}{8} = \frac{12}{8} \\ 7 & \frac{5}{8} \longrightarrow \frac{5}{8} \\ \hline 10 & \qquad\quad\ \frac{7}{8} = 10\frac{7}{8} \end{array}$$

1. $17\frac{4}{5}$
 $17\frac{2}{3}$

2. $23\frac{4}{7}$
 $15\frac{10}{21}$

3. $18\frac{1}{3}$
 $15\frac{1}{5}$

4. $26\frac{11}{16}$
 $18\frac{3}{8}$

5. $12\frac{1}{2}$
 $7\frac{4}{9}$

6. $14\frac{7}{12}$
 $6\frac{2}{5}$

7. $27\frac{4}{5}$ − $8\frac{2}{3}$

8. $13\frac{5}{8}$ − $6\frac{11}{20}$

9. $32\frac{12}{16}$ − $8\frac{5}{8}$

10. $21\frac{5}{7}$ − $13\frac{2}{3}$

11. $18\frac{2}{3}$ − $9\frac{3}{8}$

12. $23\frac{5}{16}$ − $15\frac{3}{4}$

13. $18\frac{2}{7}$ − $17\frac{1}{3}$

14. $15\frac{3}{8}$ − $8\frac{2}{3}$

15. 20 − $7\frac{5}{8}$

16. 16 − $9\frac{11}{16}$

17. $23\frac{2}{3}$ − $15\frac{3}{12}$

18. $21\frac{7}{16}$ − $8\frac{35}{64}$

19. $24\frac{1}{5}$ − $18\frac{5}{6}$

20. $27\frac{3}{12}$ − $15\frac{17}{30}$

21. $24\frac{4}{9}$ − $13\frac{27}{45}$

22. $18\frac{9}{11} - 15\frac{2}{3}$

23. $14\frac{1}{3} - 13\frac{5}{7}$

24. $24 - 23\frac{5}{18}$

APPLICATION PROBLEMS

1. A bolt of velvet contains $23\frac{3}{4}$ yards. If a clerk sells $4\frac{2}{3}$ yards, how much velvet is left in the bolt? ________

2. Mrs. Wilson bought a $12\frac{1}{2}$-pound turkey. After being roasted, the turkey weighed $8\frac{5}{8}$ pounds. How much less did the turkey weigh after roasting? ________

3. A carpenter had a board that measured $15\frac{1}{3}$ feet. He cut off a piece of board that measured $8\frac{5}{8}$ feet. How many feet of board were left? ________

4. A baker had $8\frac{1}{2}$ dozen eggs and used up $5\frac{3}{4}$ dozen. How many dozen eggs were left? ________

5. Mrs. Brown had $8\frac{1}{3}$ cups of flour and used $3\frac{4}{5}$ cups to bake a cake. How many cups of flour were left? ________

6. A slab of cheese weighed $9\frac{7}{8}$ pounds. If a clerk sold $1\frac{1}{4}$, $2\frac{3}{8}$, and $\frac{13}{16}$ pounds, how much cheese was left? ________

7. Gilbert weighed $220\frac{3}{4}$ pounds. After going on a diet, he lost the following weights: $5\frac{1}{3}$ pounds, $8\frac{3}{4}$ pounds, and $10\frac{15}{16}$ pounds. What was Gilbert's weight after dieting? ________

8. Wendy had $15\frac{1}{3}$ yards of ribbon. If she used $3\frac{1}{4}$, $2\frac{1}{3}$, and $5\frac{5}{6}$ yards, how much ribbon did she have left? ________

9. A house painter had $9\frac{1}{4}$ gallons of paint. If she used $2\frac{3}{4}$, $3\frac{7}{8}$, and $2\frac{1}{2}$ gallons, how much paint was left? ________

10. A bolt of fabric contains $43\frac{1}{3}$ meters. If a clerk sells $3\frac{2}{5}$, $5\frac{2}{3}$, and $6\frac{3}{4}$ meters, how many meters of fabric are left? ________

Review of Part VIII (Units 27–31)

In 1–9, add. Change all improper fractions to mixed numbers or whole numbers. Reduce answers to lowest terms.

1.
$$\begin{array}{r} \frac{3}{4} \\ \frac{5}{8} \\ \underline{\frac{11}{16}} \end{array}$$

2.
$$\begin{array}{r} \frac{4}{5} \\ \frac{3}{4} \\ \underline{\frac{13}{20}} \end{array}$$

3.
$$\begin{array}{r} \frac{3}{8} \\ \frac{3}{4} \\ \underline{\frac{7}{10}} \end{array}$$

4. $\frac{2}{3}$ $\frac{5}{8}$ $\frac{11}{12}$

5. $\frac{3}{5}$ $\frac{3}{4}$ $\frac{7}{20}$

6. $\frac{3}{4}$ $\frac{5}{8}$ $\frac{3}{5}$

7. $12\frac{2}{3}$ $15\frac{4}{5}$ $19\frac{8}{15}$

8. $24\frac{4}{5}$ $23\frac{13}{15}$ $36\frac{2}{3}$

9. $32\frac{3}{4}$ $27\frac{5}{6}$ $42\frac{7}{9}$

In 10–21, subtract. Reduce answers to lowest terms.

10. $\frac{3}{4}$ $\frac{5}{8}$

11. $\frac{2}{3}$ $\frac{3}{5}$

12. $\frac{4}{5}$ $\frac{3}{4}$

13. $\frac{7}{8}$ $\frac{4}{5}$

14. $\frac{9}{16}$ $\frac{3}{8}$

15. $\frac{5}{6}$ $\frac{1}{4}$

16. $12\frac{1}{3}$ $7\frac{3}{4}$

17. $23\frac{2}{3}$ $17\frac{7}{8}$

18. $28\frac{3}{8}$ $12\frac{3}{5}$

19. $35\frac{1}{6}$ $23\frac{4}{5}$

20. $38\frac{3}{7}$ $15\frac{5}{6}$

21. $18\frac{3}{16}$ $5\frac{5}{8}$

PART IX. Multiplying Fractions

UNIT 32. Multiplication of Fractions

Every fraction is made up of a numerator (top number) and a denominator (bottom number). To multiply one fraction by another fraction, you must multiply the numerator by the numerator and the denominator by the denominator.

RULE

To multiply fractions, multiply the top numbers and multiply the bottom numbers to get a single fraction.

You can remember this rule as "multiply across." Whenever possible, reduce your product to lowest terms.

EXAMPLE 1. Multiply: $\frac{3}{4} \times \frac{2}{5}$

Solution:

Step 1: Multiply the top numbers and multiply the bottom numbers.

$$\frac{3}{4} \times \frac{2}{5} = \frac{6}{20}$$

Step 2: Reduce $\frac{6}{20}$ to lowest terms.

$$2\left|\frac{\overset{3}{\cancel{6}}}{\underset{10}{\cancel{20}}}\right. = \frac{3}{10}$$

Answer: $\frac{3}{4} \times \frac{2}{5} = \frac{3}{10}$

To simplify the multiplication of fractions, you may *divide any top number into any bottom number,* or you may *divide any bottom number into any top number.* This is called **cancelling.**

EXAMPLE 2. Multiply: $\frac{3}{8} \times \frac{4}{5}$

Solution:

Step 1: Notice that 4 divides evenly into 8. Before multiplying, divide the *top number,* 4, into the *bottom number,* 8.

$$\frac{3}{\underset{2}{\cancel{8}}} \times \frac{\overset{1}{\cancel{4}}}{5}$$

Think: "4 into 4 goes 1, and 4 into 8 goes 2."

Step 2: Multiply across.

$$\frac{3}{\underset{2}{\cancel{8}}} \times \frac{\overset{1}{\cancel{4}}}{5} = \frac{3}{10}$$

Answer: $\frac{3}{8} \times \frac{4}{5} = \frac{3}{10}$

Sometimes you can cancel in more than one way.

EXAMPLE 3. Multiply: $\frac{7}{8} \times \frac{4}{21}$

Solution: Notice that 7 divides evenly into 21 and that 4 divides evenly into 8. Divide *both pairs* of numbers; then multiply across.

$$\frac{\overset{1}{\cancel{7}}}{\underset{2}{\cancel{8}}} \times \frac{\overset{1}{\cancel{4}}}{\underset{3}{\cancel{21}}} = \frac{1}{6}$$

Think: "7 into 7 goes 1, and 7 into 21 goes 3. 4 into 4 goes 1, and 4 into 8 goes 2."

Answer: $\frac{7}{8} \times \frac{4}{21} = \frac{1}{6}$

Whenever *any number* divides evenly into both a top number and a bottom number, you may simplify by cancelling.

EXAMPLE 4. Multiply: $\frac{8}{9} \times \frac{7}{12}$

Solution: The number 4 divides evenly into both the top number, 8, and the bottom number, 12. Therefore, you can simplify by dividing 4 into 8 and 4 into 12.

$$\frac{\overset{2}{\cancel{8}}}{9} \times \frac{7}{\underset{3}{\cancel{12}}} = \frac{14}{27}$$

Think: "4 into 8 goes 2, and 4 into 12 goes 3."

Answer: $\frac{8}{9} \times \frac{7}{12} = \frac{14}{27}$

You have already learned (Unit 17) how to simplify division by crossing out the end zeros, one for one, in the dividend and in the divisor. When multiplying fractions, you may get rid of the same number of end zeros in a top number and in a bottom number.

EXAMPLE 5. Multiply: $\frac{99}{100} \times \frac{10}{13}$

Solution: Cross out one end zero in the top number, 10; cross out one end zero in the bottom number, 100.

$$\frac{99}{10\cancel{0}} \times \frac{1\cancel{0}}{13} = \frac{99}{130}$$

Answer: $\frac{99}{100} \times \frac{10}{13} = \frac{99}{130}$

Remember

You must cancel into a top number and a bottom number. You must *never* cancel into two top numbers or into two bottom numbers.

You can use the above rules for multiplying and for cancelling when you multiply more than two fractions.

EXAMPLE 6. Multiply: $\frac{3}{8} \times \frac{4}{9} \times \frac{3}{15}$

Solution: Cancel in as many ways as you can; then multiply across.

$$\frac{\overset{1}{\cancel{3}}}{\underset{2}{\cancel{8}}} \times \frac{\overset{1}{\cancel{4}}}{\underset{3}{\cancel{9}}} \times \frac{\overset{1}{\cancel{3}}}{\underset{5}{\cancel{15}}} = \frac{1}{30}$$

Answer: $\frac{3}{8} \times \frac{4}{9} \times \frac{3}{15} = \frac{1}{30}$

EXAMPLE 7. Find $\frac{3}{4}$ of $\frac{8}{15}$.

Solution: The word "of" tells you to multiply.

$$\frac{\overset{1}{\cancel{3}}}{\underset{1}{\cancel{4}}} \times \frac{\overset{2}{\cancel{8}}}{\underset{5}{\cancel{15}}} = \frac{2}{5}$$

Answer: $\frac{2}{5}$

EXERCISES

In 1–24, multiply. Simplify by cancelling where possible.

SAMPLE SOLUTIONS

a. $\frac{5}{6} \times \frac{7}{8} = \frac{35}{48}$

b. $\frac{2}{\underset{1}{\cancel{3}}} \times \frac{\overset{1}{\cancel{3}}}{5} = \frac{2}{5}$

c. $\frac{\overset{1}{\cancel{3}}}{\underset{2}{\cancel{4}}} \times \frac{\overset{3}{\cancel{6}}}{\underset{5}{\cancel{15}}} = \frac{3}{10}$

d. $\frac{\overset{1}{\cancel{3}}}{5} \times \frac{7}{\underset{3}{\cancel{9}}} \times \frac{\overset{1}{\cancel{4}}}{\underset{2}{\cancel{8}}} = \frac{7}{30}$

e. $\frac{\overset{1}{\cancel{5}}}{\underset{2}{\cancel{6}}} \times \frac{1}{\underset{2}{\cancel{10}}} \times \frac{\overset{3}{\cancel{9}}}{11} = \frac{3}{44}$

1. $\frac{4}{5} \times \frac{3}{4}$

2. $\frac{5}{6} \times \frac{3}{8}$

3. $\frac{5}{8} \times \frac{3}{4}$

4. $\frac{2}{5} \times \frac{5}{8}$

5. $\frac{12}{16} \times \frac{3}{5}$

6. $\frac{8}{9} \times \frac{2}{12}$

7. $\frac{5}{6} \times \frac{4}{5}$

8. $\frac{3}{4} \times \frac{4}{9}$

9. $\frac{5}{8} \times \frac{16}{25}$

10. $\frac{7}{8} \times \frac{12}{21}$

11. $\frac{3}{4} \times \frac{5}{8}$

12. $\frac{3}{8} \times \frac{4}{5}$

13. $\frac{9}{16} \times \frac{5}{6}$

14. $\frac{2}{3} \times \frac{1}{2} \times \frac{6}{7}$

15. $\frac{5}{16} \times \frac{4}{8} \times \frac{7}{10}$

16. $\frac{1}{4} \times \frac{5}{6} \times \frac{2}{5}$

17. $\frac{3}{4} \times \frac{1}{2} \times \frac{3}{5}$

18. $\frac{5}{12} \times \frac{4}{5} \times \frac{3}{16}$

19. $\frac{2}{3} \times \frac{5}{8} \times \frac{3}{10}$

20. $\frac{3}{4} \times \frac{5}{6} \times \frac{3}{5}$

21. $\frac{5}{16} \times \frac{2}{3} \times \frac{1}{7}$

22. $\frac{2}{5} \times \frac{3}{4} \times \frac{15}{16}$

23. $\frac{3}{8} \times \frac{4}{7} \times \frac{5}{6}$

24. $\frac{3}{16} \times \frac{7}{12} \times \frac{3}{4}$

APPLICATION PROBLEMS

SAMPLE SOLUTION

If the width of a river is $\frac{5}{8}$ of a mile, how far from either shore is a boat located exactly in midstream?

$\frac{1}{2}$ of $\frac{5}{8} = \frac{1}{2} \times \frac{5}{8} = \frac{5}{16}$ $\frac{5}{16}$ mi.

1. Mrs. Wilson had a ham that weighed $\frac{3}{4}$ pound. If she used $\frac{1}{4}$ of the ham, what fraction of a pound did she use? __________

2. Mrs. Brown had $\frac{1}{2}$ dozen eggs and used $\frac{1}{3}$ of them. How many eggs did she use? __________

3. Jim lives $\frac{3}{4}$ of a mile from town. If he walked $\frac{2}{3}$ of the distance, what part of a mile did he walk? __________

4. Mr. Brown spent $\frac{3}{5}$ of his vacation money during the first week of his trip. If $\frac{2}{3}$ of the money spent went for food and lodging, what part of his entire vacation money went for food and lodging? __________

5. A truck driver had a load of $\frac{7}{8}$ ton of sand. If he delivered $\frac{4}{5}$ of the sand, what part of a ton did he deliver? __________

6. If one man can mow his lawn in $\frac{3}{4}$ of an hour, how long will it take two men to do the job? (Assume that the two men work at the same speed.) __________

7. Movie star Cliff Dangle has an agent who gets $\frac{1}{10}$ of his earnings. Last year, Cliff earned $\frac{3}{4}$ of a million dollars. What fraction of a million did the agent get? __________

8. What fraction of a foot is $\frac{3}{8}$ of an inch? (*Hint:* 1 inch $= \frac{1}{12}$ of a foot.) __________

9. A *furlong* is a distance that is equal to $\frac{1}{8}$ of a mile. How long is $\frac{3}{4}$ furlong? __________

10. Janet won $\frac{3}{4}$ of a million dollars in a lottery. If she pays $\frac{1}{5}$ of the money in federal taxes, how much of her prize *will remain?* __________

UNIT 33. Multiplication by a Mixed Number

Mixed Numbers

To multiply a mixed number by a mixed number, *change the mixed numbers to improper fractions.* Then cancel where possible and multiply across.

Example 1. Multiply: $3\frac{3}{4} \times 4\frac{2}{3}$

Solution:

Step 1: Change the mixed numbers to improper fractions.

$3\frac{3}{4} = \frac{15}{4}$ Think: "4 × 3 = 12, and 12 + 3 = 15."

$4\frac{2}{3} = \frac{14}{3}$ Think: "3 × 4 = 12, and 12 + 2 = 14."

Step 2: Cancel and multiply across.

$$\frac{\overset{5}{\cancel{15}}}{\underset{2}{\cancel{4}}} \times \frac{\overset{7}{\cancel{14}}}{\underset{1}{\cancel{3}}} = \frac{35}{2} = 17\frac{1}{2}$$

Answer: $3\frac{3}{4} \times 4\frac{2}{3} = 17\frac{1}{2}$

Whole and Mixed Numbers

To multiply a whole number by a mixed number, change the mixed number to an improper fraction and also *change the whole number to an improper fraction.*

RULE

To change a whole number to an improper fraction, draw a fraction bar under the whole number and insert the number 1 as the denominator.

$2 = \frac{2}{1}$ $5 = \frac{5}{1}$ $20 = \frac{20}{1}$ etc.

After you change the whole number and the mixed number to improper fractions, cancel where possible and multiply across.

Example 2. Multiply: $6 \times 3\frac{2}{3}$

Solution:

Step 1: Change both numbers to improper fractions: $6 = \frac{6}{1}$ and $3\frac{2}{3} = \frac{11}{3}$.

Step 2: Cancel and multiply across.

$$\frac{\overset{2}{\cancel{6}}}{1} \times \frac{11}{\underset{1}{\cancel{3}}} = \frac{22}{1} = 22$$

Answer: $6 \times 3\frac{2}{3} = 22$

EXERCISES

In 1–16, multiply. Change answers that are improper fractions to whole numbers or to mixed numbers.

Sample Solutions

a. $2\frac{1}{2} \times 4\frac{2}{3} = \frac{5}{\underset{1}{\cancel{2}}} \times \frac{\overset{7}{\cancel{14}}}{3} = \frac{35}{3} = 11\frac{2}{3}$

b. $4\frac{3}{8} \times 8 = \frac{35}{\underset{1}{\cancel{8}}} \times \frac{\overset{1}{\cancel{8}}}{1} = \frac{35}{1} = 35$

1. $4\frac{1}{3} \times 3\frac{1}{5}$

2. $5\frac{1}{4} \times 2\frac{1}{3}$

3. $7\frac{3}{8} \times 9\frac{5}{6}$

4. $8\frac{5}{9} \times 4\frac{1}{11}$

5. $6\frac{5}{6} \times 3\frac{7}{8}$

6. $3\frac{4}{5} \times 5\frac{5}{8}$

7. $6\frac{2}{3} \times 4\frac{7}{8}$

8. $8\frac{4}{5} \times 3\frac{5}{9}$

9. $5 \times 3\frac{3}{4}$

10. $7 \times 4\frac{2}{3}$

11. $3\frac{7}{8} \times 7$

12. $6\frac{3}{16} \times 3$

13. $5 \times 6\frac{3}{5}$

14. $7 \times 4\frac{5}{6}$

15. $3 \times 5\frac{2}{5}$

16. $4\frac{4}{5} \times 8$

APPLICATION PROBLEMS

1. How many feet of lumber are needed to make 12 shelves, each measuring $5\frac{2}{3}$ feet? __________

2. Mary needs $3\frac{2}{5}$ yards of material to make a dress. How many yards of material will she need to make 4 dresses? __________

3. Tom earns $150 a week and saves $\frac{1}{3}$ of his income. How much does he save every week? __________

4. Velvet sells for $4 a yard. How much will $5\frac{3}{4}$ yards cost? __________

5. A ground beef patty weighs $\frac{1}{4}$ pound. How many pounds will 12 patties weigh? __________

6. Overtime pay is figured at $1\frac{1}{2}$ times the regular hourly rate of pay. If Richard earns $8 per hour, how much is his hourly overtime rate of pay? __________

7. At $1\frac{1}{2}$ times the regular rate, what will be the overtime rate on $6.75 per hour? (*Hint:* 6.75 may be written as $6\frac{3}{4}$.) __________

8. Carpeting sells at $9.25 per yard. How much will $15\frac{1}{2}$ yards cost? (*Hint:* $9.25 = 9\frac{1}{4}$.) __________

9. A carpenter needs $6\frac{1}{2}$ feet of lumber to make a shelf. If she has enough lumber to make $8\frac{2}{3}$ shelves, how much lumber does she have? __________

10. Stan's overtime rate is $8\frac{1}{2}$ dollars per hour. How much overtime pay did he earn for $4\frac{2}{3}$ hours? __________

UNIT 34. Multiplication of a Large Whole Number by a Mixed Number

Suppose you have to multiply a large whole number (a number with three or more digits) by a mixed number. The following procedure is usually easier than changing both numbers to improper fractions.

To multiply a large whole number by a mixed number, multiply the whole number by the *whole-number part of the mixed number.* Then multiply the whole number by the *fraction part of the mixed number.* Add both answers for the final result.

Example 1. Multiply: $235 \times 25\frac{3}{5}$

Solution: Multiply the whole number, 235, by the whole-number part of the mixed number, 25.

```
  235 | 3
  ×25 | -
      | 5
------+
 1 175
 4 70
------
 5,875
```

Next, multiply the whole number, 235, by the fraction part of the mixed number, $\frac{3}{5}$. Combine both answers for the final product.

$$\frac{\cancel{3}}{\cancel{5}_1} \times \frac{\overset{47}{\cancel{235}}}{1} = \frac{141}{1} = 141$$

```
  235 |
  ×25 | 3/5 × 235/1 = 141/1 = 141
------+
 1 175
 4 70
------
 5 875
 +141
------
 6,016
```

Answer: $235 \times 25\frac{3}{5} = 6{,}016$

Example 2. Multiply: $105 \times 79\frac{1}{3}$

Solution:

$$\begin{array}{r|l} 105 & \\ \times 79 & \frac{1}{\not{3}_1} \times \frac{\not{105}^{35}}{1} = \frac{35}{1} = 35 \\ \hline 945 & \\ 7\,35 & \\ \hline 8\,295 & \\ +35 & \\ \hline 8{,}330 & \end{array}$$

Answer: $105 \times 79\frac{1}{3} = 8{,}330$

Example 3. Multiply: $754 \times 82\frac{3}{4}$

Solution:

$$\begin{array}{r|l} 754 & \\ \times 82 & \frac{3}{\not{4}_2} \times \frac{\not{754}^{377}}{1} = \frac{1{,}131}{2} = 565\frac{1}{2} \\ \hline 1\,508 & \\ 60\,32 & \\ \hline 61\,828 & \\ +565\frac{1}{2} & \\ \hline 62{,}393\frac{1}{2} & \end{array}$$

Answer: $754 \times 82\frac{3}{4} = 62{,}393\frac{1}{2}$

EXERCISES

In 1–15, multiply.

1. $\begin{array}{r} 135 \\ 18\frac{2}{3} \\ \hline \end{array}$

2. $\begin{array}{r} 248 \\ 26\frac{5}{8} \\ \hline \end{array}$

3. $\begin{array}{r} 326 \\ 38\frac{1}{3} \\ \hline \end{array}$

4. $\begin{array}{r} 530 \\ 45\frac{7}{10} \\ \hline \end{array}$

5. $\begin{array}{r} 262 \\ 35\frac{3}{4} \\ \hline \end{array}$

6. $\begin{array}{r} 436 \\ 18\frac{1}{2} \\ \hline \end{array}$

7. $\begin{array}{r} 342 \\ 23\frac{3}{5} \\ \hline \end{array}$

8. $\begin{array}{r} 287 \\ 32\frac{5}{7} \\ \hline \end{array}$

9. $\begin{array}{r} 416 \\ 43\frac{9}{16} \\ \hline \end{array}$

10. $369 \times 37\frac{2}{3}$

11. $535 \times 36\frac{4}{5}$

12. $328 \times 43\frac{3}{4}$

13. $486 \times 46\frac{1}{3}$

14. $484 \times 26\frac{5}{16}$

15. $460 \times 34\frac{4}{5}$

APPLICATION PROBLEMS

1. A can of green beans weighs $4\frac{1}{2}$ ounces. How many ounces will 235 cans weigh? __________

2. A car averages $18\frac{1}{2}$ miles to a gallon of gasoline. How many miles will be traveled using 237 gallons? __________

3. A hotel needs $21\frac{3}{4}$ yards of carpeting to cover the floor of one room. How many yards of carpeting will be needed to cover the floors of 253 rooms? (All the rooms are the same size.) __________

4. A bottle of salad oil holds $12\frac{2}{3}$ ounces. How many ounces will 345 bottles hold? __________

5. A brick weighs $3\frac{1}{5}$ pounds. How much will 460 bricks weigh? __________

6. If it takes $15\frac{1}{2}$ hours to assemble one car, how many hours will it take to assemble 435 cars? __________

7. A bottle of medicine holds $23\frac{2}{3}$ cubic centimeters. How many cubic centimeters are contained in 639 bottles? __________

8. A perfume sells at \$$53\frac{3}{8}$ per ounce. How much would 136 ounces cost? __________

9. A share of stock sells for \$$34\frac{1}{4}$. What is the cost of 230 shares of stock? __________

10. A furniture manufacturer uses $23\frac{2}{3}$ yards of fabric to manufacture one living room set. How many yards of fabric are needed to make 240 sets? __________

Review of Part IX (Units 32–34)

In 1–18, multiply. Where possible, simplify by cancelling. Express all improper fractions as mixed numbers.

1. $\frac{3}{4} \times \frac{3}{5}$

2. $\frac{4}{5} \times \frac{3}{8}$

3. $\frac{2}{3} \times \frac{3}{10}$

4. $\frac{5}{8} \times \frac{12}{15}$

5. $\frac{3}{7} \times \frac{5}{6} \times \frac{21}{30}$

6. $\frac{3}{8} \times \frac{2}{5} \times \frac{5}{6}$

7. $\frac{5}{16} \times \frac{4}{15} \times \frac{3}{4}$

8. $\frac{2}{5} \times \frac{3}{4} \times \frac{5}{6}$

9. $3\frac{1}{2} \times 5\frac{3}{4}$

10. $7 \times 6\frac{4}{5}$

11. $4\frac{2}{3} \times 6$

12. $5\frac{3}{5} \times 4\frac{2}{3}$

13. $6\frac{1}{3} \times 4$

14. $3\frac{3}{8} \times 4\frac{1}{4}$

15. $563 \times 67\frac{2}{3}$

16. $848 \times 85\frac{3}{4}$

17. $470 \times 37\frac{4}{5}$

18. $650 \times 78\frac{7}{10}$

PART X. Dividing Fractions

UNIT 35. Division of Fractions

WORDS TO KNOW

When we **invert** something, we turn it upside down. When we **invert a fraction,** we turn it upside down by putting the top number on the bottom and the bottom number on top.

When a division problem with fractions is indicated with the ÷ symbol, the second fraction (the right-hand fraction) is the divisor. Thus, in the problem $\frac{1}{2} \div \frac{1}{4}$, the divisor is $\frac{1}{4}$.

To divide one fraction by another fraction, indicate the division with the ÷ symbol. Then follow this rule:

RULE

To divide fractions, *invert the divisor* (the second fraction) and change the problem to multiplication.

To **invert a fraction,** reverse the numerator and the denominator.

When you invert $\frac{2}{3}$, you get $\frac{3}{2}$.

When you invert $\frac{7}{8}$, you get $\frac{8}{7}$.

When you invert $\frac{99}{100}$, you get $\frac{100}{99}$.

EXAMPLE 1. Divide: $\frac{3}{4} \div \frac{3}{8}$

Solution:

Step 1: Bring down the first fraction exactly as it is written.

Step 2: Change the ÷ symbol to a × symbol.

Step 3: *Invert* the second fraction (the divisor).

$$\frac{3}{4} \div \frac{3}{8}$$

$$\downarrow \quad \downarrow \quad \downarrow$$

$$\frac{3}{4} \times \frac{8}{3}$$

Step 4: Solve the multiplication problem.

$$\frac{\overset{1}{\cancel{3}}}{\underset{1}{\cancel{4}}} \times \frac{\overset{2}{\cancel{8}}}{\underset{1}{\cancel{3}}} = \frac{2}{1} = 2$$

Answer: $\frac{3}{4} \div \frac{3}{8} = 2$

Recall that a division problem asks how many times the divisor is contained in the dividend. Example 1 asks, "How many $\frac{3}{8}$'s are there in $\frac{3}{4}$?" Applying the rule, the answer is 2. Does this make sense? Do two $\frac{3}{8}$'s make $\frac{3}{4}$? As you can see, they do: two $\frac{3}{8}$'s $= 2 \times \frac{3}{8} = \frac{2}{1} \times \frac{3}{8} = \frac{6}{8} = \frac{3}{4}$.

We check division problems with fractions just as we do with whole numbers: The answer (quotient) multiplied by the divisor should equal the dividend.

EXAMPLE 2. Divide and check: $\frac{7}{8} \div 4$

Solution: Rewrite the whole number, 4, as $\frac{4}{1}$. Then invert the divisor and multiply.

$$\frac{7}{8} \div \frac{4}{1}$$

$$\frac{7}{8} \times \frac{1}{4} = \frac{7}{32}$$

Check: The quotient, $\frac{7}{32}$, times the divisor, 4, should equal the dividend, $\frac{7}{8}$.

$$\frac{7}{\cancel{32}_8} \times \frac{\cancel{4}^1}{1} = \frac{7}{8} \checkmark$$

Answer: $\frac{7}{8} \div 4 = \frac{7}{32}$

EXAMPLE 3. Divide $\frac{3}{4}$ into $\frac{1}{2}$ and check.

Solution: Recall that the number doing the dividing is the divisor. Thus, $\frac{3}{4}$ is the divisor. (Mentally change "divide $\frac{3}{4}$ into $\frac{1}{2}$" to "$\frac{1}{2} \div \frac{3}{4}$.")

$$\frac{1}{2} \div \frac{3}{4}$$

$$\frac{1}{\cancel{2}_1} \times \frac{\cancel{4}^2}{3} = \frac{2}{3}$$

Check:

$$\frac{\cancel{2}^1}{\cancel{3}_1} \times \frac{\cancel{3}^1}{\cancel{4}_2} = \frac{1}{2} \checkmark$$

Answer: $\frac{3}{4}$ divided into $\frac{1}{2}$ is $\frac{2}{3}$.

EXERCISES

In 1–16, divide.

1. $\frac{3}{5} \div \frac{3}{4}$
2. $\frac{2}{7} \div \frac{4}{21}$
3. $\frac{5}{8} \div \frac{7}{8}$
4. $\frac{3}{16} \div \frac{4}{9}$
5. $\frac{3}{5} \div \frac{9}{10}$
6. $\frac{5}{6} \div \frac{7}{12}$
7. $\frac{1}{2} \div \frac{7}{16}$
8. $\frac{11}{12} \div \frac{4}{7}$
9. $\frac{2}{3} \div \frac{4}{5}$
10. $\frac{5}{6} \div \frac{7}{8}$
11. $\frac{5}{8} \div \frac{5}{12}$
12. $\frac{5}{8} \div \frac{5}{8}$
13. $\frac{7}{8} \div \frac{3}{16}$
14. $\frac{3}{4} \div \frac{4}{5}$
15. $\frac{1}{3} \div \frac{7}{8}$
16. $\frac{3}{8} \div \frac{7}{8}$

APPLICATION PROBLEMS

In 1–10, check your answers.

1. How many $\frac{1}{8}$'s of a pound are there in $\frac{3}{4}$ of a pound? ______
2. How many barrels of sand can be filled with $\frac{3}{4}$ ton of sand if each barrel can hold $\frac{1}{16}$ of a ton? ______
3. If you have $\frac{7}{8}$ gallon of alcohol, how many pint-size containers can you fill? (*Hint:* There are 8 pints to a gallon.) ______
4. A bus leaves every 3 minutes after the hour. How many buses are dispatched in $\frac{3}{4}$ of an hour? $\left(\textit{Hint: } 3 \text{ min.} = \frac{3}{60} \text{ hr.}\right)$ ______
5. How many 6-inch pieces of pipe can be cut from a length of pipe measuring $\frac{2}{3}$ yard? $\left(\textit{Hint: } 6 \text{ inches} = \frac{1}{6} \text{ of a yard.}\right)$ ______
6. How many $\frac{1}{4}$-hours are there in $\frac{5}{6}$ hour? ______
7. How many periods of 20 minutes are there in $\frac{3}{4}$ of an hour? $\left(\textit{Hint: } 20 \text{ minutes} = \frac{1}{3} \text{ hour.}\right)$ ______
8. It takes a typist an average of 6 minutes to type a letter. How many letters can be typed in $\frac{3}{4}$ hour? (*Hint:* Write 6 minutes as a fraction of an hour reduced to lowest terms.) ______
9. From $\frac{3}{4}$ yard of tape, how many 3-inch tabs can be cut? (*Hint:* Write 3 inches as a fraction of a yard reduced to lowest terms.) ______
10. Evelyn has $\frac{3}{4}$ pound of flour. If she needs 2 ounces to make a cookie, how many cookies can she make? ______

UNIT 36. Division by Mixed Numbers

MIXED NUMBERS

To divide one mixed number by another mixed number, *change the mixed numbers to improper fractions.* Then divide, as shown in the preceding unit, by inverting the divisor and multiplying.

$$\frac{11}{3} \div \frac{9}{2}$$
$$\downarrow \quad \downarrow \quad \downarrow$$
$$\frac{11}{3} \times \frac{2}{9} = \frac{22}{27}$$

Example 1. Divide: $3\frac{2}{3} \div 4\frac{1}{2}$

Solution: Change $3\frac{2}{3}$ to $\frac{11}{3}$ and change $4\frac{1}{2}$ to $\frac{9}{2}$. Then divide.

Answer: $3\frac{2}{3} \div 4\frac{1}{2} = \frac{22}{27}$

WHOLE AND MIXED NUMBERS

To divide a whole number by a mixed number, *change both the mixed number and the whole number to improper fractions.* Then divide.

EXAMPLE 2. Divide: $5 \div 1\frac{1}{2}$

Solution: Change 5 to $\frac{5}{1}$ and change $1\frac{1}{2}$ to $\frac{3}{2}$.

$$\frac{5}{1} \div \frac{3}{2}$$

$$\frac{5}{1} \times \frac{2}{3} = \frac{10}{3} = 3\frac{1}{3}$$

Answer: $5 \div 1\frac{1}{2} = 3\frac{1}{3}$

EXERCISES

In 1–14, divide.

1. $4\frac{3}{4} \div 5\frac{1}{2}$

2. $3\frac{2}{3} \div 6\frac{3}{5}$

3. $5\frac{3}{16} \div 2\frac{1}{2}$

4. $2\frac{1}{12} \div 3\frac{3}{4}$

5. $4\frac{2}{3} \div 4\frac{4}{6}$

6. $6\frac{2}{3} \div 3\frac{1}{5}$

7. $3\frac{3}{4} \div 2\frac{2}{5}$

8. $14\frac{1}{2} \div 3\frac{5}{8}$

9. $2\frac{2}{3} \div 7\frac{1}{2}$

10. $8 \div \frac{3}{4}$

11. $3\frac{4}{5} \div 8$

12. $5\frac{3}{4} \div 5$

13. $8 \div 2\frac{2}{5}$

14. $6 \div 7\frac{7}{8}$

APPLICATION PROBLEMS

In 1–10, check your answers.

1. If a dress requires $3\frac{1}{5}$ yards of material, how many dresses can be made with 32 yards of material? ________

2. A plane flies 845 miles in $1\frac{2}{3}$ hours. How many miles does the plane average in one hour? ________

3. Tom drove 153 miles in $3\frac{1}{2}$ hours. How many miles did he average in one hour? ________

4. If $1\frac{3}{4}$ ounces of butter are used to make a box of cookies, how many boxes of cookies can be made with 7 pounds of butter? ________

5. If a board $10\frac{1}{4}$ feet long is cut into 6 equal pieces, how long will each piece be? ________

6. If it takes $\frac{3}{4}$ yard of fabric to make a child's blouse, how many blouses can be made with $6\frac{3}{4}$ yards of fabric? ________

7. A length of lumber measures $4\frac{4}{5}$ meters. How many pieces of lumber measuring $1\frac{1}{5}$ meters can the length be cut into? ________

8. How many pieces of ribbon each measuring 15 inches can be cut from a length of ribbon measuring $6\frac{1}{4}$ feet? ________

9. A bag contains $11\frac{2}{3}$ pounds of raisins. How many 20-ounce boxes can be filled with the raisins? ________

10. A piece of fabric measures $6\frac{2}{3}$ yards. If 4 feet of fabric are used to make a skirt, how many skirts can be made from the piece of fabric? ________

UNIT 37. Simplifying Complex Fractions

WORDS TO KNOW

So far, in your work with fractions, numerators and denominators have always been whole numbers. It is possible for the numerator or the denominator, or both, to be a fraction or a mixed number. When this happens, the fraction is called a **complex fraction.**

The process of division can be indicated in three ways. Thus, "3 divided by 4" can be shown by:

(1) the division box, $4\overline{)3}$.

(2) the division symbol, $3 \div 4$.

(3) the fraction bar, $\frac{3}{4}$.

The fraction bar, between the numerator and the denominator, can be read as "divided by." You will use this fact to simplify **complex fractions,** which are fractions that have a fraction in the numerator or in the denominator, or in both.

To "simplify" a complex fraction means to change it to a proper fraction or a mixed number that has the same value.

EXAMPLE 1. Simplify: $\dfrac{\frac{1}{2}}{\frac{2}{3}}$

Solution: Instead of reading this complex fraction as "$\frac{1}{2}$ over $\frac{2}{3}$," read it as "$\frac{1}{2}$ divided by $\frac{2}{3}$." In symbols:

$$\dfrac{\frac{1}{2}}{\frac{2}{3}} = \frac{1}{2} \div \frac{2}{3}$$

To simplify the complex fraction, do the indicated division.

$$\frac{1}{2} \div \frac{2}{3} = \frac{1}{2} \times \frac{3}{2} = \frac{3}{4}$$

Answer: $\dfrac{\frac{1}{2}}{\frac{2}{3}} = \frac{3}{4}$

RULE

To simplify a complex fraction, rewrite the fraction as a division problem with the fraction in the numerator divided by the fraction in the denominator. Then do the division.

EXAMPLE 2. Simplify: $\dfrac{3\frac{1}{4}}{5}$

Solution: Rewrite the complex fraction as a division problem.

$$\dfrac{3\frac{1}{4}}{5} = 3\frac{1}{4} \div 5 = \frac{13}{4} \div \frac{5}{1}$$

Do the division.

$$\frac{13}{4} \div \frac{5}{1} = \frac{13}{4} \times \frac{1}{5} = \frac{13}{20}$$

Answer: $\dfrac{3\frac{1}{4}}{5} = \frac{13}{20}$

EXERCISES

In 1–12, simplify the complex fraction.

1. $\dfrac{\frac{3}{8}}{\frac{2}{3}}$ **2.** $\dfrac{\frac{3}{4}}{\frac{7}{8}}$ **3.** $\dfrac{\frac{3}{5}}{5}$

4. $\dfrac{16\frac{2}{3}}{100}$ **5.** $\dfrac{6}{\frac{3}{4}}$ **6.** $\dfrac{\frac{3}{8}}{\frac{5}{16}}$

7. $\dfrac{3\frac{2}{3}}{4\frac{4}{5}}$

8. $\dfrac{27\frac{1}{2}}{36\frac{4}{7}}$

9. $\dfrac{37\frac{1}{2}}{100}$

10. $\dfrac{12\frac{5}{6}}{8\frac{2}{3}}$

11. $\dfrac{12\frac{1}{2}}{100}$

12. $\dfrac{\frac{3}{5}}{4\frac{1}{3}}$

UNIT 38. Finding a Number When a Fractional Part of It Is Known

Suppose that 5 students in a class are absent, and this represents $\frac{1}{5}$ of the number of students in the class. How would you find the total number of students?

The above facts are that 5 students are $\frac{1}{5}$ of the total number, that is, $5 = \frac{1}{5} \times ?$.

Since division is the opposite of multiplication, and so will undo multiplication, you can use division to find the unknown factor. The multiplication problem $5 = \frac{1}{5} \times ?$ can be changed to the division problem $5 \div \frac{1}{5} = ?$.

Therefore, to find a number when a fractional part is known, use the following rule:

RULE

To find a number when you know a fractional part of it, divide the known part by the fraction. The quotient will be the unknown number.

Applying the rule, we divide 5 by $\frac{1}{5}$:

$$5 \div \frac{1}{5} = \frac{5}{1} \times \frac{5}{1} = \frac{25}{1} = 25$$

There are 25 students in the class.

Example 1. Clint works after school, and saves $50 of his earnings each week. If this is $\frac{2}{5}$ of his earnings, what is his total income?

Solution: Since you know that $50 is $\frac{2}{5}$ of the income, divide the $50 by $\frac{2}{5}$ to find the total.

$$50 \div \frac{2}{5} = \frac{\overset{25}{\cancel{50}}}{1} \times \frac{5}{\underset{1}{\cancel{2}}} = 125$$

Check: If the answer is correct, then $\frac{2}{5}$ of $125 should equal $50.

$$\frac{2}{\underset{1}{\cancel{5}}} \times \frac{\overset{25}{\cancel{125}}}{1} = 50 \checkmark$$

Answer: His total income is $125.

Remember

In a word problem, always check your answer against the *given facts.*

Example 2. If $\frac{3}{16}$ of an unknown number is 12, find the number.

Solution: To find the unknown number, divide the known part, 12, by the fraction, $\frac{3}{16}$.

$$12 \div \frac{3}{16} = \frac{\overset{4}{\cancel{12}}}{1} \times \frac{16}{\underset{1}{\cancel{3}}} = 64$$

Check: Is $\frac{3}{16}$ of 64 equal to 12?

$$\frac{3}{\underset{1}{\cancel{16}}} \times \frac{\overset{4}{\cancel{64}}}{1} = 12 \checkmark$$

Answer: 64

The $\frac{\text{IS}}{\text{OF}}$ Fraction

In solving word problems that deal with fractional parts of an unknown number, many students have difficulty remembering which given number should be divided into which. A good way to remember which number becomes the divisor and which number becomes the dividend is to think of the "fraction" $\frac{\text{IS}}{\text{OF}}$.

The number in the problem related to IS is written as the numerator, and will be the dividend.

The number related to OF is written as the denominator, and will be the divisor.

The given facts in Example 1 can be expressed as: "($50 is) ($\frac{2}{5}$ of) what amount?" Using the $\frac{\text{IS}}{\text{OF}}$ method, write a fraction using the 50 (IS) as the numerator and the $\frac{2}{5}$ (OF) as the denominator. You get $\frac{50}{\frac{2}{5}}$. Carry out the division.

$$\frac{\text{IS}}{\text{OF}} = \frac{50}{\frac{2}{5}} = 50 \div \frac{2}{5}$$

$$= \frac{\overset{25}{\cancel{50}}}{1} \times \frac{5}{\underset{1}{\cancel{2}}} = 125$$

The given facts in Example 2 may be stated as: "(12 is) ($\frac{3}{16}$ of) what number?" Write the fraction 12 (IS) over $\frac{3}{16}$ (OF) to get $\frac{12}{\frac{3}{16}}$. Carry out the division.

$$\frac{\text{IS}}{\text{OF}} = \frac{12}{\frac{3}{16}} = 12 \div \frac{3}{16}$$

$$= \frac{\overset{4}{\cancel{12}}}{1} \times \frac{16}{\underset{1}{\cancel{3}}} = 64$$

No matter what the actual wording may be, all problems about fractional parts of unknown numbers can be reworded so that one number is related to IS and the other is related to OF. You can then use the fraction $\frac{\text{IS}}{\text{OF}}$ to write a numerical fraction that will solve the problem.

Exercises

In 1–10, find the unknown number.

Sample Solutions

a. ($\frac{1}{2}$ of) what number (is 12)?

$$\frac{IS}{OF} = \frac{12}{\frac{1}{2}} = \frac{12}{1} \times \frac{2}{1} = \frac{24}{1} = 24$$

b. (16 is) ($\frac{1}{2}$ of) what number?

$$\frac{IS}{OF} = \frac{16}{\frac{1}{2}} = \frac{16}{1} \times \frac{2}{1} = \frac{32}{1} = 32$$

1. $\frac{3}{4}$ of what number is 26?

2. $\frac{2}{3}$ of what number is 14?

3. $\frac{5}{8}$ of what number is 20?

4. $\frac{2}{5}$ of what number is 42?

5. $\frac{3}{8}$ of what number is 12?

6. 35 is $\frac{1}{3}$ of what number?

7. 30 is $\frac{5}{8}$ of what number?

8. 27 is $\frac{2}{3}$ of what number?

9. 23 is $\frac{1}{4}$ of what number?

10. 40 is $\frac{1}{12}$ of what number?

APPLICATION PROBLEMS

Sample Solution

Of the Newton High School cheerleaders, $\frac{1}{6}$ are sophomores.

If 4 are sophomores, find the total number of cheerleaders.

Use $\frac{\text{IS}}{\text{OF}}$: (4 is) ($\frac{1}{6}$ of) what number?

$$\frac{IS}{OF} = \frac{4}{\frac{1}{6}} = \frac{4}{1} \times \frac{6}{1} = \frac{24}{1} = 24$$

24

1. Tom received 976 votes for Class President. If this represents $\frac{4}{5}$ of all the votes, how many students voted in the election? __________

2. The school baseball team won 22 games. If they won $\frac{2}{3}$ of the games played, what was the total number of games played? __________

3. A stadium has 20,088 general admission seats. Find the total number of seats if this represents $\frac{2}{3}$ of the stadium seats. __________

4. John saves $45 a week. If he is saving $\frac{5}{8}$ of his weekly income, how much does he earn in a week? ______

5. Mary spent $35, which is $\frac{1}{6}$ of her weekly income, on food. What is her weekly income? ______

6. The price of a coat was reduced by $43 during a $\frac{1}{3}$-off sale. What was the original price of the coat? ______

7. Frank was sick for 2 days and worked only 3 days last week. If his pay for the 3 days was $222, what is his regular pay for a 5-day week? (*Hint:* What fraction of the week did he work?) ______

8. A salesperson earned $131 commission on the sale of a refrigerator. If his commission was $\frac{1}{5}$ of the selling price, what was the selling price of the refrigerator? ______

9. The price of a dress was reduced by $24. If the reduction was $\frac{2}{5}$ of the original price, what was the original price of the dress? ______

10. Michael bought a car and made a $\frac{1}{3}$ down payment. If the balance is $6,842, what was the full price of the car? ______

UNIT 39. Finding What Fractional Part One Number Is of Another Number

RULE

To find what fractional part one number is of another number, write a fraction with the *total amount as the denominator* and the *partial amount as the numerator.*

EXAMPLE 1. If you had $25 and you spent $15, what fractional part of your money did you spend?

Solution: You spent $15 out of a total of $25. Therefore, write a fraction with 15 as the numerator and 25 as the denominator. Then reduce the fraction.

$$\frac{\text{amount spent}}{\text{total}} = \frac{\overset{3}{\cancel{15}}}{\underset{5}{\cancel{25}}} = \frac{3}{5}$$

Check: Does $\frac{3}{5}$ of $25 equal $15?

$$\frac{3}{\underset{1}{\cancel{5}}} \times \frac{\overset{5}{\cancel{25}}}{1} = \frac{15}{1} = 15 \checkmark$$

Answer: You spent $\frac{3}{5}$ of your money.

Alternate Solution: You may also use the $\frac{\text{IS}}{\text{OF}}$ method to solve this example. The problem asks: "(15 is) what fractional part (of 25)?" Write the fraction $\frac{\text{IS}}{\text{OF}} = \frac{15}{25}$ and reduce to lowest terms, $\frac{3}{5}$.

EXAMPLE 2. What part of 49 is 42?

Solution: Since 42 is the "part" and 49 is the "total," form a fraction with 42 as the numerator and 49 as the denominator. Reduce.

$$7\Big|\frac{\overset{6}{\cancel{42}}}{\underset{7}{\cancel{49}}} = \frac{6}{7}$$

Check: Does $\frac{6}{7}$ of 49 equal 42?

$$\frac{6}{\underset{1}{\cancel{7}}} \times \frac{\overset{7}{\cancel{49}}}{1} = \frac{42}{1} = 42 \checkmark$$

Answer: $\frac{6}{7}$

Alternate Solution: You may also use the $\frac{\text{IS}}{\text{OF}}$ method to solve this example. The problem asks: "(42 is) what part (of 49)?" Write the fraction $\frac{\text{IS}}{\text{OF}} = \frac{42}{49}$ and reduce to lowest terms, $\frac{6}{7}$.

Note: As you did in other word problems, look for *key words* to help you. When you are finding what fractional part one number is of another number, words such as *total, complete,* and *all* indicate the larger number (that will form the denominator). Words such as *part, portion,* and *amount spent* indicate the smaller number (that will form the numerator).

EXERCISES

In 1–14, find the fractional part.

SAMPLE SOLUTIONS

a. (6 is) what part (of 12)?

$\frac{IS}{OF} = \frac{6}{12} = \frac{1}{2}$

b. What part (of 18) (is 3)?

$\frac{IS}{OF} = \frac{3}{18} = \frac{1}{6}$

1. 12 is what part of 18?

2. 16 is what part of 20?

3. 28 is what part of 42?

4. 49 is what part of 56?

5. 75 is what part of 100?

6. 35 is what part of 85?

7. 24 is what part of 64?

8. What part of 64 is 16?

9. What part of 75 is 25?

10. What part of 80 is 10?

11. What part of 72 is 30?

12. What part of 70 is 21?

13. What part of 144 is 36?

14. What part of 150 is 75?

APPLICATION PROBLEMS

SAMPLE SOLUTION

Of 500 seniors who entered the 12th grade at Washington High School, 475 received diplomas in June. What fraction of the seniors graduated?

(475 is) what part (of 500)?

$\frac{IS}{OF} = \frac{475}{500}$ $\quad 25 \bigg| \frac{\overset{19}{\cancel{475}}}{\underset{20}{\cancel{500}}} = \frac{19}{20}$

$\frac{19}{20}$

1. There are 20 girls in a class of 36 students. What part of the class is girls? ________
2. A car radiator holds 20 quarts. If it contains 8 quarts of antifreeze, what part of the contents is antifreeze? ________
3. There are 8 students absent in a class of 32 students. What fractional part of the class is absent? ________
4. Tom bought a television set for $125 and made a down payment of $50. What fractional part of the price is the down payment? ________
5. Paula is 24 years old. If she spent 4 years in college, what part of her life did she spend in college? ________
6. A coat originally selling for $60 was reduced by $15. What fraction of the original price was the reduction? ________
7. A radio originally selling for $96 was reduced to $60. What fraction of the original price is the new price? ________
8. The Wilson family bought a house for $65,000 and made a down payment of $13,000. What fraction of the price was the down payment? ________
9. A retailer had 250 blouses in stock. If 100 blouses were sold, what fraction of the stock remains? ________
10. A jacket originally selling for $39 was reduced by $12. What fraction of the original price will the new price be? ________

Review of Part X (Units 35–39)

In 1–12, divide.

1. $\frac{3}{5} \div \frac{3}{10}$

2. $\frac{5}{6} \div \frac{2}{3}$

3. $\frac{3}{16} \div \frac{2}{7}$

4. $\frac{3}{4} \div \frac{2}{3}$

5. $\frac{4}{5} \div \frac{8}{15}$

6. $\frac{5}{8} \div \frac{10}{15}$

7. $3\frac{1}{3} \div 4\frac{3}{5}$

8. $5\frac{3}{4} \div 6$

9. $4 \div 5\frac{7}{8}$

10. $2\frac{2}{3} \div 1\frac{4}{5}$

11. $5 \div 2\frac{3}{16}$

12. $6\frac{1}{2} \div 4$

In 13–16, simplify the complex fraction.

13. $\dfrac{\frac{2}{3}}{\frac{4}{5}}$

14. $\dfrac{2\frac{1}{5}}{3\frac{1}{4}}$

15. $\dfrac{12\frac{1}{2}}{100}$

16. $\dfrac{5\frac{5}{8}}{6}$

In 17–20, find the unknown number.

17. $\frac{5}{8}$ of what number is 36?

18. $\frac{3}{5}$ of what number is 35?

19. 40 is $\frac{2}{3}$ of what number?

20. 32 is $\frac{5}{6}$ of what number?

In 21–24, find the fractional part.

21. 16 is what part of 48?

22. 24 is what part of 72?

23. What part of 35 is 7?

24. What part of 85 is 25?

Cumulative Review of Parts VII–X

In 1–5, represent the given facts as fractions.

1. A book has 280 pages and you read 140 pages. __________
2. A down payment of $50 was made on a radio that sells for $225. __________
3. A light is on during twenty minutes of each hour. __________
4. Six eggs were broken out of two dozen. __________
5. It is cold for three months out of the year. __________

In 6–9, reduce each fraction to lowest terms.

6. $\frac{240}{300}$ **7.** $\frac{1{,}000}{8{,}000}$ **8.** $\frac{1{,}500}{4{,}500}$ **9.** $\frac{468}{824}$

In 10–13, change each fraction to a higher equivalent that has the given denominator.

10. $\frac{13}{25} = \frac{\quad}{100}$ **11.** $\frac{12}{16} = \frac{\quad}{64}$ **12.** $\frac{2}{3} = \frac{\quad}{60}$ **13.** $\frac{4}{5} = \frac{\quad}{75}$

In 14–17, circle the larger fraction.

14. $\frac{3}{4}$ $\frac{7}{8}$ **15.** $\frac{3}{5}$ $\frac{2}{3}$ **16.** $\frac{7}{8}$ $\frac{13}{16}$ **17.** $\frac{4}{5}$ $\frac{5}{7}$

In 18–21, arrange each set of fractions in descending order of value (largest first).

18. $\frac{3}{4}$ $\frac{7}{8}$ $\frac{13}{16}$ **19.** $\frac{2}{3}$ $\frac{7}{9}$ $\frac{5}{6}$ **20.** $\frac{3}{10}$ $\frac{4}{15}$ $\frac{2}{5}$ **21.** $\frac{11}{14}$ $\frac{5}{7}$ $\frac{23}{28}$

In 22–25, change each improper fraction to a mixed number.

22. $\frac{25}{6}$ **23.** $\frac{13}{3}$ **24.** $\frac{17}{16}$ **25.** $\frac{35}{8}$

In 26–29, change each mixed number to an improper fraction.

26. $2\frac{3}{5}$ **27.** $4\frac{2}{3}$ **28.** $9\frac{1}{7}$ **29.** $10\frac{5}{6}$

In 30–35, add. Change improper fractions to mixed numbers or to whole numbers.

30. $\frac{19}{25} + \frac{21}{25} + \frac{18}{25} + \frac{23}{25}$

31. $\frac{29}{64} + \frac{37}{64} + \frac{45}{64} + \frac{39}{64}$

32. $\frac{3}{4}$, $\frac{4}{5}$, $\frac{7}{10}$

33. $\frac{2}{3}$, $\frac{5}{6}$, $\frac{3}{5}$

34. $\frac{7}{8}$, $\frac{5}{6}$, $\frac{11}{12}$

35. $\frac{5}{7}$, $\frac{3}{4}$, $\frac{11}{14}$

In 36–39, subtract. Reduce answers to lowest terms.

36. $\frac{5}{9} - \frac{7}{18}$

37. $\frac{13}{16} - \frac{3}{8}$

38. $\frac{11}{15} - \frac{3}{10}$

39. $\frac{5}{7} - \frac{1}{3}$

In 40–47, multiply. Simplify by cancelling where possible. Reduce answers to lowest terms.

40. $\frac{5}{16} \times \frac{2}{3} \times \frac{3}{5}$

41. $\frac{14}{15} \times \frac{3}{7} \times \frac{1}{2}$

42. $\frac{3}{4} \times \frac{8}{9} \times \frac{2}{3}$

43. $\frac{4}{5} \times \frac{10}{18} \times \frac{2}{3}$

44. $3\frac{3}{4} \times 2\frac{1}{2}$

45. $4\frac{1}{3} \times 9$

46. $5\frac{3}{5} \times 3\frac{1}{3}$

47. $5\frac{5}{8} \times 6\frac{2}{3}$

In 48–51, divide. Reduce answers to lowest terms.

48. $\frac{2}{3} \div \frac{5}{8}$

49. $\frac{9}{15} \div \frac{3}{5}$

50. $\frac{5}{16} \div \frac{3}{3}$

51. $\frac{7}{8} \div \frac{3}{4}$

In 52–67, perform the indicated operations. Answers that are improper fractions should be expressed as mixed numbers. Reduce all answers to lowest terms.

52. $21\frac{1}{3}$, 19, $+24\frac{11}{12}$

53. $19 - 18\frac{5}{14}$

54. $7\frac{1}{3} \div 5\frac{1}{2}$

55. 575, $\times 28\frac{4}{5}$

56. $6\frac{3}{4} \div 2\frac{1}{4}$

57. 20, $18\frac{4}{5}$, $+34\frac{3}{4}$

58. 63, $\times 34\frac{2}{3}$

59. $30 - 15\frac{3}{4}$

60. $\begin{array}{r} 964 \\ \times 56\frac{3}{4} \\ \hline \end{array}$

61. $3\frac{3}{4} \div 4\frac{3}{8}$

62. $14\frac{12}{16} - 10\frac{3}{4}$

63. $\begin{array}{r} 14\frac{7}{9} \\ 18\frac{5}{6} \\ +19 \\ \hline \end{array}$

64. $35\frac{3}{5} - 26\frac{7}{8}$

65. $\begin{array}{r} 24 \\ 23\frac{5}{8} \\ +22\frac{2}{3} \\ \hline \end{array}$

66. $8\frac{4}{7} \div 6$

67. $\begin{array}{r} 760 \\ \times 86\frac{3}{5} \\ \hline \end{array}$

In 68–71, simplify the complex fraction.

68. $\dfrac{\frac{3}{8}}{\frac{3}{4}}$

69. $\dfrac{9}{\frac{3}{5}}$

70. $\dfrac{5\frac{1}{3}}{2}$

71. $\dfrac{3\frac{1}{2}}{4\frac{1}{2}}$

In 72–75, find the unknown number.

72. $\frac{4}{5}$ of what number is 132?

73. $\frac{2}{3}$ of what number is 120?

74. 135 is $\frac{3}{7}$ of what number?

75. 114 is $\frac{2}{3}$ of what number?

In 76–79, find the fractional part.

76. 48 is what part of 60?

77. 75 is what part of 90?

78. What part of 72 is 45?

79. What part of 35 is 28?

80. If $\frac{1}{2}$ of a given number is 12, what is $\frac{1}{3}$ of the given number?

81. If $\frac{1}{5}$ of a given number is 20, what is $\frac{1}{4}$ of the given number?

82. If $\frac{2}{3}$ of a given number is 40, what is $\frac{2}{5}$ of the given number?

PART XI. Decimal Fractions

UNIT 40. Understanding Decimal Fractions

WORDS TO KNOW

Proper fractions, improper fractions, and mixed numbers are all written with a fraction bar. Fractions with a fraction bar are called **common fractions.**

When a common fraction has a denominator of a power of 10, it can be expressed as a **decimal fraction.** In decimal fractions, the fraction bar is replaced with the **decimal point.**

A mixed fraction consists of a whole number and a common fraction. Similarly, a **mixed decimal** consists of a whole number and a decimal fraction.

Perhaps you have noticed that amounts of money on personal checks are often written as **common fractions.** Nine dollars and fifty cents is written as $\$9\frac{50}{100}$. When you write the amount $\$9\frac{50}{100}$ as \$9.50, you express the common fraction $\frac{50}{100}$ as the **decimal fraction** .50. Also, you express the mixed fraction $9\frac{50}{100}$ as the **mixed decimal** 9.50. In either example, the fraction bar is replaced by the **decimal point.**

A decimal fraction is similar to a common fraction, because it represents a part of a unit. A common fraction can have any number as its denominator, but a decimal fraction must have a denominator that is a power of 10. Powers of 10, you recall, are 10, 100, 1,000, 10,000, and so on.

When you write a decimal fraction, you do not write the denominator. *The denominator of a decimal fraction is never written down.* Rather, it is determined by the number of digits to the right of the decimal point.

Here are some examples of common fractions written as decimal fractions:

$\frac{7}{10} = .7$ Read ".7" as "seven *tenths.*" Note that the 7 is *one place* to the right of the decimal point and the number *10* has *one zero.*

$\frac{13}{100} = .13$ Read ".13" as "thirteen *hundredths.*" Note that the right-hand digit is *2 places* to the right of the decimal point and the number *100* has *two zeros.*

$\frac{135}{1,000} = .135$ Read ".135" as "one hundred thirty-five *thousandths.*" Note that the right-hand digit is *3 places* to the right of the decimal point and the number *1,000* has *three zeros.*

As you move to the right of the decimal point, the value of each digit is divided by 10. Thus, a value of *tenths* is divided by 10 to get a value of *hundredths;* a value of hundredths is divided by 10 to get a value of *thousandths;* etc.

Remember

When you read the name of a decimal fraction, you *say the denominator* even though the denominator is not actually written down.

The following table will enable you to read the names of decimal fractions:

Table II: Names of Decimal Fractions

Number of Digits to the Right of the Decimal Point	Name of the Decimal Fraction	Example	Name of the Example
1 digit	tenths	.5	five tenths
2 digits	hundredths	.18	eighteen hundredths
3 digits	thousandths	.101	one hundred one thousandths
4 digits	ten-thousandths	.0004	four ten-thousandths
5 digits	hundred-thousandths	.00022	twenty-two hundred-thousandths
6 digits	millionths	.000001	one millionth

In addition to indicating the denominators of decimal fractions, the decimal point separates the whole unit from the decimal fraction in a mixed decimal. Mixed decimals are read by adding the names of decimal fractions to the names of whole numbers. The decimal point is read as the word "and." For example, "5.06" is read "five and six hundredths," and "99.44" is read "ninety-nine and forty-four hundredths."

EXERCISES

1. Read the following decimal fractions and mixed decimals out loud and write them as word phrases.

a. .4 ____________________

b. 2.07 ____________________

c. .15 ____________________

d. .125 ____________________

e. 7.025 ____________________

f. .0623 ____________________

g. 5.50 ____________________

h. .070 ____________________

i. .045 ____________________

j. 1.0063 ____________________

k. 16.24 ____________________

l. .08 ____________________

2. Write each of the following word phrases as a decimal fraction.

a. six hundredths ____________________

b. eight tenths ____________________

c. twenty-three hundredths ____________________

d. two hundred twenty-three thousandths ____________________

e. thirty hundredths ____________________

f. four and forty-seven thousandths ____________________

g. three thousand two hundred twenty-seven ten-thousandths ____________________

h. five hundred twenty-eight ten-thousandths ____________________ *i.* one and nine thousandths ______________

j. eight hundredths ______________

k. seven tenths ______________

l. ten and twenty-three ten-thousandths ______________

3. Change each of the following common fractions to a decimal fraction.

a. $\frac{5}{100}$ *b.* $\frac{15}{1,000}$ *c.* $\frac{4}{10}$ *d.* $\frac{40}{100}$ *e.* $\frac{45}{100}$

f. $\frac{99}{1,000}$ *g.* $\frac{125}{1,000}$ *h.* $\frac{100}{1,000}$ *i.* $\frac{453}{10,000}$ *j.* $\frac{53}{1,000}$

4. Change each of the following decimal fractions to a common fraction.

a. .061 *b.* .5 *c.* .85 *d.* .005 *e.* .0352

f. .06 *g.* .0305 *h.* .535 *i.* .75 *j.* .075

APPLICATION PROBLEMS

SAMPLE SOLUTION

John earns $100 each week and has a deduction of $5 for medical insurance. Write the deduction as a decimal fraction of his earnings.

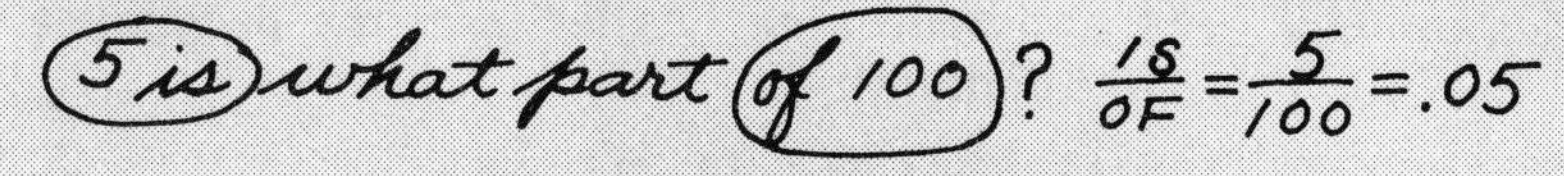

.05

1. Tom had $10 and spent $7. Write as a decimal fraction the part of the money he spent. ______________

2. A coat selling for $100 was reduced by $20. Write the reduction as a decimal fraction of the price. ______________

3. A salesperson earns $15 for every $100 of sales. Write the earnings as a decimal fraction of the sales. ______________

4. A retailer received a shipment of 1,000 drinking glasses and found 115 of them to be broken. Write the number of broken glasses as a decimal fraction of the whole shipment. ______________

5. Janet bought a car for $10,000 and made a down payment of $3,000. Write the down payment as a decimal fraction of the cost of the car. ______________

6. A real estate tax is $8 for every $100 of assessed value. Write the tax as a decimal fraction of assessed value. ______

7. A sales tax is 7¢ for every $1 of purchase. Write the sales tax as a decimal fraction of the purchase. (*Hint:* $1 = 100¢.) ______

8. In a survey on toothpaste, 6 out of 10 people preferred a certain brand. Write the ratio as a decimal fraction. ______

9. A radio originally selling for $300 was reduced by $60. Write the reduction as a decimal fraction of the original price. (*Hint:* Reduce the common fraction to a fraction with the denominator 10.) ______

10. The Goodwin family bought a house for $60,000 and made a down payment of $12,000. Write the down payment as a decimal fraction of the cost of the house. ______

UNIT 41. Comparing Decimals

WORDS TO KNOW

In our discussions of decimal fractions, we will use the word **decimal** to mean "decimal fraction." Thus, "comparing decimals" means "comparing decimal fractions."

When you compared the values of common fractions (Unit 25), you found a *common denominator.*

To compare the values of **decimals** that have different numbers of decimal places, you must make them alike by changing them to decimals that have the *same number of decimal places.* When two decimals have the same number of decimal places, they have a common denominator. Thus, .07 and .67 have a common denominator of 100. The decimals .018 and .180 have a common denominator of 1,000.

To compare decimals:

Step 1: Change the given decimals to decimals that have the same number of decimal places. To do this, *place zeros to the right of the last digit* until the desired number of decimal places is reached.

Step 2: Compare the resulting numbers. The higher number is the decimal with the larger value.

EXAMPLE 1. Which is larger, .5 or .45?

Solution: To compare .5 and .45, change them to 2-place decimals: .5 = .50 and .45 = .45. The decimal .50 (50 hundredths) is larger than the decimal .45 (45 hundredths).

Answer: .5 is larger than .45.

EXAMPLE 2. Which is smaller, .4 or .04?

Solution: Change .4 and .04 to 2-place decimals: .4 = .40 and .04 = .04. The decimal .04 (4 hundredths) is smaller than the decimal .40 (40 hundredths).

Answer: .04 is smaller than .4.

EXAMPLE 3. Arrange the following decimals in order of descending values (largest first):

.01 .1 .001

Solution: Change the given numbers to 3-place decimals: .01 = .010, .1 = .100, and .001 = .001. The largest decimal is .100 (100 thousandths), followed by .010 (10 thousandths), followed by .001 (1 thousandth).

Answer: .1 .01 .001

EXERCISES

In 1–6, circle the largest decimal.

1. .7 .07 .070
2. .013 .0115 .12
3. .05 .50 .0050
4. .60 .8 .900
5. .45 .045 .405
6. .4 .05 .008

In 7–10, underline the decimals that have the same value.

7. .04 .040 .0040
8. .8 .08 .80
9. .03 .003 .0300
10. .17 .170 .0170

In 11–14, arrange the decimals in ascending order of size (smallest value first).

SAMPLE SOLUTIONS

a. .400 .150 .235   .15   .235   .4

b. .0300 .0230 .4000 .0415   .023   .03   .0415   .4

11. .125 .1259 .2 .02 ____ ____ ____ ____
12. .6 .605 .006 .07 ____ ____ ____ ____
13. .025 .205 .250 .0025 ____ ____ ____ ____
14. .75 .075 .0780 .8 ____ ____ ____ ____

APPLICATION PROBLEMS

1. John walked .6 mile and Tom walked .58 mile. Who walked farther? ____________
2. Which weighs more, .64 pound or .5 pound? ____________
3. Two brands of coffee are priced at $1.15 a can. If one can weighs 1.25 pounds and the other weighs 1.2 pounds, which brand is the better buy? ____________
4. Billy's height is 34.55 centimeters, and Bobby's height is 34.6 centimeters. Which boy is taller? ____________

5. A rubber washer is .094 inch thick, and a zinc washer is .0945 inch thick. Which washer has the greater thickness? ____________

6. Which weighs more, .063 kilogram or .63 kilogram? ____________

7. Heidi jogged 2.50 miles, and Jay jogged 2.05 miles. Who jogged the greater distance? ____________

8. A bottle of Brand X orange drink contains .8 liter, and a bottle of Brand Y contains .75 liter. Which brand contains more orange drink? ____________

9. The diameters of 3 tulip bulbs are 5.04 centimeters, 5.4 centimeters, and 5.045 centimeters. Which is the largest? ____________

10. Four lengths of lumber measure 2.85 meters, 2.09 meters, 2.908 meters, and 2.91 meters. Arrange the lengths from largest to smallest. ____________

UNIT 42. Rounding Off Decimals

When a whole number is *rounded off,* its actual value is changed to an approximate value. In Unit 2, you learned how to round off whole numbers to any given place, such as the tens place or the hundreds place.

To round off decimals, you follow the same steps you used to round off whole numbers. However, in the final step, you omit all digits inside the parentheses.

After deciding on the value to be rounded off to, do the following steps:

Step 1: Place parentheses around all digits to the right of this value.

Step 2: A. If the left-hand digit inside the parentheses is 5 or more, add "1" to the part outside the parentheses. *Note:* In this step, the actual value of "1" will vary. If you round off to *tenths,* then "1" will have a value of .1; if you round off to *thousandths,* then "1" will have a value of .001; etc.

B. If the left-hand digit inside the parentheses is less than 5, do not add "1."

Step 3: Omit all digits inside the parentheses.

EXAMPLE 1. Round off .2645 to the nearest *tenth.*

Step 1: Place parentheses around all digits to the right of the *tenths* place.

.2(645)

Step 2: Since the left-hand digit inside the parentheses is greater than 5, add "1" to the part outside the parentheses. (In this example, "1" is one *tenth.*)

```
 .2(645)
+.1
 .3(645)
```

Step 3: Omit all digits inside the parentheses.

.3(~~645~~)

Answer: .2645 = .3, to the nearest tenth.

EXAMPLE 2. Round off .2645 to the nearest *hundredth.*

Step 1: Place parentheses around all digits to the right of the *hundredths* place.

.26(45)

Steps 2 and 3: Since the left-hand digit inside the parentheses is less than 5, do not add "1." Then omit all digits inside the parentheses.

```
 .26(45)
  +0
 .26(45)
```

(the 45 in the result is crossed out)

Answer: .2645 = .26, to the nearest hundredth.

EXAMPLE 3. Round off .2645 to the nearest *thousandth.*

Step 1: Place parentheses around all digits to the right of the *thousandths* place.

.264(5)

Steps 2 and 3: Since the digit inside the parentheses is 5, add "1" to the part outside the parentheses. (In this example, "1" is one *thousandth.*) Then omit the digit inside the parentheses.

$$\begin{array}{r} .264(5) \\ +.001 \\ \hline .265(\not{5}) \end{array}$$

Answer: .2645 = .265, to the nearest thousandth.

Sometimes more than one digit of the part outside the parentheses will be changed.

EXAMPLE 4. Round off 3.1974 to the nearest *hundredth.*

Step 1: Place parentheses around all digits to the right of the *hundredths* place.

3.19(74)

Steps 2 and 3: Since the left-hand digit inside the parentheses is greater than 5, add "1" to the part outside the parentheses. (In this example, "1" is one *hundredth.*) Then omit the digits inside the parentheses.

$$\begin{array}{r} 3.19(74) \\ +.01 \\ \hline 3.20(\not{74}) \end{array}$$

Answer: 3.1974 = 3.20, to the nearest hundredth.

EXERCISES

In 1–8, round off to the nearest *tenth.*

1. .37 **2.** .43 **3.** .55 **4.** .86

5. .158 **6.** .638 **7.** .456 **8.** .076

In 9–16, round off to the nearest *hundredth.*

9. .344 **10.** .248 **11.** .295 **12.** .305

13. .125 **14.** .238 **15.** .3264 **16.** .3485

In 17–24, round off to the nearest *thousandth.*

17. .2452 **18.** .0348 **19.** .1153 **20.** .00085

21. .0042 **22.** .23965 **23.** .0346 **24.** .2508

APPLICATION PROBLEMS

1. The diameter of a rod measures .678 of a centimeter. What is the measurement to the nearest *hundredth*? __________

2. A package of meat weighs 2.53 pounds. Rounded off to the nearest *tenth*, what is the weight of the package? __________

3. The thickness of a disc measures .1256 of an inch. Round off the measurement to the nearest *thousandth*. __________

4. A metal bar measures 5.347 inches. Round off the measurement to the nearest *hundredth*. __________

5. Round off 3.25 billion to the nearest *tenth* of a billion. __________

6. Round off 3.63 to the nearest *tenth*. What is the decimal value of the digit omitted? __________

7. What is the approximate weight of 2.053 ounces of gold rounded off to the nearest *tenth*? __________

8. What is the approximate weight of .65 of a pound rounded off to the nearest *pound*? __________

9. Round off 10.4507 to the nearest *hundredth*. __________

10. A recipe calls for 2.05 grams of cinnamon. Round off the quantity to the nearest *tenth*. __________

UNIT 43. Changing Decimals to Common Fractions

Decimals have denominators of powers of 10.

To change a decimal to a common fraction, *rewrite the decimal as a fraction that has the equivalent denominator in a power of 10.* Reduce this common fraction to lowest terms.

A quick way of determining the denominator is to remember that the *number of digits* in the decimal is the same as the *number of zeros* in the denominator. For example:

A *one-digit* decimal changes to a fraction with *one zero* in the denominator: .8 changes to $\frac{8}{10}$.

A *two-digit* decimal changes to a fraction with *two zeros* in the denominator: .35 changes to $\frac{35}{100}$.

A decimal with *three digits* changes to a fraction with *three zeros* in the denominator: .415 changes to $\frac{415}{1,000}$.

A decimal with *four digits* changes to a fraction with *four zeros* in the denominator: .0356 changes to $\frac{356}{10,000}$.

EXAMPLE 1. Change .04 to a common fraction.

Solution: Rewrite the decimal *four hundredths* as the common fraction $\frac{4}{100}$ and reduce to lowest terms. (Notice that there are *2 digits* in the decimal .04 and *2 zeros* in the fraction $\frac{4}{100}$.)

$$.04 = \frac{4}{100} \qquad 4\Big|\frac{\overset{1}{\cancel{4}}}{\underset{25}{\cancel{100}}} = \frac{1}{25}$$

Answer: $.04 = \frac{1}{25}$

EXAMPLE 2. Change .625 to a common fraction.

Solution: Rewrite the decimal *six hundred twenty-five thousandths* as the common fraction $\frac{625}{1,000}$ and reduce to lowest terms. (Notice that there are *3 digits* in .625 and *3 zeros* in $\frac{625}{1,000}$.)

$$.625 = \frac{625}{1{,}000}$$

$$25\left|\frac{\overset{25}{\cancel{625}}}{\underset{40}{\cancel{1{,}000}}}\right. = \frac{25}{40} \qquad 5\left|\frac{\overset{5}{\cancel{25}}}{\underset{8}{\cancel{40}}}\right. = \frac{5}{8}$$

Answer: $.625 = \frac{5}{8}$

EXAMPLE 3. Change $.12\frac{1}{2}$ to a common fraction.

Solution: Rewrite the decimal *twelve and one-half hundredths* as $\frac{12\frac{1}{2}}{100}$ and simplify the complex fraction.

$$.12\frac{1}{2} = \frac{12\frac{1}{2}}{100} = 12\frac{1}{2} \div 100$$

$$= \frac{25}{2} \div \frac{100}{1}$$

(*Reminder:* To divide fractions, invert the divisor and multiply.)

$$\frac{25}{2} \div \frac{100}{1} = \frac{\overset{1}{\cancel{25}}}{2} \times \frac{1}{\underset{4}{\cancel{100}}} = \frac{1}{8}$$

Answer: $.12\frac{1}{2} = \frac{1}{8}$

EXERCISES

In 1–12, change the decimal to a common fraction and reduce to lowest terms.

1. .3 **2.** .05 **3.** .23

4. .008 **5.** .119 **6.** .013

7. .60 **8.** .030 **9.** .235

10. .010 **11.** .255 **12.** .75

In 13–18, change the decimal to a complex fraction and simplify.

13. $.33\frac{1}{3}$ **14.** $.14\frac{2}{7}$ **15.** $.16\frac{2}{3}$

16. $.37\frac{1}{2}$ **17.** $.66\frac{2}{3}$ **18.** $.66\frac{1}{2}$

APPLICATION PROBLEMS

1. The thickness of a nail is .25 of an inch. Write the thickness as a common fraction. ______
2. Jane bought $.87\frac{1}{2}$ of a pound of meat. What common fraction of a pound did she buy? ______
3. Write $.75 as a common fraction of a dollar. ______
4. The decimal fraction .60 of a mile is equal to what common fraction of a mile? ______
5. Write .2 of a second as a common fraction. ______
6. A package of ground beef weighs 2.5 pounds. Write the weight as a common mixed fraction. ______
7. Cheryl bought $.37\frac{1}{2}$ yard of ribbon. What common fraction of a yard did she buy? ______
8. Elaine jogs 3.80 miles every day. Write the distance as a common mixed fraction. ______
9. Richard works .35 mile from his home. Write the distance as a common fraction of a mile. ______
10. What common fraction of a dozen is $.83\frac{1}{3}$ dozen? ______

Review of Part XI (Units 40–43)

In 1–5, write each decimal as a word phrase.

1. .6 ______
2. .03 ______
3. .45 ______
4. .115 ______
5. .007 ______

In 6–10, write each word phrase as a decimal.

6. nine tenths ______
7. eighty-two hundredths ______
8. sixty-five thousandths ______
9. seven ten-thousandths ______
10. one and four hundredths ______

In 11–16, circle the largest decimal.

11. .8 .65 .08 .080

12. .5 .40 .050 .05

13. .12 .025 .0125 .02

14. .30 .03 .030 .0300

15. .45 .050 .05 .405

16. .705 .75 .075 .0750

In 17–22, round off to the nearest *hundredth.*

17. .048 **18.** .263 **19.** .005 **20.** .506 **21.** .123 **22.** .018

In 23–27, round off to the nearest *thousandth.*

23. .1255 **24.** .0008 **25.** .0634 **26.** .0038 **27.** .2365

In 28–36, change each decimal fraction to a common fraction and reduce to lowest terms.

28. .75 **29.** .07 **30.** .1

31. .20 **32.** .125 **33.** .375

34. .020 **35.** 6.002 **36.** 8.325

In 37–38, change the decimals to common fractions and circle the fraction with the greatest value.

37. .1 .01 .001 .0001

38. .1 .10 .100 .1000

PART XII. Adding, Subtracting, and Multiplying Decimals

UNIT 44. Addition of Decimals and Mixed Decimals

Adding decimals is similar to adding whole numbers. However, there are two things to remember when writing a column of decimals and mixed decimals to be added:

1. You must line up the decimal points, one under the other, in a straight line.
2. You must line up the digits to the right and to the left of each decimal point, one under the other, according to their value.

EXAMPLE 1. Add: 3.5 + 10.23 + .235

Solution: Line up the digits and decimal points correctly, one under the other.

```
   3.5
  10.23
    .235
  ------
```

Bring down the decimal point to where your answer will be.

```
   3.5
  10.23
    .235
  ------
     .
```

Add as with whole numbers.

```
   3.5
  10.23
    .235
  ------
  13.965
```

Answer: 3.5 + 10.23 + .235 = 13.965

EXAMPLE 2. Add: 8.97 + 89.7 + 897

Solution:

```
  121
    8.97
   89.7
  897
  ------
  995.67
```

Answer: 8.97 + 89.7 + 897 = 995.67

EXAMPLE 3. Add: $7 + $.59 + 27¢

Solution:

```
    1
  $7.00
    .59
    .27
  -----
  $7.86
```

Answer: $7 + $.59 + 27¢ = $7.86

EXERCISES

In 1–8, add.

1. 4.3 + .235 + 5.05

2. .35 + .175 + .62 + .4

3. 22.53 + 12.05 + 9.275

4. 16.84 + 23.246 + 34.5

5. 18.63 + 24.075 + 237.8

6. 115.45 + 230.125 + 452.864

7. 357.643 + 562.48 + 763.2875

8. 642.687 + 578.9 + 785.2576

APPLICATION PROBLEMS

1. Tom bought a coat for $97.50, a suit for $65.95, and a shirt for $3.75. How much money did he spend? __________

2. Janice had $237.64 in the bank, and made a deposit of $63.58. How much does she have in the bank? __________

3. Harry earned $185.65 one week and his wife earned $97.45. What were their combined earnings? __________

4. Find the total of the following distances: 5.6 miles, 12.1 miles, 108.4 miles, and 15.6 miles. __________

5. James earned the following commissions: $33.65, $27.83, $53.42, and $43.58. Find his total commission. __________

6. A secretary made the following bank deposit: cash, $347.95; checks, $373.43, $528.75, and $487.62. What is the total amount of the deposit? __________

7. Alice bought three packages of meat weighing 2.63 kilograms, .43 kilogram, and .97 kilogram. What is the total weight of the three packages of meat? __________

8. The base price of Pedro's new car is $8,650.85. He has the following optional equipment: automatic transmission, $875.68; power steering, $589.83; stereo, $368.47; and tinted glass, $97.39. What is the total cost of the car? __________

9. Alicia had the following deductions from her weekly pay check: federal withholding tax, \$68.37; FICA tax, \$28.73; and state tax, \$14.64. Find the total deductions. ____________

10. A salesperson sold the following lengths of fabric: 2.43 yards, 3.5 yards, 4.54 yards, and 3.07 yards. How many yards of fabric were sold? ____________

For 11–12, use a calculator.

11. A traveling sales representative submitted the expense account below. Find the totals for each expense item, the daily totals, and the grand total for the week.

Expense Item	DATES 5/6–5/10					Expense Item Totals
	Mon.	Tues.	Wed.	Thurs.	Fri.	
Motel	\$48.00	52.00	46.00	45.00	55.00	
Breakfast	2.87	3.05	3.58	3.25	3.68	
Lunch	5.62	5.85	6.35	5.43	6.85	
Dinner	12.90	11.63	12.86	13.75	11.87	
Taxi	9.45	8.60	7.50	8.75	9.65	
Car Rental	33.00	38.00	39.00	36.00	42.00	
Daily Totals						

Grand Total

12. In the following record of tax deductions, find the total deductions for each employee and the total for each type of tax. Check for accuracy by finding the grand total of columns and rows.

Employee	DEDUCTIONS				Total Deductions per Employee
	Federal Tax	FICA Tax	State Tax	City Tax	
Alsop	\$97.52	23.65	11.23	9.35	
Burns	83.21	19.32	8.78	8.42	
Cortez	94.48	21.67	10.36	9.15	
Davidson	98.37	23.86	11.49	9.63	
Eames	86.51	18.87	9.79	8.73	
Franco	97.15	23.36	11.20	9.28	
Tax Totals					

Grand Total

UNIT 45. Subtraction of Decimals and Mixed Decimals

To subtract decimals and mixed decimals, set up the subtraction problem as you did for addition:

1. Line up the decimal points, one under the other, in a straight line.
2. Line up the digits correctly, one under the other, according to their values.

Then follow the same procedure as for subtracting whole numbers.

Example 1. Subtract: 35.68 − 15.25

Solution: Line up the digits and decimal points correctly.

$$\begin{array}{r} 35.68 \\ -15.25 \\ \hline \end{array}$$

Bring down the decimal point to where your answer will be.

$$\begin{array}{r} 35.68 \\ -15.25 \\ \hline . \end{array}$$

Subtract as with whole numbers.

$$\begin{array}{r} 35.68 \\ -15.25 \\ \hline 20.43 \end{array}$$

Check your answer.

$$\begin{array}{r} 20.43 \\ +15.25 \\ \hline 35.68 \end{array} \checkmark$$

Answer: 35.68 − 15.25 = 20.43

Example 2. Subtract: 125.07 − 83.48

Solution:

$$\begin{array}{r} 125.07 \\ -83.48 \\ \hline 41.59 \end{array}$$

Check:

$$\begin{array}{r} 41.59 \\ +83.48 \\ \hline 125.07 \end{array} \checkmark$$

Answer: 125.07 − 83.48 = 41.59

Example 3. Subtract: 7.6 − 4.582

Solution: Write 7.6 as 7.600, so that it has as many places as the subtrahend.

$$\begin{array}{r} 7.600 \\ -4.582 \\ \hline 3.018 \end{array}$$

Check:

$$\begin{array}{r} 3.018 \\ +4.582 \\ \hline 7.600 \end{array} \checkmark$$

Answer: 7.6 − 4.582 = 3.018

EXERCISES

In 1–21, subtract.

1. $\begin{array}{r} .36 \\ .18 \\ \hline \end{array}$ **2.** $\begin{array}{r} .44 \\ .27 \\ \hline \end{array}$ **3.** $\begin{array}{r} .27 \\ .08 \\ \hline \end{array}$ **4.** $\begin{array}{r} .237 \\ .025 \\ \hline \end{array}$

5. $\begin{array}{r} .49 \\ .17 \\ \hline \end{array}$ **6.** $\begin{array}{r} .235 \\ .123 \\ \hline \end{array}$ **7.** $\begin{array}{r} .423 \\ .347 \\ \hline \end{array}$ **8.** $\begin{array}{r} 27.05 \\ 23.46 \\ \hline \end{array}$

9. 60.07
42.38

10. 47.163
21.687

11. 32.6
14.732

12. 28.52
13.756

13. .63 − .423

14. .523 − .35

15. .067 − .0048

16. 23.5 − 15.67

17. 47.8 − 22.9

18. 19.05 − 9.25

19. 35 − 15.63

20. 43 − 9.316

21. 1.3 − .5432

In 22–31, subtract:

22. 12.62 from 25.06

23. 27.5 from 46.628

24. 14.03 from 32.005

25. 19.325 from 28

26. 12.8 from 35.05

27. 18.005 from 23

28. $37.48 from $249.63

29. $346.35 from $740

30. $473.67 from $3,060.05

31. $3,638.47 from $4,005

APPLICATION PROBLEMS

1. A suit selling for $73.84 is reduced to $69.95. What is the amount of the reduction? ______

2. Mrs. O'Connor earned $9,628.87 last year. This year her income is $10,745.65. How much more did she earn this year? ______

3. Tom ran the 100-yard dash in 14.30 seconds, and later ran the same distance in 12.07 seconds. By how much did he improve his time? ______

4. A dress selling for $78.35 was reduced by $28.68. What is the new price of the dress? ______

5. Harry bought a hat for $18.65, a sports jacket for $57.68, and a tie for $7.89. If he had $98.65 before he bought the clothes, how much does he have left? ______

6. John earned $195.50 last week and had the following deductions taken from his pay check: federal income tax, $37.60; social security tax, $9.37; union dues, $1.93. Find John's net (take-home) pay. ______

7. Carol bought a television set for $345. She made a down payment of $73.48 and two installment payments of $53.65 each. How much does she still owe? ______

8. Andrea bought a blouse for $19.95, a skirt for $35.75, and a pair of shoes for $29.95. If she paid with a $100 bill, how much change did she get? ______

9. A bolt of fabric has 45 yards. A clerk sold 4.35 yards, 5.7 yards, and 3.33 yards. How many yards of fabric remain? ______

10. Jim's checking account had a balance of $735.25. If he wrote checks for $135.60, $87.56, $235, and $47.98, what is his new balance? ______

For 11–15, use a calculator.

11. Samantha earns $19,685 a year. Her friend earns $1,450 less. How much does Samantha's friend earn a year? ______

12. Rachel's gross salary last week was $496.75. There were the following deductions: federal tax, $96.53; state tax, $53.42; FICA tax, $26.28. What was her take-home pay (the net amount of her salary)? ______

13. Joel's net salary was $388.75. He paid the following amounts: credit card bill, $43.80; groceries, $56.85; telephone bill, $47.67. How much of his salary was left? ______

14. Sara bought a stereo set for $788.99 on the layaway plan and made three payments of $38.62, $45.50, and $72.90. How much does Sara still owe for the stereo? ______

15. The Santos family left on a car trip from New York to Florida, a distance of 2,570 kilometers. In four days, they traveled the following distances: 429.9 kilometers, 450.8 kilometers, 412.2 kilometers, and 439.5 kilometers. How much of the distance remained? ______

UNIT 46. Multiplication of Whole Numbers by Decimals

The multiplication of whole numbers by decimals or by mixed decimals is similar to the multiplication of whole numbers by whole numbers. However, you must determine where to insert the decimal point in the answer.

RULE

To multiply decimals and whole numbers:

1. Set up the problem and multiply as with whole numbers.
2. Count the number of decimal places to the right of the decimal point in the problem (the number of decimal places in the multiplicand or the number of decimal places in the multiplier). Moving from right to left, count as many decimal places in the product as there are decimal places in the problem, and insert the decimal point there.

Example 1. Multiply: 23.43 × 25

Solution: Multiply as with whole numbers.

```
 12 1
 23.43
   ×25
 11715
 4686
 58575
```

Count the number of decimal places to the right of the decimal point in the problem (in the multiplicand).

```
23.43  → (2) decimal places
  ×25
```

Since the multiplicand has 2 decimal places to the right of its decimal point, the product must have 2 decimal places to the right of its decimal point.

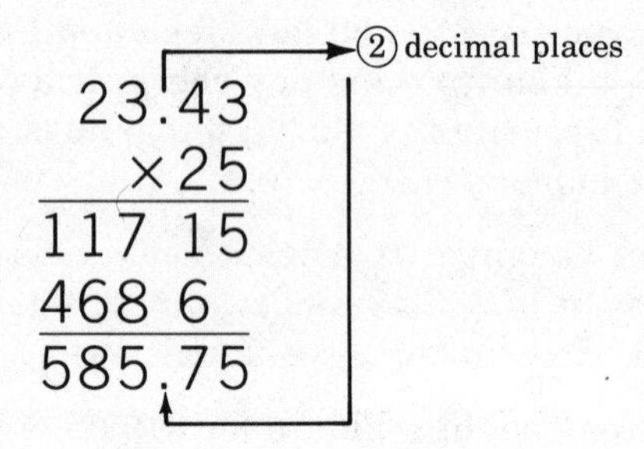

Answer: 23.43 × 25 = 585.75

Example 2. Multiply: 174 × .083

Solution:

```
  53
  21
  174
 ×.083   → (3) decimal places
   522
 13 92
 14.442
```

Answer: 174 × .083 = 14.442

EXERCISES

In 1–25, multiply.

1. 123 3.5	**2.** 23.28 14	**3.** 235 .25	**4.** 1,473 .24	**5.** 46.05 32

6. 562
   .07

7. 2,474
   3.6

8. 5,433
   2.15

9. 634.24
   36

10. 2,364
    .19

11. 4,247
    .34

12. 56,235
    .236

13. 23.47
    235

14. 5,628
    .14

15. 7,365
    4.6

16. 63.46
    145

17. 3,468
    3.47

18. 53.52
    235

19. 243.6
    235

20. 4,368
    3.06

21. $35.73
    15

22. $19.65
    35

23. $247.98
    27

24. $149.95
    46

25. $653.75
    87

APPLICATION PROBLEMS

1. How much will 135 floor tiles cost at $.23 each? ______
2. Tom earns $153.50 a week. How much does he earn in a year? ______
3. How much will 53 yards of carpeting cost at $7.65 a yard? ______
4. If a package of sugar weighs 2.125 pounds, how much would 335 packages weigh? ______
5. The Chadwick family spends an average of $87.68 a week on food. How much does the Chadwick family spend on food in one year? ______
6. A jar of instant coffee weighs 226.8 grams. How much will 48 jars weigh? ______
7. Valerie bought a color television set, to be paid for by a down payment of $75 and 16 installments of $21.75 each. What is the total cost of the television set? ______

8. Harry bought 3 shirts at $14.75 each, 8 pairs of socks at $1.79 a pair, and 4 ties at $6.85 each. How much did Harry spend? ____________

9. Flo's Food Stop received a delivery of two dozen 14.5-ounce packages of cheddar cheese and one dozen 6.25-ounce packages of Swiss cheese. What was the total weight of the cheese? ____________

10. The following supplies were purchased for an office: 250 memo pads at 19¢ each, 75 typewriter ribbons at $1.29 each, 125 reams of paper at $3.95 each, and 65 dozen ballpoint pens at 97¢ per dozen. What was the total cost of the purchase? ____________

UNIT 47. Multiplication of Decimals by Decimals

To multiply two decimals or mixed decimals, follow the rule explained in the preceding unit. However, since both the multiplicand *and* the multiplier have decimal places to the right of the decimal point, you must count the *total* number of decimal places in the problem. Add the number of decimal places in the top number (the multiplicand) to the number of decimal places in the bottom number (the multiplier). Moving from right to left, count as many places in the product as there are total decimal places in the problem. Insert the decimal point there.

EXAMPLE 1. Multiply: $348.25 × .16

Solution: Multiply as with whole numbers.

```
 348.25
   ×.16
 208950
 34825
 557200
```

Count the number of decimal places to the right of the decimal point in the top number. Count the number of decimal places to the right of the decimal point in the bottom number. Add these two numbers.

```
348.25  → 2
  ×.16  → 2
          (4) decimal places
```

To place the decimal point in your product, count 4 decimal places, moving from right to left.

```
 348.25  → 2
   ×.16  → 2
           (4) decimal places
 20 8950
 34 825
 55.7200
```

Since the answer is in dollars and cents, cross out the last two zeros, because they have no value.

$55.72~~00~~

Answer: $348.25 × .16 = $55.72

EXAMPLE 2. Multiply: 53.6 × 2.02

Solution:

```
   53.6  → 1
  ×2.02  → 2
           (3) decimal places
  1 072
 107 20
108.272
```

Answer: 53.6 × 2.02 = 108.272

In placing a decimal point in your product, it may be necessary to use zeros to the left as placeholders.

EXAMPLE 3. Multiply: .25 × .07

Solution:

```
 .25      → 2
× .07     → 2
                (4) decimal places
.(0)175
```

Answer: .25 × .07 = .0175

EXERCISES

In 1–20, multiply.

1. 43.26 × 5.4 **2.** 46.3 × .025 **3.** 3.27 × .35 **4.** .456 × .34 **5.** 1.64 × .015

6. .182 × .11 **7.** 245.6 × .045 **8.** 38.65 × .86 **9.** 5.638 × 24.5 **10.** 253.48 × .65

11. 437.05 × 3.47 **12.** 568.34 × .605 **13.** 720.62 × .458 **14.** 649.42 × 23.08 **15.** 482.63 × .008

16. 574.65 × 20.07 **17.** 867.38 × 52.7 **18.** 5,642.8 × .539 **19.** 738.36 × .075 **20.** 65.84 × .427

APPLICATION PROBLEMS

1. Brand X gasoline sells for $1.06 a gallon. What is the cost of 18.5 gallons? ________
2. How much would 4.65 yards of velvet cost if one yard sells for $4.62? ________
3. The average rainfall in a city is 3.425 inches per month. How much rain will fall in 4.5 months? (Since your answer will be approximate, round it off to the nearest *hundredth.*) ________
4. A car rental company charges $105 a week plus $.06 per mile. If you travel 1,625.5 miles in seven days, how much would the car cost you? ________
5. An acre of farmland produces 34.25 bushels of wheat. How many bushels will be produced by a 20.8-acre farm? ________
6. James drives an average of 50.75 miles per hour. How many miles will he travel in 6.4 hours? ________
7. If a liter of water from Mr. Earp's well contains .85 gram of minerals, what is the weight, to the nearest *hundredth* of a gram, of the minerals in 270.25 liters of water? ________
8. On a trip to Europe, Joan drove 3,724.8 kilometers. If one kilometer is .62 mile, how many miles did she drive? ________
9. A fruit-nut mix contained 5.25 pounds of walnuts at $1.52 a pound, 4.5 pounds of raisins at $.88 a pound, and 2.4 pounds of coconut at $1.40 a pound. What was the total cost of the mix? ________
10. A cab ride in Winter Falls costs $.80 for the first .25 mile, and $.15 for every .25 mile after that. If Ms. Cross lives 1.75 miles from the train station, what will a ride to the station cost her? ________

UNIT 48. Rounding Off to the Nearest Penny in Multiplication

When you obtain a product that must be expressed in dollars and cents, it is necessary to round off to the nearest *hundredth.* Since a penny is a hundredth of a dollar, you must "round off to the nearest penny."

In Example 1 of the preceding unit, you rounded off the product $55.7200 to $55.72. Most of the time, however, the digits to be rounded off will not be zeros.

EXAMPLE 1. Multiply: $362.66 × .36

Solution: Perform the multiplication and place the decimal point in the product by counting four decimal places, from right to left.

```
   $362.66
      ×.36
   -------
   21 7596
  108 798
  --------
 $130.5576
```

In working with dollars and cents, your final answer must have only *two places* to the right of the decimal point. (Cents are hundredths of a dollar, and the value of the second digit to the right of the decimal point is hundredths.) Therefore, you must round off 130.5576 to the nearest hundredth. Follow the steps explained in the unit on rounding off decimals (Unit 42):

Step 1: Place parentheses around all digits to the right of the hundredths place.

$$130.55(76)$$

Step 2: Since the left-hand digit inside the parentheses is greater than 5, add "1" to the part outside the parentheses. Omit all digits inside the parentheses. In working with dollars and cents, the "1" will always be "1 penny," written ".01."

$$\begin{array}{r} 130.55(76) \\ +.01 \\ \hline 130.56(\not{7}\not{6}) \end{array}$$

Answer: $362.66 × .36 = $130.56

Since you have practiced rounding off whole numbers (Unit 2) and decimals (Unit 42), you should be able to round off a product to the nearest penny *mentally*.

Remember

When rounding off to the nearest penny, if the third digit to the right of the decimal point is 5 or more, add a penny to your answer. If the third digit is less than 5, do not add a penny.

Example 2. Multiply: $235.45 × .25

Solution:

$$\begin{array}{r} \$235.45 \\ \times .25 \\ \hline 11\ 77\ 25 \\ 47\ 09\ 0 \\ \hline \$58.86(\not{2}\not{5}) \end{array} = \$58.86$$

Answer: $235.45 × .25 = $58.86

Example 3. Find, to the nearest penny, the cost of 1.42 grams of a chemical if the price is (*a*) $2.12 per gram and (*b*) $1.90 per gram.

Solutions:

(*a*)

$$\begin{array}{r} 1.42 \\ \times \$2.12 \\ \hline 2\ 84 \\ 14\ 2 \\ 2\ 84 \\ \hline \$3.01(\not{0}\not{4}) \end{array} = \$3.01$$

Answer: $3.01

(*b*)

$$\begin{array}{r} 1.42 \\ \times \$1.90 \\ \hline 1\ 27\ 80 \\ 1\ 42 \\ \hline \$2.69(\not{8}\not{0}) \end{array} = \$2.70$$

Answer: $2.70

EXERCISES

In 1–20, multiply and round off to the nearest penny.

1. $\begin{array}{r} \$34.42 \\ .25 \\ \hline \end{array}$

2. $\begin{array}{r} \$46.37 \\ .045 \\ \hline \end{array}$

3. $\begin{array}{r} \$27.43 \\ .07 \\ \hline \end{array}$

4. $\begin{array}{r} \$53.64 \\ .34 \\ \hline \end{array}$

5. $\begin{array}{r} \$242.65 \\ .125 \\ \hline \end{array}$

6. $\begin{array}{r} \$256.43 \\ .68 \\ \hline \end{array}$

7. $\begin{array}{r} \$347.67 \\ .036 \\ \hline \end{array}$

8. $\begin{array}{r} \$362.49 \\ .43 \\ \hline \end{array}$

9. $465.55
   .38

10. $478.64
    .46

11. $538.78
    .53

12. $673.15
    .64

13. $2,435.23
    .25

14. $2,536.28
    .47

15. $3,648.47
    .54

16. $4,765.83
    .057

17. $4,267.56
    .48

18. $5,658.72
    .47

19. $5,748.68
    .54

20. $5,658.76
    .57

APPLICATION PROBLEMS

1. What is the cost of .75 liter of a juice drink that sells at $.68 per liter? ________

2. To find the total cost of an item, including sales tax, Sandy multiplies the price of the item by 1.065. What is the total cost of a toaster if the price is $14.50? ________

3. Josh worked as a stock clerk during the summer, earning $4.35 an hour. If he worked 32.6 hours one week, how much did he earn that week? ________

4. How much would 87.9 gallons of heating fuel cost at $1.17 a gallon? ________

5. A ream of paper costs $14.75. What is the cost of .5 ream? ________

6. How much would 4.73 pounds of meat cost at $3.65 per pound? ________

7. Coal sells for $93.85 per ton. How much will 5.43 tons of coal cost? ________

8. If Sal bought gasoline at $1.13 a gallon, what did he pay for 15.7 gallons? ________

9. A real estate tax rate is $.075 for every dollar of assessed value. What would be the real estate tax on a house assessed at $27,695? ________

10. A retailer estimates that out of every dollar of sales, $.23 is his gross profit. What is his gross profit on a sale of $1,347.85? ________

For 11–12, use a calculator.

11. Charlene bought the following spices for her cooking class: 1.25 ounces of paprika at $1.03 an ounce, 1.38 ounces of ginger at $1.27 an ounce, 1.40 ounces of cinnamon at $.85 an ounce, 1.45 ounces of cloves at $2.31 an ounce, and 1.35 ounces of nutmeg at $1.13 an ounce. What was the total cost of her purchase? ________________

12. The Acme Paper Company employs six part-time workers. The chart shows the number of hours each one worked during the first full week in April, and the hourly rate paid each worker. Complete the table to find the amount paid each worker, and find the total amount of the payroll.

	Number of Hours					Total Number of Hours	Hourly Rate	Earnings
Employee	M	T	W	Th	F			
Leslie	5.75		6.25		4.5		$4.25	
Brian	7	6.5		5.75	5		3.85	
Jose		3.25	6.75	7	4.5		4.15	
Fran	5.5	5.5	5.5	7	5.5		3.90	
Xavier		4.5	4.75		5.25		4.35	
Gerry	6.75			6.25	7		3.95	
							Total	

UNIT 49. Multiplication of Whole Numbers, Decimals, and Mixed Decimals by Powers of 10

WHOLE NUMBERS

To multiply a whole number by 10, place one zero to the right of the number. Thus,

$$23 \times 10 = 23(0) = 230$$

To multiply a whole number by 100, place two zeros to the right of the number. Thus,

$$23 \times 100 = 23(00) = 2{,}300$$

To multiply a whole number by 10, 100, 1,000, or any power of 10: Place as many 0's to the right of the given number (the multiplicand) as there are zeros in the multiplier.

EXAMPLE 1. Multiply: $253 \times 100{,}000$

Solution: Since there are five zeros in 100,000, place five zeros to the right of 253:

$$253(00000)$$

Answer: $253 \times 100{,}000 = 25{,}300{,}000$

DECIMALS

To multiply a decimal or a mixed decimal by a power of 10, move the decimal point the same number of places to the right as there are zeros in the multiplier.

EXAMPLE 2. Multiply: $12.35 × 10

Solution: Move the decimal point one place to the right.

$$\$12.35 \times 10 = \$12.3.5 = \$123.5$$

Because this problem deals with dollars and cents, and since cents are hundredths of a dollar, place a zero after the 5 (5 tenths) to make it 50 cents (50 hundredths).

Answer: $12.35 × 10 = $123.50

EXAMPLE 3. Multiply: 15.47 × 100

Solution: Move the decimal point two places to the right.

$$15.47 \times 100 = 15.47. = 1{,}547.$$

When there are no digits to the right of a decimal point, it is not necessary to include the decimal point.

Answer: 15.47 × 100 = 1,547

EXAMPLE 4. Multiply: 25.87 × 1,000

Solution: Move the decimal point three places to the right.

$$25.87 \times 1{,}000 = 25.87(\ ).$$

Since there are only two digits to the right of the decimal point, and since you moved the decimal point three places, insert a zero after the 7 as a placeholder: 2587(0).

Answer: 25.87 × 1,000 = 25,870

EXAMPLE 5. Multiply: .006 × 100

Solution: Move the decimal point two places to the right.

$$.006 \times 100 = .00.6 = 00.6$$

Since the zeros to the left of the decimal point have no value, and since they are not needed as placeholders, omit them.

Answer: .006 × 100 = .6

Remember

When multiplying whole numbers or decimals by powers of 10:

1. For whole numbers, place as many zeros to the right of the given number as there are zeros in the multiplier.
2. For decimals, move the decimal point the same number of places to the right as there are zeros in the multiplier.

EXERCISES

In 1–24, multiply. Perform the multiplications *mentally*.

1. 235 × 10
2. 470 × 10
3. 236 × 1,000
4. 140 × 100
5. 537 × 10,000
6. 325 × 10
7. 236 × 100
8. 307 × 10,000
9. 215 × 10
10. $13.05 × 10
11. $15.23 × 100
12. $12.63 × 1,000

13. $.63 × 100 **14.** $29.07 × 10 **15.** $.059 × 100

16. $.003 × 100 **17.** $123.00 × 10 **18.** $35.75 × 1,000

19. $.005 × 10,000 **20.** 45.61 × 10 **21.** 27.35 × 100

22. 15.08 × 10 **23.** 27.63 × 100 **24.** 43.37 × 1,000

APPLICATION PROBLEMS

1. Find the price of 100 floor tiles at $.57 each. ________
2. Molly bought 10 yards of fabric at $3.75 a yard. Find the total amount she paid. ________
3. Mrs. Atkins is buying a refrigerator on the installment plan, and has 10 payments left. If each payment is $17.53, how much does she still owe for the refrigerator? ________
4. A certain drug costs $.42 a capsule. How much would 100 capsules cost? ________
5. Wallpaper sells for $6.85 a roll. How much would 10 rolls cost? ________
6. A company allows its salespeople $1.27 per mile as car expenses. How much will be allowed for 1,000 miles? ________
7. How much will 100 calendars cost at $.63 each? ________
8. Frank earns $7.45 per hour. How much will he earn in 10 hours? ________
9. A company bought 10 calculators at $128.95 each. What was the total cost of the calculators? ________
10. How much will 100 ballpoint pens cost at $.125 each? ________

Review of Part XII (Units 44–49)

In 1–4, add.

1. 125.53 + 25.065 + 1.255 + .05

2. .05 + .125 + .008 + .8

3. 15.3 + 128.75 + 8.075 + .3075

4. 35.005 + 28.15 + 18.1 + 5.03

In 5–10, subtract.

5. 157.25 − 65.06

6. 24.4 − 12.53

7. 65 − 43.030

8. 27.05 − 14.165

9. 38.5 − 15.125

10. 150.3 − 85.5

In 11–22, multiply and round off to the nearest penny.

11. $2.85 × 15

12. $38.25 × 36

13. $265 × .35

14. $347 × .06

15. $127.50 × 4.6

16. $353 × 8.5

17. $87.08 × .27

18. $385 × .10

19. $465.27 × 6.8

20. $362.87 × .25

21. $305.05 × .06

22. $27.58 × 3.8

In 23–30, multiply *mentally*.

23. 14.5 × 10

24. 506 × 100

25. 35.8 × 100

26. 1.045 × 1,000

27. $2.50 × 10

28. $3.87 × 100

29. $12.95 × 1,000

30. $245 × 100

For 31–33, use a calculator.

31. Julio was hoping to get his time for the 100-yard dash down to 11.50 seconds. His best times, in seconds, in one week's practice were: Monday, 13.72; Tuesday, 12.58; Wednesday, 12.07; Thursday, 13.81; Friday, 11.94.

a. For each day, find by how much he exceeded his 11.50-second goal. ______ ______ ______ ______ ______

b. By how much did his Friday time beat his Monday time? ____________

32. Find the total cost of Mr. Linsdale's grain purchases if he bought 60 bushels of corn at $2.54 per bushel, 85 bushels of wheat at $3.56 per bushel, 52 bushels of oats at $1.46 per bushel, 78 bushels of barley at $2.42 per bushel, and 50 bushels of soybeans at $5.92 per bushel. ____________

33. For a health project, Teresa kept a record, as shown in the table, of her intake of riboflavin for one day.

a. Calculate the totals in the table.

b. At Teresa's age, the recommended daily intake of riboflavin is 1.8 milligrams. Compare her total intake with the recommended amount. Was it more or less, and by how much? ________________

Food	Quantity	Milligrams of Riboflavin	Total per Item
Orange juice	1 cup	.03 mg per cup	
Scrambled eggs	2 eggs	.16 mg per egg	
Toast	2 slices	.06 mg per slice	
Milk	3 cups	.4 mg per cup	
Chili	1 cup	.26 mg per cup	
Potato chips	20 chips	.01 mg per 10 chips	
Milk shake	1 cup	.67 mg per cup	
Tomato juice	.75 cup	.07 mg per cup	
Beef stew	1.5 cups	.17 mg per cup	
Lettuce	1 cup	.03 mg per cup	
Blue cheese dressing	2 tablespoons	.02 mg per tablespoon	
Cherry pie	1 slice	.12 mg per slice	
Ice cream	.5 cup	.33 mg per cup	
Pizza	2 slices	.18 mg per slice	
Cola	1 cup	0 mg	
		Total Intake for the Day	

PART XIII. Dividing Decimals

UNIT 50. Division When the Dividend Is a Decimal

To divide a whole number (the divisor) into a decimal or a mixed decimal (the dividend), you proceed as in the division of whole numbers. However, you must know where to put the decimal point in your answer (the quotient).

RULE

When the dividend is a decimal, place the decimal point in your answer *directly above the decimal point in the dividend.* Then divide as with whole numbers.

EXAMPLE. Divide: 150.75 ÷ 25

Solution: Use the division box to set up the problem, as with whole numbers. Then place a decimal point directly above the decimal point in the dividend.

$$25\overline{)150.75}$$

Now divide as with whole numbers.

$$\begin{array}{r} 6.03 \\ 25\overline{)150.75} \\ \underline{150} \\ 75 \\ \underline{75} \end{array}$$

Answer: 150.75 ÷ 25 = 6.03

EXERCISES

In 1–12, divide. Round off each answer to the nearest *tenth.*

1. $6\overline{).989}$
2. $7\overline{).538}$
3. $27\overline{)46.23}$
4. $35\overline{)250.60}$
5. $32\overline{)803.07}$
6. $43\overline{)135.63}$
7. $52\overline{)460.85}$
8. $45\overline{)605.50}$
9. $37\overline{)475.82}$

10. $63\overline{)1{,}472.87}$

11. $65\overline{)3{,}417.07}$

12. $72\overline{)3{,}053.17}$

In 13–18, divide. Give answers in dollars and cents.

13. $1,185.75 ÷ 45

14. $3,498.45 ÷ 83

15. $3,325.64 ÷ 142

16. $8,014.50 ÷ 225

17. $10,942.25 ÷ 253

18. $13,125.40 ÷ 310

APPLICATION PROBLEMS

1. Mr. Garcia bought a car for $9,880.56. If the car is to be paid for in 36 equal installments, how much will each payment be? ________
2. Mrs. Kelly bought 43 yards of velvet for $378.83. What was the price of the velvet per yard? ________
3. The Wilsons spent $1,968.75 on their 25-day vacation. What was the average cost per day? ________
4. Tom paid $2,317.64 in federal income tax last year. On the average, how much income tax did he pay each week? ________
5. Mr. Adams paid $3,424.56 in rent last year. Find his monthly rent. ________
6. Jane earns $22,139 a year. What is her weekly salary? ________
7. A real estate office bought 12 model A typewriters for $5,711.40. How much did each typewriter cost? ________

8. A department store bought 250 sleeveless blouses for $1,687.50. What is the cost of each blouse? ______________

9. Edward earned $349.20 for a 40-hour week. What is his hourly rate of pay? ______________

10. Robert bought a dozen pairs of socks for $22.68. How much is one pair? ______________

UNIT 51. Division of Whole Numbers, Decimals, and Mixed Decimals by Powers of 10

WHOLE NUMBERS

Every whole number has a decimal point after the right-hand digit, whether or not the decimal point is actually written. Thus, "6" may be written "6." and "27" may be written "27."

RULE

To divide a whole number by 10, by 100, or by any power of 10, move the decimal point in the dividend to the left as many places as there are zeros in the divisor.

EXAMPLE 1. Divide: 762 ÷ 10

Solution: Move the decimal point 1 place to the left:

$$762. \div 10 = 76.2. = 76.2$$

Answer: 762 ÷ 10 = 76.2

EXAMPLE 2. Divide: 762 ÷ 100

Solution: Move the decimal point 2 places to the left:

$$762. \div 100 = 7.62. = 7.62$$

Answer: 762 ÷ 100 = 7.62

EXAMPLE 3. Divide: 762 ÷ 1,000

Solution: Move the decimal point 3 places to the left:

$$762. \div 1{,}000 = .762. = .762$$

Answer: 762 ÷ 1,000 = .762

EXAMPLE 4. Divide: 762 ÷ 10,000

Solution: Move the decimal point 4 places to the left and place a zero to the right of the decimal point as a placeholder:

$$762. \div 10{,}000 = .(\ )762. = .0762$$

Answer: 762 ÷ 10,000 = .0762

DECIMALS

To divide a decimal or a mixed decimal by a power of 10, follow the preceding rule: *Move the decimal point in the dividend to the left as many places as there are zeros in the divisor.* Include zeros as placeholders when necessary.

EXAMPLE 5. Divide: 1.72 ÷ 100

Solution: Move the decimal point 2 places to the left:

$$1.72 \div 100 = .(\ )1.72 = .0172$$

Answer: 1.72 ÷ 100 = .0172

EXAMPLE 6. Divide: .06 ÷ 100

Solution: Move the decimal point 2 places to the left:

$$.06 \div 100 = .(\ )(\ ).06 = .0006$$

Answer: .06 ÷ 100 = .0006

EXERCISES

In this set of exercises, you should perform the divisions *mentally.*

In 1–12, divide each number by 10.

1. 85	**2.** 78	**3.** 57	**4.** 230	**5.** 460	**6.** 650
7. 32.5	**8.** 47.8	**9.** 4.63	**10.** 23.37	**11.** 3.2	**12.** .478

In 13–24, divide each number by 100.

13. 56	**14.** 80	**15.** 240	**16.** 34.8
17. 542	**18.** 63.8	**19.** 4.53	**20.** 67.39
21. 5,600	**22.** 548.7	**23.** 65.93	**24.** .768

In 25–36, divide each number by 1,000.

25. 3,478	**26.** 400	**27.** 2,620	**28.** 628
29. 264.5	**30.** 8.25	**31.** 5,620	**32.** 675.9
33. 520	**34.** 3,570	**35.** 89.2	**36.** .72

In 37–44, divide each number by 10,000.

37. 4,620	**38.** 5,729	**39.** 15,625	**40.** 24,700
41. 43,234	**42.** 275.07	**43.** 6,842.8	**44.** 852

APPLICATION PROBLEMS

1. Matt's old car averages 10 miles per gallon of gasoline. How many gallons will Matt need to travel 847 miles? __________

2. For the senior prom, 100 seniors must pay $1,360 for the hotel hall. How much will each senior have to pay? (Each senior pays the same amount.) __________

3. A 1,000-member congregation must raise $22,640 to repair the church. If every member pays the same amount, how much must each member contribute? __________

4. At a benefit concert attended by 10,000 people, the total receipts were $58,500. What was the average price per ticket? __________

5. A construction company is building an apartment house and needs 100 floor tiles for each kitchen. How many kitchens can be tiled with 5,600 tiles? ________

6. Paper clips sell at $62 per 100 boxes. What is the price of one box of paper clips? ________

7. A box of 10 artists' paint brushes costs $8.90. What is the average cost of one brush? ________

8. On a vacation trip, the Cullen family traveled 1,000 miles. If they spent $60 on gasoline, what was the average cost per mile of the gasoline? ________

9. One kilogram is equal to 1,000 grams. How many kilograms are there in 3,460 grams? ________

10. One kilogram of instant coffee sells for $20. How much do 495 grams cost? (*Hint:* See Problem 9. How much does 1 gram cost?) ________

UNIT 52. Division When the Divisor Is a Decimal

Never try to divide when the divisor contains a decimal point. If there is a decimal point in your divisor, *change the divisor to a whole number.*

To do this, move the decimal point in the divisor to the right until it is "outside of the number." If the divisor is 1.62, change it to the whole number 162. by moving the decimal point *two* places to the right. If the divisor is .162, change it to the whole number 162. by moving the decimal point *three* places to the right. Recall that every whole number has a decimal point to the right of its right-hand digit.

Since you have moved the decimal point in the divisor to the right, you must now move the decimal point in the dividend to the right *the same number of places* you moved the decimal point in the divisor.

After moving the decimal points, divide as if the divisor were a whole number.

EXAMPLE 1. Divide: $.12\overline{)24.36}$

Solution: Move the decimal point in the divisor two places to the right.

$$.12.\overline{)24.36}$$

Next, move the decimal point in the dividend two places to the right.

$$.12.\overline{)24.36.}$$

What you have done is the same as multiplying both the divisor and the dividend by 100 in order to change the divisor, .12, to a whole number: $.12 \times 100 = 12.$ and $24.36 \times 100 = 2{,}436.$

Now you divide as with whole numbers.

$$\begin{array}{r} 203 \\ 12\overline{)2{,}436} \\ 2\,4 \\ \hline 36 \\ 36 \\ \hline \end{array}$$

Check: Does your quotient, 203, multiplied by the original divisor, .12, equal the original dividend, 24.36?

$$\begin{array}{rl} 203 & \leftarrow \text{quotient} \\ \times .12 & \leftarrow \text{divisor} \\ \hline 4\,06 & \\ 20\,3 & \\ \hline \checkmark\ 24.36 & \leftarrow \text{dividend} \end{array}$$

Answer: $.12\overline{)24.36} = 203$

EXAMPLE 2. Divide: $1.5\overline{)225}$

Solution: Move the decimal point in the divisor one place to the right (1.5 × 10 = 15).

$$1.5.\overline{)225}$$

Next, move the decimal point in the dividend one place to the right.

$$1.5.\overline{)225.(\ ).}$$

Since there are no digits to the right of the decimal point in the dividend, fill in the empty space with a zero as a placeholder (225 × 10 = 2,250).

$$1.5.\overline{)225.0.}$$

Divide as with whole numbers.

$$\begin{array}{r} 150 \\ 15\overline{)2{,}250} \\ 15 \\ \hline 75 \\ 75 \\ \hline \end{array}$$

Check:

$$\begin{array}{rl} 150 & \leftarrow \text{quotient} \\ \times 1.5 & \leftarrow \text{divisor} \\ \hline 750 & \\ 150 & \\ \hline ✓\ 225.0 & \leftarrow \text{dividend} \end{array}$$

Answer: $1.5\overline{)225} = 150$

EXAMPLE 3. Divide: .1848 ÷ .14

Solution: Set up the problem and move the decimal point two places to the right in both the divisor and the dividend. Then divide as if the divisor were a whole number. After you move the decimal point in your dividend, you place the decimal point in your answer directly above the decimal point in your dividend.

$$\begin{array}{r} 1.32 \\ .14.\overline{).18.48} \\ 14 \\ \hline 4\,4 \\ 4\,2 \\ \hline 28 \\ 28 \\ \hline \end{array}$$

Check:

$$\begin{array}{rl} 1.32 & \leftarrow \text{quotient} \\ \times .14 & \leftarrow \text{divisor} \\ \hline 528 & \\ 132 & \\ \hline ✓\ .1848 & \leftarrow \text{dividend} \end{array}$$

Answer: .1848 ÷ .14 = 1.32

Remember

To perform division when there is a decimal point in the divisor:

1. **Change** the divisor to a whole number by moving the decimal point to the right. (The divisor is a whole number when the decimal point is "outside," or to the right, of the divisor.)
2. **Move** the decimal point in the dividend to the right the same number of places as you moved the decimal point in the divisor.
3. **Divide** as if the divisor were a whole number, placing the decimal point in your answer directly above the decimal point in your dividend.

EXERCISES

In 1–15, divide.

SAMPLE SOLUTIONS

a.
$$\begin{array}{r} 291. = 291 \\ .15.\overline{)43.65.} \\ 30 \\ \hline 136 \\ 135 \\ \hline 15 \\ 15 \\ \hline \end{array}$$

b.
$$\begin{array}{r} 700. = 700 \\ .32.\overline{)224.00.} \\ 224 \\ \hline \end{array}$$

1. $.25\overline{)37.75}$
2. $.08\overline{)27.20}$
3. $3.8\overline{)132.24}$
4. $.055\overline{)267.3}$
5. $.042\overline{)470.4}$
6. $5.6\overline{)651.84}$
7. $.15\overline{)625.5}$
8. $.075\overline{)427.5}$
9. $.27\overline{)2.5839}$
10. $.23\overline{)621}$
11. $.40\overline{)9.26}$
12. $.08\overline{)25.22}$
13. $204.75 ÷ .875
14. $15.97 ÷ .025
15. $95.04 ÷ .375

APPLICATION PROBLEMS

1. Oranges sell at $1.45 a dozen. How many dozen can you buy with $13.05? __________
2. A wholesale butcher charged $30.52 for an order of beef. If the price of the beef was $1.09 a pound, how much did the beef weigh? __________
3. An airplane flew 997.15 miles in 2.75 hours. What was its average speed in miles per hour? __________

4. Gasoline sells at $1.45 a gallon. How many gallons can you buy with $17.40? ______

5. Mrs. Rennie paid $7.20 for a sack of potatoes. If the potatoes sold at $.36 per pound, how many pounds of potatoes did the sack hold? ______

6. If ham sells for $.35 per ounce, how many ounces can you buy with $4.20? ______

7. If 2.52 kilograms of cheese cost $9.45, what is the price per kilogram? ______

8. Mrs. Lugo paid $548.25 for carpeting. If the price per square yard is $12.75, how many square yards did she buy? ______

9. Hugo's gross pay last week was $288.75. If his hourly rate is $8.25, how many hours did he work? ______

10. Linda bought a stereo on the installment plan for $2,826. Her monthly payments will be $78.50. For how many months will she be making payments? ______

UNIT 53. Rounding Off to the Nearest Penny in Division

In business problems, it is often necessary to divide amounts of money into smaller amounts. Sometimes the answer will contain a fraction of a penny. In actual business transactions, however, you cannot pay out or collect a fraction of a penny. Hence, in working with dollars and cents in division problems, you cannot keep a fraction of a penny in your answer. You either have to raise the fraction to a whole penny or drop the fraction out.

Example 1. Divide $415.75 into 20 equal parts.

Solution: Set up the division problem and divide. You get a remainder of 15.

```
      $20.78
20 )$415.75
     40
     ----
      15 7
      14 0
      ----
       1 75
       1 60
       ----
         15 ← remainder
```

When dividing dollars and cents, use the following rule for rounding off:

RULE

If the remainder is half or more than half of the divisor, add a penny to your answer. If the remainder is less than half of the divisor, do not add a penny to your answer.

In this example, half of the divisor is 10. Since the remainder is 15, add a penny to your answer.

```
                       9
                  $20.78 = $20.79
             20 )$415.75
half of the
divisor → (10)    40
                  ----
                   15 7
                   14 0
                   ----
                    1 75
                    1 60
                    ----
                    (15) ← remainder
```

Answer: $415.75 ÷ 20 = $20.79

EXAMPLE 2. Divide: 30)$536.53

Solution: Divide as usual, and obtain a remainder of 13.

```
                $17.88 = $17.88
            30 )$536.53
half          30
of the
divisor → (15)
              236
              210
               26 5
               24 0
                2 53
                2 40
                  (13) ← remainder
```

Because the remainder is less than half of the divisor, do not add a penny to your answer.

Answer: 30)$536.53 = $17.88

Alternate Method: Carry out the division to *three* decimal places. If the *third digit* to the right of the decimal point is *5 or more*, *add* a penny to your answer. If the digit is *less than 5*, do *not* add a penny.

This is the same rule you learned in Unit 48.

```
          $17.88(4) = $17.88
     30 )$536.53(0)
          30
          236
          210
           26 5
           24 0
            2 53
            2 40
              130
              120
```

EXERCISES

In 1–12, divide and round off to the nearest penny.

1. 23)$695.83
2. 25)$1,032.30
3. 15)$230.52
4. 34)$690.00
5. 22)$749.43
6. 19)$820.88
7. 36)$739.20
8. 18)$578.85
9. 24)$590.52
10. 18)$547.98
11. 36)$1,153.63
12. 31)$1,240.07

APPLICATION PROBLEMS

In 1–10, round off your answer to the nearest penny.

1. At the A and Z Furniture Store, according to sales records, the average installment purchase is $843.71 and is paid for in 24 equal payments. What is the amount of the average payment? ______
2. Last year, Stanley spent $12.68 on postage. On the average, how much did he spend each month? ______
3. Meg has saved $505.58 for her two-week vacation. How much can she afford to spend on each day of her vacation? (Assume that the vacation is exactly 14 days.) ______
4. Harry spent $139.52 last year on newspapers. On the average, how much did he spend each day? (Use a 365-day year.) ______
5. After adding up the food costs and medical expenses, Shirley found that her pet cat cost $185.25 last year. How much does Shirley spend per month on her cat? ______
6. Dennis earns $23,575 a year. What is his weekly salary? ______
7. Last year, Jack was able to save $2,890 from his salary. What was the average amount he saved each week? ______
8. Three people won a lottery of $1,600,000. How much will each one get? ______
9. A 21-day vacation trip to London, Paris, and Rome costs $1,990. What is the average cost per day? ______
10. Curtis bought 3 pairs of shoes for $65. What is the average price per pair? ______

For 11–12, use a calculator.

11. Mr. Tasco received a shipment of supplies for his stationery store. The bill gave the total cost of each item, and he wanted the unit cost. Fill in the table.

Item	Quantity Received	Total Cost	Unit Cost
Composition Books	4 dozen	$36.50	
Calendars	2 dozen	52.50	
Marking Pens	$3\frac{1}{2}$ dozen	26.00	
Paperback Books	20 books	45.00	
Pencils	1 gross (12 dozen)	11.75	
Looseleaf Binders	$2\frac{1}{2}$ dozen	27.50	

12. Rachel was comparing the price per ounce of breakfast cereals, to see which was the best buy. Complete the chart.

Cereal	Price per Box	Number of Ounces in Box	Price per Ounce
Whizzos	$1.49	10	
Rice Pops	1.79	14.5	
Oaties	1.59	12	
Corn Nips	1.89	14	
Superwheats	1.55	11.5	
Crunchies	1.69	12.8	

UNIT 54. Division of a Smaller Whole Number by a Larger Whole Number

There are times when you must divide a smaller number by a larger number. For example, if 5 boys want to share 2 dollars, how much should each boy get?

Because the number of dollars is smaller than the number of boys, the answer will not be in whole dollars. Rather, the answer will be in cents, which are the decimal fractions of dollars. *Cents,* you recall, are *hundredths* of dollars.

Example 1. Divide: $2 ÷ 5 Round off your answer to the nearest penny.

Solution: Place two zeros after the decimal point because you want the answer in hundredths of dollars. Then place a decimal point in your answer directly above the decimal point in the dividend.

5) $2.00

Now you divide as with whole numbers.

```
    $.40
5 ) $2.00
     2 0
```

Answer: Each boy will get $.40.

The same method is used to divide any smaller number by a larger number. Also, the dividend can be an amount of something, such as miles, pounds, hours, etc. To divide a smaller whole number by a larger whole number, use this rule:

RULE

To divide a smaller whole number by a larger whole number, place a decimal point to the right of the dividend. Then:

1. For an answer to the nearest *tenth,* place *one zero* to the right of the decimal point.
2. For an answer to the nearest *hundredth,* place *two zeros* to the right of the decimal point.
3. For an answer to the nearest *thousandth,* place *three zeros* to the right of the decimal point.

And so on.

Then divide and round off the remainder.

Since dividing a smaller number by a larger number often does not come out even, be sure you know how to round off your decimal answer. To round off a decimal answer to a given value such as tenths, hundredths, or thousandths, follow the procedure explained in the preceding unit. However, instead of rounding off to the nearest penny, round off to the given value.

Example 2. Divide: 7 ÷ 12 Round off the answer to the nearest *tenth.*

Solution: Place a decimal point and one zero after the 7. Divide and round off.

$$\begin{array}{r} 6 \\ .\cancel{5} = .6 \\ 12\overline{)7.0} \\ 6\ 0 \\ \hline 1\ 0 \end{array}$$

half of the divisor → (6) (1 0) ← remainder

Since the remainder is larger than half of the divisor, you add 1 unit to the quotient. In this example, the unit is tenths, so you add .1 and .5, getting .6.

Answer: 7 ÷ 12 = .6, to the nearest tenth.

Example 3. Divide: 7 ÷ 12 Round off the answer to the nearest *hundredth.*

Solution: Place a decimal point and two zeros after the 7. Divide and round off.

$$\begin{array}{r} .58 = .58 \\ 12\overline{)7.00} \\ 6\ 0 \\ \hline 1\ 00 \\ 96 \\ \hline 4 \end{array}$$

half of the divisor → (6) (4) ← remainder

Answer:
7 ÷ 12 = .58, to the nearest hundredth.

Example 4. Divide: 7 ÷ 12 Round off the answer to the nearest *thousandth.*

Solution: Place a decimal point and three zeros after the 7. Divide and round off.

$$\begin{array}{r} .583 = .583 \\ 12\overline{)7.000} \\ 6\ 0 \\ \hline 1\ 00 \\ 96 \\ \hline 40 \\ 36 \\ \hline 4 \end{array}$$

half of the divisor → (6) (4) ← remainder

Answer:
7 ÷ 12 = .583, to the nearest thousandth.

In word problems involving division, you may have to round off the answers. In problems dealing with cost or price, *always round off to the nearest penny.* When other units are involved, such as yards, pounds, or hours, you must be told whether the answer is to be in whole units, tenths, hundredths, etc. If you are not told, *always round off to the nearest whole unit.*

Example 5. A snail travels 5 inches in 7 hours. Find, to the nearest *tenth* of an inch, how far it travels in one hour.

Solution: To find the distance traveled in *one* hour, divide the entire distance, 5 inches, by the total time, 7 hours.

$$\begin{array}{r} .7 \\ 7\overline{)5.0} \\ 4\ 9 \\ \hline 1 \end{array}$$

half of the divisor → ($3\frac{1}{2}$) (1) ← remainder is less than half of the divisor

Do not add .1 to the quotient.

Answer: .7 of an inch

EXERCISES

In 1–5, divide. Round off the quotient to the nearest *tenth.*

1. $7\overline{)4}$ **2.** $15\overline{)11}$ **3.** $8\overline{)5}$ **4.** $13\overline{)10}$ **5.** $23\overline{)19}$

In 6–10, divide. Round off to the nearest *hundredth.*

6. $12\overline{)8}$ **7.** $18\overline{)7}$ **8.** $28\overline{)19}$ **9.** $13\overline{)5}$ **10.** $26\overline{)23}$

In 11–15, divide. Round off to the nearest *thousandth*.

11. $16\overline{)5}$ **12.** $35\overline{)23}$ **13.** $27\overline{)15}$ **14.** $32\overline{)28}$ **15.** $23\overline{)21}$

APPLICATION PROBLEMS

SAMPLE SOLUTION

For an experiment, Sandy must measure equal amounts of a chemical into 7 test tubes. If he must use 2 grams of the chemical, how much goes into each test tube? Round off your answer to the nearest *hundredth* of a gram. *Each tube gets $\frac{1}{7}$ of 2, or $\frac{2}{7}$ gram.*

$$\begin{array}{r} .28\overset{9}{} = .29 \\ 7\overline{)2.00} \\ \underline{14} \\ 60 \\ \underline{56} \\ 4 \end{array}$$

.29 gram

1. Tee shirts sell at 5 for $16. How much is one shirt? ________

2. Mary needs 6 yards of material to make 7 blouses. To the nearest *tenth* of a yard, how much material is used to make one blouse? ________

3. What is the price of one pair of socks if they sell at $9.55 a dozen? ________

4. If an electricity bill for one month is $41.10, what is the average cost of electricity for one day? Use a 30-day month. ________

5. Sara earns $8 an hour. How much does she earn for 1 minute of work? ________

6. If it takes 3 hours to assemble 5 TV sets, what decimal fraction of an hour is needed to assemble 1 TV set? ________

7. Two feet is what decimal fraction of a yard? (*Hint:* 3 feet = 1 yard.) ________

8. A package of 3 pens sells for $2. What is the cost of 1 pen? ________

9. Oranges sell at 6 for $1. How much does 1 orange cost? ________

10. Ninety seconds is what fraction of an hour? Answer with a decimal fraction to the nearest *thousandth*. (*Hint:* 90 seconds is 1.5 minutes. You can use the $\frac{\text{IS}}{\text{OF}}$ fraction.) ________

UNIT 55. Changing Common Fractions to Decimal Fractions

Recall that a fraction is a "part of a unit." A *common fraction* is a part of its denominator. Thus, the fraction $\frac{3}{4}$ is "three parts out of four" or "three fourths." A *decimal fraction* is also a part of its denominator, even though the denominator is not actually written down. Thus, the fraction .75 is "75 parts out of a hundred" or "75 hundredths." Notice that both the common fraction $\frac{3}{4}$ and the decimal fraction .75 have the same value: $\frac{3}{4} = .75$. In money, 3 quarters is the same as $.75.

You can say that you have "half a dollar" or that you have "$.50." Both fractions have the same value, because $\frac{1}{2} = .50$. The common fraction $\frac{1}{2}$ of a dollar means that 1 dollar has been divided by 2. The fraction $\frac{3}{4}$ means that the number 3 has been divided by 4.

RULE

To change a common fraction to a decimal fraction, divide the numerator by the denominator.

Follow the rule for dividing a smaller number by a larger number that was explained in the preceding unit.

EXAMPLE 1. Change $\frac{4}{5}$ to a decimal fraction.

Solution: $\frac{4}{5}$ means 4 divided by 5.

$$\begin{array}{r} .8 = .8 \\ 5\overline{)4.0} \\ 4\,0 \\ \hline \end{array}$$

Answer: $\frac{4}{5} = .8$

Alternate Method: Draw a division box next to the denominator and let the numerator "fall in." (You used this method in your work with improper fractions in Unit 26.) Then divide as usual.

$$\begin{array}{r} 4 \searrow \\ 5\overline{)4} \end{array}$$

EXAMPLE 2. Change $\frac{2}{3}$ to a decimal fraction. Round off your answer to the nearest *hundredth*.

Solution: $\frac{2}{3}$ means 2 divided by 3. Or, $\frac{2}{3\overline{)2}}$.

$$\begin{array}{r} 7 \\ .6\not{6} = .67 \\ 3\overline{)2.00} \\ 1\,8 \\ \hline 20 \\ 18 \\ \hline 2 \end{array}$$

half of the divisor → $1\frac{1}{2}$

(2) ← remainder

Answer: $\frac{2}{3} = .67$, to the nearest hundredth.

EXERCISES

In 1–15, change the common fraction to a decimal fraction. Round off the answer to the nearest *hundredth*.

1. $\frac{5}{16}$ **2.** $\frac{3}{16}$ **3.** $\frac{9}{20}$ **4.** $\frac{1}{6}$ **5.** $\frac{3}{7}$

6. $\frac{5}{9}$ 7. $\frac{5}{12}$ 8. $\frac{7}{11}$ 9. $\frac{15}{24}$ 10. $\frac{21}{28}$

11. $\frac{13}{15}$ 12. $\frac{27}{30}$ 13. $\frac{15}{40}$ 14. $\frac{24}{36}$ 15. $\frac{12}{32}$

APPLICATION PROBLEMS

1. A nut measures $\frac{9}{32}$ of an inch in diameter. What is the measurement of the nut in *thousandths* of an inch? ______

2. John ran $\frac{4}{7}$ of a mile. How many *hundredths* of a mile did he run? ______

3. A moon rock weighs $\frac{7}{16}$ of a pound. How many *hundredths* of a pound does it weigh? ______

4. For an experiment, a chemist needs $\frac{5}{9}$ of a gallon of a certain chemical. How many *thousandths* of a gallon does she need? ______

5. The time of flight of an experimental rocket is $\frac{5}{12}$ of an hour. What is the time of flight in *thousandths* of an hour? ______

6. The outer diameter of a bushing is $\frac{7}{8}$ inch. What is the measurement in *thousandths* of an inch? ______

7. A "$\frac{1}{3}$-off" sale means that $\frac{1}{3}$ of every dollar of the original price is discounted. How many cents per dollar are discounted? ______

8. The Browns bought $\frac{7}{8}$ ton of coal. How many *hundredths* of a ton did they buy? ______

9. During a "$\frac{1}{4}$-off" sale, how many cents out of every dollar are discounted? ________________

10. A package of ham weighs 5 ounces. How many *thousandths* of a pound does it weigh? (*Hint:* 16 oz. = 1 lb.) ________________

UNIT 56. Changing the Fraction in a Mixed Number to a Decimal to Simplify Multiplication and Division

When you must multiply or divide a whole number by a mixed number, the problem can often be simplified by changing the common fraction to a decimal fraction.

EXAMPLE 1. Multiply: $345 \times 23\frac{1}{4}$

Solution: You may find it easier to multiply if you change $23\frac{1}{4}$ to 23.25.

$$\begin{array}{r} 23.25 \\ \times 345 \\ \hline 116\ 25 \\ 930\ 0 \\ 6\ 975 \\ \hline 8{,}021.25 \end{array}$$

Answer: $345 \times 23\frac{1}{4} = 8{,}021.25$

Note: To multiply using common fractions, first change $23\frac{1}{4}$ to an improper fraction.

$$345 \times 23\frac{1}{4} = \frac{345}{1} \times \frac{93}{4}$$

Then, you must multiply the numerators, 345×93. Finally, you must divide this product by 4. In general, there is less chance for error in performing one operation than in performing several separate operations.

EXAMPLE 2. Divide: $125 \div 25\frac{1}{2}$

Solution: Change $25\frac{1}{2}$ to 25.5 and divide.

$$\begin{array}{r} 4.90 \\ 25.5\overline{)125.0.00} \\ 102\ 0 \\ \hline 23\ 00 \\ 22\ 95 \\ \hline 50 \end{array} = 4.90$$

half of the divisor → $127\frac{1}{2}$

50 ← remainder

Answer: $125 \div 25\frac{1}{2} = 4.90$

EXERCISES

In 1–6, change the mixed fraction to a mixed decimal and multiply.

1. $247 \times 35\frac{3}{4}$
2. $527 \times 47\frac{1}{5}$
3. $345\frac{1}{2} \times 37$

4. $638\frac{3}{5} \times 43$

5. $528 \times 124\frac{1}{4}$

6. $728 \times 46\frac{2}{8}$

In 7–12, change the mixed fraction to a mixed decimal and divide. If the quotient does not come out even, round off the answer to the nearest *hundredth*.

7. $215 \div 12\frac{1}{2}$

8. $354\frac{4}{5} \div 38$

9. $527\frac{1}{5} \div 43$

10. $1,356\frac{3}{4} \div 120$

11. $2,435 \div 15\frac{4}{5}$

12. $3,528\frac{1}{2} \div 28$

APPLICATION PROBLEMS

1. If a cinder block weighs $15\frac{1}{2}$ pounds, how many pounds will 358 cinder blocks weigh? __________

2. A clothing manufacturer uses $4\frac{3}{4}$ yards of material to make a suit. How many suits can he make with 475 yards of material? __________

3. A sewing machine weighs $23\frac{4}{5}$ pounds. How many pounds would 324 sewing machines weigh? __________

4. A fruit picker can pick $12\frac{1}{2}$ bushels of tomatoes in one hour. How many bushels can he pick in 138 hours? __________

5. Gasoline sells for $\$1.38\frac{2}{5}$ a gallon. How many gallons can be bought with \$16? (Answer to the nearest *tenth* of a gallon.) __________

6. A furnace filter costs $\$.37\frac{1}{2}$. How much will 250 filters cost? ________

7. The Wilsons bought $43\frac{3}{4}$ square yards of carpeting for \$371.88. What is the price per square yard of the carpeting? ________

8. James received \$63 for working $8\frac{3}{4}$ hours overtime. Find his hourly overtime rate. ________

9. Paper clips sell at $\$4.12\frac{1}{5}$ per 1,000. How much will 4,500 paper clips cost? (*Hint:* Find how many 1,000's are contained in 4,500.) ________

10. Business cards are printed at $\$1.45\frac{1}{2}$ per 100. How much would it cost to print 1,250 cards? (*Hint:* How many 100's are there in 1,250?) ________

Review of Part XIII (Units 50–56)

In 1–12, divide and round off to the nearest *penny.*

1. 5) \$2,605.85	**2.** 24) \$3,263.50	**3.** 42) \$368.75
4. 4.5) \$372.08	**5.** .18) \$680	**6.** 85) \$568.52
7. .06) \$853	**8.** 27) \$4,321.48	**9.** .05) \$475.68
10. 6.3) \$248.95	**11.** .38) \$874.70	**12.** .53) \$654.80

In 13–18, divide. Round off to the nearest *hundredth.*

13. 12) 3	**14.** 15) 6	**15.** 23) 18
16. 65) 42	**17.** 128) 85	**18.** 350) 275

In 19–24, change each fraction to a decimal. Round off to the nearest *hundredth.*

19. $\frac{2}{3}$	**20.** $\frac{5}{7}$	**21.** $\frac{12}{15}$
22. $\frac{25}{63}$	**23.** $\frac{19}{45}$	**24.** $\frac{48}{72}$

In 25–26, change the mixed number to a mixed decimal and multiply.

25. $9\frac{99}{100} \times 87$

26. $847 \times 25\frac{2}{5}$

In 27–28, change the mixed number to a mixed decimal and divide.

27. $7{,}128 \div 9\frac{9}{10}$

28. $54.2 \div 3\frac{4}{5}$

For 29–36, use a calculator.

In 29–34, change each mixed number to a mixed decimal, and compute. Round answers to the nearest *thousandth.*

29. $12\frac{5}{8} \times 2\frac{3}{4}$

30. $43\frac{13}{15} + 8\frac{3}{7}$

31. $143\frac{1}{4} \div 17$

32. $61 - 18\frac{5}{9}$

33. $258 \times 12\frac{2}{3}$

34. $97\frac{3}{8} \div 42\frac{1}{16}$

35. Ms. Swenson recorded her students' test grades as common fractions, which she then changed to decimal fractions rounded to the nearest hundredth. For example, if Ben Smith got 15 out of 18 problems correct, his grade was $\frac{15}{18}$, or .83. Find, to the nearest *hundredth,* the grades for students getting the following numbers of questions correct:

a. 14 out of 17 __________

b. 23 out of 25 __________

c. 12 out of 19 __________

d. 21 out of 30 __________

e. 7 out of 12 __________

36. Some neighbors on Elm Street were comparing the fuel economy of their automobiles. They kept a week's record of mileage and gasoline consumption, as shown in the table. Find the number of miles per gallon for each. Round answers to the nearest *tenth.*

	Miles Traveled	Gallons Used	Miles per Gallon
Alice	103.6	9.4	
Carl	47.8	3.9	
Dan	242.1	17.9	
Lucy	95.7	7.5	
Walt	168.3	11.8	

PART XIV. Percents

UNIT 57. Understanding Percents

WORDS TO KNOW

You know that there are 100 *cents* in a dollar and 100 years in a *century*. This makes it easy to remember that **percent** means *per hundred.*

THE MEANING OF %

The word **percent** can be read "hundredths" or "out of 100." The symbol **%** is often used in place of the word **percent.** Note what the % symbol means in the following examples:

Salespeople who earn a commission of "15% on sales" earn $15 for every $100 worth of merchandise they sell. They earn "$15 out of $100."

A "7% sales tax" means that you are taxed $.07 every time you spend $1.00; that is, the tax is "7¢ out of 100¢."

When a school has a student population that is "55% boys," then "55 students out of every 100 students are boys."

If the amount of spoilage in a crate of oranges is found to be "10%," this means that "10 oranges out of every 100 oranges" are spoiled.

PERCENTS WITH FRACTIONS

When a common fraction or a decimal fraction is involved in a percent problem, you usually keep the fraction in its original form. Note how the following expressions are written with the % symbol:

"$66\frac{2}{3}$ out of 100" may be expressed as "$66\frac{2}{3}$%."

"87.5 per hundred" means "87.5%."

"$\frac{1}{3}$ hundredths" means "$\frac{1}{3}$ of 1 percent" or "$\frac{1}{3}$%."

THE MEANING OF 100%

We use the expression "100%" to mean "*all* of something" or "*the whole group.*" Note what "100%" means in each of the following:

A can labeled "100% pure juice" contains no water or coloring. The contents are "*all* pure juice."

When a theater is filled to "100% capacity," you know that "*all* the seats" are filled.

If "100% of the sophomore class" attended a game, then the "*whole class*" attended.

GREATER THAN 100%

When you mean *more* than the whole group or *more* than all of something, you use expressions such as "150%" or "200%."

If your income this year is "150%" of last year's income, it is 50% *greater* than it was last year. "150%" means *all* of something (100%), plus 50% more.

If the cost of something is "200%" of the old price, the new price is *twice* the old price. "200%" means two times 100%.

The expression "475%" means four times 100%, plus 75% more.

UNKNOWN PERCENTS

Since you know that *100%* means *all*, you can figure out percents that are not given in a problem. You do this by subtracting the given percent from 100%. For example:

The school that has a student population of 55% boys must have a population of 45% girls, since 100% − 55% = 45%.

The crate that has 10% of its oranges spoiled has 100% − 10% = 90% of its oranges not spoiled.

Common Fractions

Since percents mean *hundredths*, you can write any percent as a common fraction that has a denominator of 100. For example:

6% means "6 out of 100," or $\frac{6}{100}$.

$$25\% = \frac{25}{100}$$

$$2.5\% = \frac{2.5}{100}$$

$$66\frac{2}{3}\% = \frac{66\frac{2}{3}}{100}$$

$$250\% = \frac{250}{100}$$

Decimal Fractions

Finally, you can write any percent as a decimal fraction. Simply rewrite the common fraction as a decimal fraction. For example:

$$6\% = \frac{6}{100} = .06$$

$$25\% = \frac{25}{100} = .25$$

$$2.5\% = \frac{2.5}{100} = .025$$

$$66\frac{2}{3}\% = \frac{66\frac{2}{3}}{100} = .66\frac{2}{3}$$

$$250\% = \frac{250}{100} = 2.5$$

Exercises

In 1–29, fill in the blanks.

1. 25 out of 100 means ________ %

2. 5 out of 100 means ________ %

3. 87 out of 100 means ________ %

4. 1 out of 100 means ________ %

5. $12\frac{1}{2}$ out of 100 means ________ %

6. 37.5 out of 100 means ________ %

7. $4\frac{1}{2}$ out of 100 means ________ %

8. 95 out of 100 means ________ %

9. $33\frac{1}{3}$ out of 100 means ________ %

10. 100 out of 100 means ________ %

11. 8 hundredths means ________ %

12. 75 hundredths means ________ %

13. 100 hundredths means ________ %

14. 350 hundredths means ________ %

15. $\frac{1}{4}$ hundredths means ________ %

16. $5\frac{1}{2}$ hundredths means ________ %

17. $33\frac{1}{3}$ hundredths means ________ %

18. $179\frac{1}{4}$ hundredths means ________ %

19. 32.7 hundredths means ________ %

20. 125.3 hundredths means ________ %

21. $\frac{5}{100}$ = ________ %

22. $\frac{27}{100}$ = ________ %

23. $\frac{12\frac{1}{2}}{100}$ = ________ %

24. $\frac{80}{100}$ = ________ %

25. $\frac{37.5}{100}$ = ________ %

26. $\frac{33\frac{1}{3}}{100}$ = ________ %

27. $\frac{125}{100}$ = ________ %

28. $\frac{175.5}{100}$ = ________ %

29. $\frac{100}{100}$ = ________ %

APPLICATION PROBLEMS

Sample Solution

When Ocean Township had an election for Town Councillor, votes were cast by 72% of the registered voters. Of every hundred registered voters, how many voted?

72% means 72 out of 100. — *72 voted.*

Of every hundred registered voters, how many did not vote?

If 72% voted, then 100% − 72%, or 28%, did not vote. 28% means 28 out of 100. — *28 did not vote.*

1. A school has 48% girls. How many students out of every 100 are girls? ________

   How many students out of every 100 are boys? ________

2. If there is a sales tax of 7%, how many dollars tax will you pay for every purchase of $100? ________

   How many dollars for a purchase of $200? ________

3. If 15% of a class is absent, what is the percent of students present? ________

4. A family spends 25% of its income on rent and 15% on food. What percent of the income is left for other expenses? ________

5. The price of a watch was reduced by 35%. What percent of the price remains? ________

6. A dress selling for $100 was reduced by 20%. What is the new price of the dress? ________

7. Vincent got a 13% increase in salary. What percent of the old salary is the new salary? ________

8. Write 8% as cents of a dollar. ________

9. Write 25¢ as a percent of a dollar. ________

10. In fifteen years, the price of gasoline tripled. What percent of the original price was the new price? ________

UNIT 58. Changing Percents to Decimals and Decimals to Percents

Percents to Decimals

In many problems, you must perform multiplication and division with percents. Before you can do this, you must always change the percent to a decimal. In Unit 57, you learned how to change a percent to a decimal by first changing the percent to a common fraction. You can also change a percent to a decimal *directly.*

RULE

> To change a percent to a decimal, rewrite the percent without the % sign. Then move the decimal point two places to the *left*.

Examples

$$30\% = .30. = .30$$

(Recall that every whole number has a decimal point, even though the decimal point is not actually written.)

$$5\% = .()5. = .05$$

(Recall the use of zeros as placeholders.)

$$250\% = 2.50. = 2.50$$

(When a percent exceeds 100%, it is rewritten as a mixed decimal.)

$$4\frac{1}{2}\% = .()4.\frac{1}{2} = .04\frac{1}{2}$$

(In a mixed number, the decimal point is between the whole number and the common fraction, even though it is not actually written.)

When a percent is a mixed number, it is sometimes convenient to eliminate the common fraction from the decimal equivalent. Do this by first changing the mixed number to a mixed decimal. Then follow the above rule.

Examples

$$4\frac{1}{2}\% = 4.5\% = .()4.5 = .045$$

(Notice that .045 and $.04\frac{1}{2}$ have the same value.)

$$99\frac{44}{100}\% = 99.44\% = .99.44$$

$$= .9944$$

You may find it helpful to think of percents as the hundred cents contained in a dollar. To change a percent to a decimal, simply *rewrite the percent as cents of a dollar.* For example, write 25% as .25, just as you write 25¢ as $.25. Also, 7% = .07 (7¢ = $.07) and 150% = 1.50 (150¢ = $1.50).

Decimals to Percents

When a problem requires a percent as the answer, you will often have to change your decimal answer to a percent. To change a given decimal to a percent, you reverse the preceding rule.

RULE

> To change a decimal to a percent, rewrite the decimal with the % sign. Then move the decimal point two places to the *right*.

Examples

$$.65 = .65. = 65\%$$

$$.03 = .03. = 3\%$$

(Since the zero to the left of the 3 is not needed as a placeholder, you do *not* write "03%.")

$$2.25 = 2.25. = 225\%$$

$$.875 = .87.5 = 87.5\% \text{ or } 87\frac{1}{2}\%$$

$$.008 = .00.8 = .8\%$$

$$.12\frac{1}{2} = .12.\frac{1}{2} = 12\frac{1}{2}\% \text{ or } 12.5\%$$

$$1.2 = 1.2(). = 120\%$$

(A mixed decimal changes to a percent that is greater than 100%.)

EXERCISES

In 1–20, change each percent to a decimal. When the percent is a mixed number, keep the common fraction.

1. 28%
2. 6%
3. 1%
4. 35%
5. 100%
6. $12\frac{1}{2}\%$
7. $5\frac{3}{4}\%$
8. 23.8%
9. 8%
10. 15.9%
11. $33\frac{1}{3}\%$
12. .001%
13. 125%
14. 300%
15. .008%
16. .4%
17. 105%
18. .5%
19. 3.25%
20. 6.7%

In 21–40, change each decimal to a percent. When the decimal contains a common fraction, keep it.

21. .50
22. .28
23. .07
24. .45
25. .15
26. 3.15
27. $.25\frac{1}{2}$
28. $.05\frac{1}{2}$
29. 1.35
30. .005
31. .4
32. 1.00
33. $.12\frac{3}{4}$
34. .325
35. .001
36. .9
37. $.68\frac{4}{5}$
38. .025
39. .008
40. .08

APPLICATION PROBLEMS

1. "The Cost-of-Living Index increased by 1.7%." Express the percent as a decimal. ______
2. During the season, a pitcher won 16 games out of 20 for an average of .800. What percent of games pitched did he win? ______

   What percent did he lose? ______
3. If there is a sales tax of 7%, how many cents do you pay on a $1 purchase? ______
4. How many cents is 75% of a dollar? ______
5. A ball player has a batting average of .350. What is the percent of hits for times at bat? ______
6. If a sales tax is $8\frac{1}{4}$%, how many cents are paid for every $1 of sale? ______

   How much tax is paid for every $100 of sale? ______
7. Charles took a Civil Service test and answered correctly 85 questions out of the 100 questions on the test. If all the questions have the same value, what is his percent score? ______
8. A department store is having a "45%-off" sale. What is the reduction on an item selling for $1? ______

   On an item selling for $10? ______

   On an item selling for $100? ______
9. A bank pays 8.75% on certificates of deposit. How much interest will $100 earn in one year? ______
10. Albert had a $1,000 certificate of deposit. If he earned $96.50 in interest in one year, what is the percent interest paid by the bank? ______

UNIT 59. Changing Percents to Fractions and Fractions to Percents

PERCENTS TO FRACTIONS

Many problems dealing with percents can be solved by using fractions. As you learned in Unit 57, since *percent* means *hundredth*, you can write any percent as a fraction that has a denominator of 100. You will now use this skill to change percents to fractions.

RULE

To change a percent to a fraction, write the percent as a fraction that has a denominator of 100. Then reduce to lowest terms.

Examples

$$20\% = \frac{20}{100} \quad 20\big| \frac{\overset{1}{\cancel{20}}}{\underset{5}{\cancel{100}}} = \frac{1}{5}$$

$$75\% = \frac{75}{100} \quad 25\big| \frac{\overset{3}{\cancel{75}}}{\underset{4}{\cancel{100}}} = \frac{3}{4}$$

$$12\tfrac{1}{2}\% = \frac{12\frac{1}{2}}{100} = 12\tfrac{1}{2} \div \frac{100}{1}$$

$$= \frac{\overset{1}{\cancel{25}}}{2} \times \frac{1}{\underset{4}{\cancel{100}}} = \frac{1}{8}$$

Fractions to Percents

There are two ways of changing a fraction to a percent.

First, if you can raise a fraction to an equivalent fraction that has a denominator of 100, then you can change the resulting fraction to a percent.

Example 1. Change $\frac{1}{5}$ to a percent.

Solution: Since 5 divides evenly into 100, you can raise $\frac{1}{5}$ to an equivalent fraction that has a denominator of 100.

$$\frac{1}{5} = \frac{\quad}{100}$$

$$\frac{1}{5} \rightarrow \frac{20}{100}$$

Think: "5 into 100 goes 20, and 20 × 1 = 20."

Answer: $\frac{1}{5} = \frac{20}{100} = 20\%$

Second, when the denominator of a fraction does not divide evenly into 100, use the following rule:

Rule

> When a fraction cannot be raised to an equivalent fraction that has a denominator of 100, divide the numerator by the denominator. Then carry your answer to three places and change the decimal to a percent.

Example 2. Change $\frac{3}{8}$ to a percent.

Solution: Since 8 will not divide evenly into 100, you must use the rule. Divide 8 into 3, and carry your answer to three places.

```
    .375
8 )3.000
   2 4
     60
     56
      40
      40
```

$.375 = .37.5 = 37.5\%$

Answer: $\frac{3}{8} = 37.5\%$

Example 3. Change $\frac{6}{7}$ to a percent.

```
    .857 = .85.7 = 85.7%
7 )6.000
   5 6
     40
     35
      50
      49
       1
```

Reminder: You may write the division box next to the denominator and let the numerator "drop in."

```
   6↘
7 )6.00
```

Answer: $\frac{6}{7} = 85.7\%$

Table III shows some common fractions expressed as 3-place decimal equivalents and as percents.

Table III: Fraction, Decimal, and Percent Equivalents

Fraction	Decimal	Percent	Fraction	Decimal	Percent
$\frac{1}{8}$	.125	12.5%	$\frac{3}{5}$	.600	60%
$\frac{1}{6}$	.167	16.7%	$\frac{5}{8}$	.625	62.5%
$\frac{1}{5}$	.200	20%	$\frac{2}{3}$	.667	66.7%
$\frac{1}{4}$	.250	25%	$\frac{3}{4}$	.750	75%
$\frac{1}{3}$	.333	33.3%	$\frac{4}{5}$	.800	80%
$\frac{3}{8}$	.375	37.5%	$\frac{5}{6}$	.833	83.3%
$\frac{2}{5}$	.400	40%	$\frac{7}{8}$	.875	87.5%
$\frac{1}{2}$	.500	50%			

EXERCISES

In 1–16, change each percent to a fraction. Reduce the fraction to lowest terms.

1. 10% **2.** 30% **3.** 90% **4.** 85%

5. 5% **6.** 32% **7.** 64% **8.** 14%

9. 38% **10.** 76% **11.** 4% **12.** $7\frac{1}{2}\%$

13. $6\frac{1}{4}\%$ **14.** $8\frac{1}{3}\%$ **15.** $42\frac{6}{7}\%$ **16.** $55\frac{5}{9}\%$

In 17–31, change each fraction to a percent.

17. $\frac{7}{10}$ 18. $\frac{3}{50}$ 19. $\frac{15}{25}$ 20. $\frac{17}{20}$ 21. $\frac{25}{75}$

22. $\frac{20}{80}$ 23. $\frac{12}{48}$ 24. $\frac{14}{35}$ 25. $\frac{7}{28}$ 26. $\frac{15}{18}$

27. $\frac{17}{22}$ 28. $\frac{23}{45}$ 29. $\frac{3}{7}$ 30. $\frac{32}{55}$ 31. $\frac{42}{85}$

APPLICATION PROBLEMS

Reduce all fractions to lowest terms.

1. Last term, 20% of a class failed a test. What fraction of the class failed the test? ________

   What fraction of the class passed? ________

2. A living room set is reduced by $\frac{1}{3}$ off the regular price. What is the percent of reduction? ________

   What percent of the original price is the sales price? ________

3. A dress is on sale at 25% off the regular price. What fraction of the regular price is the sale price? ________

4. Tom bought a refrigerator and gave $\frac{1}{5}$ of the full price as a down payment. What percent of the full price was the down payment? ________

5. Serita received $\frac{4}{5}$ of the votes cast for class president. What percent of the votes did she receive? ________

6. A lamp is reduced by 35%. What fraction of the price is the reduction. ________

7. It is estimated that 5 out of 8 registered voters vote on Election Day. What percent of the registered voters vote? ________

8. A fruit punch contains 60% water. What fraction of the contents is water? ________

9. Edward bought a coat on the layaway plan, to be paid for in 8 equal payments. If he made 7 payments, what percent of the cost of the coat did he pay? ________

10. For every $3 of earnings, Alice pays $1 in federal income tax. What percent of her income is paid in federal taxes? ________

UNIT 60. Fractional Parts of 1%

Percents are sometimes expressed as *fractions* of 1%. A percent such as "$\frac{1}{4}$%" means "$\frac{1}{4}$ of 1%" and "$\frac{1}{2}$%" means "$\frac{1}{2}$ of 1%," etc. When performing calculations, it is usually simpler to change fractions of percents to decimals.

RULE

To change a fraction of a percent to a decimal:

1. Change the "common-fraction percent" to a "decimal-fraction percent" by rewriting the common fraction as a decimal fraction.
2. Change the resulting percent to a decimal by moving the decimal point two places to the left and dropping the % sign.

EXAMPLE 1. Change $\frac{1}{4}$% to a decimal.

Solution: Since $\frac{1}{4}$ = .25, change $\frac{1}{4}$% to .25%. Then drop the percent sign and move the decimal point two places to the left:

$$\frac{1}{4}\% = .25\% = .00.25 = .0025$$

Answer: $\frac{1}{4}$% = .0025

EXAMPLE 2. Change $\frac{7}{8}$% to a decimal.

Solution: Change $\frac{7}{8}$ to its decimal equivalent. Use Table III (page 202), or divide the denominator into the numerator.

```
     .875
  8 )7.000
     6 4
      60
      56
       40
       40
```

Change the percent to a decimal.

$$\frac{7}{8}\% = .875\% = .00.875$$

$$= .00875$$

Answer: $\frac{7}{8}$% = .00875

Some fractions, when changed to decimals, do not come out even. In cases like these, you must round off the decimal equivalent to a given number of places.

EXAMPLE 3. Change $\frac{1}{3}$% to a 5-place decimal.

Solution: Change $\frac{1}{3}$ to a 3-place decimal equivalent. (Since, in changing a percent to a decimal, we move the decimal point two places to the left, we need only three places in the decimal equivalent.) Use Table III (page 202), or divide the denominator into the numerator.

```
         .333
 1/3 = 3 )1.000
           9
           10
            9
           10
            9
            1  ← less than half
                 the divisor
```

$$\frac{1}{3}\% = .333\%$$

$$= .00.333$$

$$= .00333$$

Answer: $\frac{1}{3}$% = .00333

EXERCISES

In 1–18, change each fraction of a percent to a 5-place decimal. In 1–5, use Table III (page 202).

1. $\frac{1}{4}\%$ **2.** $\frac{2}{3}\%$ **3.** $\frac{3}{8}\%$ **4.** $\frac{5}{6}\%$ **5.** $\frac{4}{5}\%$ **6.** $\frac{3}{10}\%$

7. $\frac{7}{10}\%$ **8.** $\frac{44}{100}\%$ **9.** $\frac{2}{7}\%$ **10.** $\frac{4}{9}\%$ **11.** $\frac{2}{11}\%$ **12.** $\frac{13}{15}\%$

13. $\frac{8}{9}\%$ **14.** $\frac{19}{20}\%$ **15.** $\frac{18}{21}\%$ **16.** $\frac{11}{16}\%$ **17.** $\frac{18}{25}\%$ **18.** $\frac{25}{32}\%$

APPLICATION PROBLEMS

SAMPLE SOLUTION

If 1% of a given number is 40, how much will $\frac{1}{4}\%$ be? *$\frac{1}{4}\%$ is $\frac{1}{4}$ of 1%. Thus, $\frac{1}{4}\%$ of the number is $\frac{1}{4}$ of $40 = \frac{1}{\not{4}_1} \times \frac{\not{40}^{10}}{1} = 10$.* 10

1. Which is a larger percent: $\frac{16}{17}\%$ or .9%? __________

2. If $\frac{1}{5}\%$ of a given number is equal to 25, how much is 1% of the given number? __________

3. Is $\frac{9}{10}\%$ larger than 1%? __________

4. If 1% of a given number is 50, how much is $\frac{1}{2}\%$ of the given number? __________

5. "Last month the Cost-of-Living Index rose by three-tenths of one percent." Write the percent as a decimal. __________

6. If Joe Stubb's batting average is .320, what percent of his times at bat did he get a hit? __________

7. What fraction of a dollar is $\frac{1}{4}\%$? __________

8. A bank pays 10.875% on some certificates of deposit. What fraction of a percent is the .875% equal to? __________

9. Bank A pays 10.6% interest and Bank B pays $10\frac{5}{8}\%$. Which bank pays the higher interest? __________

10. What fraction of a percent is .375%? __________

UNIT 61. Finding a Percent of a Given Amount

A common problem in everyday life is figuring out the amount of sales tax that must be paid for a given purchase. To do this, you must know the percent of the tax, and you must know the amount of the purchase.

A sales tax of 7% (7% means 7¢ per 100¢) tells you that $.07 must be paid for every $1.00 of purchase. A purchase of $2.00 would be taxed $.14, a purchase of $3.00 would be taxed $.21, etc.

In general, if you know an amount, you can find any percent of that amount by multiplying.

RULE

To find a percent of an amount, change the percent to a decimal. Then multiply the amount by the decimal equivalent of the percent. (If the given amount is in dollars and cents, round off to the nearest penny.)

EXAMPLE 1. Find $4\frac{1}{2}\%$ of $5,000.

Solution:

$$4\frac{1}{2}\% = 4.5\% = .04.5 = .045$$

$$\begin{array}{r} \$5{,}000 \\ \times .045 \\ \hline \$225.000 \end{array}$$

Answer: $225.00

EXAMPLE 2. Find the amount of sales tax on a purchase of $43.30 if the tax is 7%.

Solution: Since $7\% = \frac{7}{100} = .07$, multiply $43.30 by .07 and round off.

$$\begin{array}{r} \$43.30 \\ \times .07 \\ \hline \$3.03(10) \end{array} = \$3.03$$

Answer: The tax is $3.03.

You can often save money by buying merchandise during special sales when stores sell goods at advertised **discounts** from regular prices. The expression "10% discount" means that the original price has been reduced by 10%; at a "50%-discount sale" or a "half-off sale," the original price has been reduced by 50%; etc.

EXAMPLE 3. What is the sale price of a coat at a "35%-discount sale," if the coat regularly sells for $95.75?

Solution:

Step 1: Find the amount of discount by multiplying the original price, $95.75, by the discount percent, 35%.

$$\begin{array}{r} \$95.75 \\ \times .35 \\ \hline 4\ 78\ 75 \\ 28\ 72\ 5 \\ \hline \$33.51(25) \end{array} = \$33.51$$

Step 2: Find the sale price by subtracting the amount of discount, $33.51, from the original price, $95.75.

$$\begin{array}{r} \$95.75 \\ -33.51 \\ \hline \$62.24 \end{array}$$

Answer: The sale price is $62.24.

EXERCISES

In 1–15, find:

1. 5% of 58
2. 15% of 72
3. 6% of 125
4. 25% of $12.63
5. 21% of $47.93
6. 38% of 74.38
7. $22\frac{1}{2}$% of $135.23
8. $6\frac{1}{4}$% of $24.87
9. 6.25% of 630
10. 4.7% of $77.85
11. .25% of 6,432
12. $25\frac{1}{5}$% of $372.42
13. $\frac{1}{4}$% of 375
14. $\frac{4}{5}$% of 462
15. .6% of 32.16

16. Use a calculator for this problem.

From the given information, find the amount of discount and the new price of each of the following items. As a sample solution, this information has been computed for the first item.

	Original Price	Discount Percent	Amount of Discount	New Price
	$87.95	35%	$30.78	$57.17
a.	235.80	33%		
b.	347.90	25%		
c.	185.75	30%		

$87.95
× .35
$30.78(25) = $30.78

$87.95
− 30.78
$57.17

If your calculator has a percent key, you may use that key instead of first converting the percent to a decimal.

APPLICATION PROBLEMS

1. A dress selling for $37.50 was reduced by 25%.

What is the amount of reduction? ______

What is the new price? ______

If there is a 5% sales tax, find the total cost of the dress. ______

2. The Browns bought a house for $58,575 and made a down payment of 20%. What is the amount of the down payment? ______

3. Because of a heavy snowfall, 30% of the students were absent. If the total number of students in the school is 1,250, how many students were absent? ______

4. The total number of students in a school is 2,340. If 45% of the students are girls, how many students are boys? ______

5. The Nelson family income is $29,780 a year. They plan to use the following budget: food, 20%; shelter, 25%; clothing, 15%; savings, 5%; all other expenses, 35%. How much will they allow for each item?

food: ______

shelter: ______

clothing: ______

savings: ______

other: ______

6. A salesperson earns $12\frac{1}{2}$% commission. What is the commission earned on a sale of $15,875.95? ______

7. A real estate tax rate is 6.75% per year. If a house is assessed at $23,450, what is the yearly tax paid? ______

8. Nancy got a 15% increase in salary. If her salary last year was $19,750, what is her new salary? ______

9. Malcolm has a savings account of $2,475.90. If the bank pays $5\frac{3}{4}$% interest and he makes no deposits or withdrawals, how much money will he have at the end of one year? ______

10. A bedroom set regularly sells for $1,678.75.

If it is reduced by 45%, what is the new price of the set? ______

If there is a sales tax of 8%, what is the total amount a customer will pay? ______

For 11–12, use a calculator.

11. Carrie's job for a mail order company is to check the accuracy of calculation on customer orders. She found some errors in the Total Price column of the following order.

a. Correct the Total Price column.

b. Check, and correct if necessary, the merchandise total, sales tax, shipping charge, and order total.

Description	Code No.	Color	Size	Number of Items	Price Each	Total Price	*Corrections*
Shirt	S2	*Blue*	*M*	2	$14.99	$29.98	
Pillow	P17	—	*King*	4	16.95	65.80	
Lamp	L9	—	—	1	39.95	39.95	
Sheet	SH5	*Beige*	*Full*	2	8.59	19.18	
				Merchandise Total		164.81	
				7% State Sales Tax		10.54	
				Shipping Charge (65¢ *per item*)		5.85	
				ORDER TOTAL		$182.20	

12. The Stangers went shopping at the Depaul Department Store's spring housewares sale. Complete the record of their purchases.

Item	Original Price	Discount Percent	Amount of Discount	Sale Price
Dinnerware	$125.00	55%		
Toaster	18.95	35%		
Electric Can Opener	14.25	15%		
Canister Set	12.89	25%		
Steak Knives	16.50	65%		
Pitcher	15.00	20%		
			Subtotal	
			8.25% Sales Tax	
			Delivery Charge	4.00
			Total	

UNIT 62. Finding What Percent One Number Is of Another Number

It is sometimes necessary to find what percent one number is of another number. To do this, start by writing a fraction. In Unit 59, you learned how to express a fraction as a percent.

Example 1. Express "30 out of 70" as a percent (to the nearest *whole percent*).

Solution: "30 out of 70" is $\frac{30}{70} = \frac{3}{7}$.

$$7\overline{)3.00} = .42\tfrac{6}{7} \approx .43$$

```
      3
    .42 = .43
 7 )3.00
    2 8
     20
     14
      6 ← remainder is larger
          than half of the divisor
```

Answer: "30 out of 70" is 43%.

Many word problems that deal with percents can be solved by writing the given information as a fraction. This fraction can then be changed to a percent.

Example 2. In a class of 25 students, 5 students are absent. Find the percent of students absent.

Solution: Write a fraction comparing the number absent to the total number of students.

$$\frac{\text{number of students absent}}{\text{total number of students}} = \frac{5}{25}$$

If you have difficulty setting up the fraction, you can use the $\frac{\text{IS}}{\text{OF}}$ method. The facts in this problem can be stated as: "(5 is) what percent (of 25)?"

$$\frac{\text{IS}}{\text{OF}} = \frac{5}{25}$$

Change the fraction to a percent.

$$\frac{5}{25} \rightarrow \frac{20}{100}$$

Think: "25 into 100 goes 4, and 4 × 5 = 20."

Answer: 20%

Example 3. What percent of 1,776 is 373? Answer to the nearest *whole percent.*

Solution: "(373 is) what % (of 1,776)?"

$$\frac{\text{IS}}{\text{OF}} = \frac{373}{1,776}$$

```
          .21 = .21
1,776 )373.00
       355 2
        17 80
        17 76
            4 ← remainder is less than
                half the divisor
```

Answer: 21%

EXERCISES

In 1–6, write the indicated ratio as a fraction. Then change the fraction to the nearest *whole percent.*

1. 28 out of 32 **2.** 30 out of 54 **3.** 12 out of 60

4. 32 out of 58 **5.** 50 out of 125 **6.** 64 out of 96

In 7–15, solve to the nearest *whole percent.*

7. 32 is what % of 64?

8. 49 is what % of 70?

9. 35 is what % of 120?

10. 43 is what % of 78?

11. 68 is what % of 134?

12. 123 is what % of 436?

13. 215 is what % of 430?

14. 75 is what % of 240?

15. 235 is what % of 638?

APPLICATION PROBLEMS

1. Tom had $165 and spent $89.10. What percent of his money did he spend? __________

2. Pierre earned $13,500 last year. If he saved $1,215, what percent of his income did he save? __________

3. There are 18 girls and 27 boys in class. What percent of the class is boys? __________

4. A coat regularly selling for $175 was reduced to $140. What is the percent of the reduction? __________

5. The Wilsons obtained a $13,535 mortgage on their house. If the first year's interest is $947.45, what percent of the mortgage is this interest? __________

6. Ellen earned $166.25 interest in one year. If her original deposit was $2,375, what rate of interest did the bank pay? __________

7. If $4.48 in sales tax is paid on an item priced at $89.50, what is the rate of the sales tax? __________

8. The DeMato family bought a house for $56,000 and made a down payment of $14,000. What is the percent of the down payment? __________

9. A retailer had 340 blouses in stock and sold 153 blouses during a special sale. What percent of the stock was sold? __________

10. In an election district, there are 3 registered Democrats for every 4 registered Republicans. What percent of the registered voters are Republicans? __________

For 11–12, use a calculator.

11. Last week, Jimmy's take-home pay was $342.28. He spent $52.53 for food, $38.45 for clothing, $21.60 for fare, and $12.50 for entertainment. What percent of his pay did he spend for each? What percent of his week's pay is left? Write answers to the nearest *percent.*

Food: __________

Clothing: __________

Fare: __________

Entertainment: __________

Percent left: __________

12. The total land area in all the National Parks is about 16,000,000 acres. Some of the larger parks are listed in the table. Find, to the nearest *percent,* what percent of the total is in each of these parks.

National Park	Number of Acres	Percent of 16,000,000
Big Bend	708,118	
Everglades	1,398,800	
Glacier	1,013,595	
Grand Canyon	1,218,375	
Great Smoky Mountains	517,368	
Isle Royale	571,796	
Mount McKinley	1,939,493	
North Cascades	504,780	
Olympic	908,720	
Yellowstone	2,219,823	
Yosemite	760,917	

UNIT 63. Finding a Number When a Percent of It Is Known

You can find a number when you know a percent of it and you know the number that the percent represents.

RULE

> To find a number when you know a percent of it and you know the number that the percent represents, divide the known number by the percent.

EXAMPLE 1. If a 30% reduction in price is equal to $15.00, find the original price.

Solution: Using the rule, divide the reduction in dollars by the reduction percent. The reduction is $15 and the reduction percent is 30%. Therefore, divide: $15 ÷ 30%.

$$\frac{\$15}{30\%} = \frac{\$15}{.30} \qquad .30.\overline{)\$15.00.} = \$50.$$

$$\begin{array}{r} 15\,0 \\ \hline \end{array}$$

Check: If $50 is the correct answer, then 30% of $50 should equal the given reduction of $15.

$$\begin{array}{r} \$50 \\ \times .30 \\ \hline \$15.00 \end{array} \checkmark$$

This method is easily remembered if you use the $\frac{\text{IS}}{\text{OF}}$ fraction: "($15 is) (30% of) what price?"

$$\frac{\text{IS}}{\text{OF}} = \frac{\$15}{30\%} = \frac{\$15}{.30}$$

Perform the indicated division, as shown above.

Answer: The original price is $50.00.

EXAMPLE 2. $87\frac{1}{2}\%$ of what number is 56?

Solution: Divide the number by the percent $\left(87\frac{1}{2}\% = 87.5\% = .875\right)$.

$$\begin{array}{r} 64. \\ .875.\overline{)56.000.} \\ 52\,50 \\ \hline 3\,500 \\ 3\,500 \\ \hline \end{array}$$

By the $\frac{\text{IS}}{\text{OF}}$ fraction:

"(56 is) ($87\frac{1}{2}\%$ of) what number?"

$$\frac{\text{IS}}{\text{OF}} = \frac{56}{87\frac{1}{2}\%} = \frac{56}{.875}$$

Perform the indicated division, as shown above.

Check: Does $87\frac{1}{2}\%$ of $64 = 56$?

$$\begin{array}{r} 64 \\ \times .87 \\ \hline 4\,48 \\ 51\,2 \\ \hline 55\,68 \\ +32 \\ \hline 56.00 \end{array} \checkmark \qquad \frac{1}{2} \times \frac{64}{1} = 32$$

Answer: 64

Note that this method is really the same one that you used in Unit 38 to find a number when you knew a *fractional* part of it. In Example 2, if you know that $87\frac{1}{2}\%$ is the same as $\frac{7}{8}$, you can find the answer by dividing $\frac{7}{8}$ into 56.

$$56 \div \frac{7}{8} = \frac{\overset{8}{\cancel{56}}}{1} \times \frac{8}{\underset{1}{\cancel{7}}} = 64$$

EXERCISES

In 1–10, find the unknown number.

SAMPLE SOLUTIONS

a. 45% of what number is 90?

$\frac{IS}{OF} = \frac{90}{45\%} = \frac{90}{.45}$

.45)90.00 = 200

200

b. \$4.75 is $\frac{1}{5}$% of what amount?

$\frac{IS}{OF} = \frac{\$4.75}{\frac{1}{5}\%} = \frac{\$4.75}{.2\%} = \frac{\$4.75}{.002}$

.002)4.750 = 2,375

4 6
15
14
10
10

\$2,375

1. 16% of what number is 48?

2. 65% of what number is 260?

3. 31% of what number is 279?

4. $3\frac{1}{4}$% of what number is 65.5?

5. 2.5% of what number is 5.65?

6. $33\frac{1}{3}$% of what number is 78?

7. \$105.93 is 45% of what amount?

8. \$6.77 is 5% of what amount?

9. \$141.75 is 35% of what amount?

10. \$234.75 is $37\frac{1}{2}$% of what amount?

APPLICATION PROBLEMS

1. Isidore bought a radio at a sale where all prices were reduced by 30%. If the amount of the reduction was $45.75, what was the original price of the radio? ____________

   What was the new price? ____________

2. How much money must be invested at 8% to earn $1,000 in interest in a year? ____________

3. If 55% of the students in a school are boys, and there are 858 boys, find the total number of students. ____________

4. A television set was reduced by $32.48. If this represents a reduction of 35%, find the original price. ____________

5. The 25% depreciation (loss of value) of a car is $735. Find the original cost of the car. ____________

6. Linda bought a coat and paid $88.29. If the amount included an 8% sales tax, what was the price of the coat without the sales tax? (*Hint:* $88.29 is 108% of the price of the coat.) ____________

7. The Perkins family bought a house and made a 25% down payment of $19,625. Find the price of the house. ____________

8. Paul got a raise of $3,525, which represented a 15% increase in his yearly salary. What was his salary before the increase? ____________

9. Elmira earns a commission of $12\frac{1}{2}$% of sales. If she earned $167.75 in commissions, find the amount of her sales. ____________

10. Edward earned $258.75 in interest for a year on a certificate of deposit. If the rate of interest is 10.35%, what was the initial amount of the certificate? ____________

For 11–12, use a calculator.

11. Mr. Hardy made up four different math tests. As shown in the table, he recorded the number of questions a student should be able to answer correctly on each test to earn a passing grade of 65%. Calculate the total number of questions on each test. Round off your answers to whole numbers. (*Hint:* 39 is 65% of what number?)

Test	Number of Correct Answers Needed for a Grade of 65%	Total Number of Questions on Test
A	39	
B	26	
C	14	
D	8	

12. Ms. Stark received statements of interest earned on three accounts. Use the amount of interest and the rate of interest to find how much money she has in each account.

Account	Interest Earned	Rate of Interest	Amount in the Account
AMB	$38.49	9.5%	
WS&L	75.16	8.4%	
PRT	14.92	5.6%	

UNIT 64. Finding the Percent of Decrease or Increase

To find the percent of decrease or increase, set up a fraction with the *amount of decrease or increase* as the numerator and the *original number* as the denominator.

EXAMPLE 1. A radio that was selling for $50.00 is reduced to $37.50. What is the percent of the reduction?

Solution: Since you have to find the reduction percent, you first find the reduction *in dollars* by subtracting the new price from the original price.

$$\begin{array}{rl} \$50.00 & \leftarrow \text{original price} \\ -37.50 & \leftarrow \text{new price} \\ \hline \$12.50 & \leftarrow \text{reduction} \end{array}$$

The amount of reduction is $12.50. Now you find what percent $12.50 is of the original price, $50.00. Write a fraction with $12.50 as the numerator and $50 as the denominator. Then change the fraction to a percent.

$$\frac{12.50}{50} \rightarrow \frac{25}{100}$$ Think: "50 into 100 goes 2, and 2 × 12.50 = 25."

Using the $\frac{\text{IS}}{\text{OF}}$ fraction:

"($12.50 is) what % (of $50)?"

$$\frac{\text{IS}}{\text{OF}} = \frac{\$12.50}{\$50} = \frac{25}{100} = 25\%$$

Answer: The reduction is 25%.

EXAMPLE 2. If a salary is increased from $120 to $138, what is the percent of increase?

Solution: Find the amount of increase.

$$\begin{array}{r} \$138.00 \\ -120.00 \\ \hline \$18.00 \end{array}$$

Write a fraction with the increase as the numerator and the original salary as the denominator.

$$\frac{\$18}{\$120} = 120\overline{)18.00} = .15 = 15\%$$

$$\begin{array}{r} 12\ 0 \\ \hline 6\ 00 \\ 6\ 00 \\ \hline \end{array}$$

Using the $\frac{\text{IS}}{\text{OF}}$ fraction: "($18 is) what % (of $120)?"

$$\frac{\text{IS}}{\text{OF}} = \frac{\$18}{\$120}$$

Perform the indicated division, as shown above.

Answer: The increase is 15%.

Remember

When figuring percents of increase or decrease, the *original amount* is always the denominator.

EXERCISES

Answer all exercises to the nearest *whole percent.*

In 1–6, find the percent of decrease if:

1. $235 is decreased to $185. **2.** $325 is decreased to $128. **3.** 327 is decreased to 295.

4. $387 is decreased to $234.

5. 436 is decreased to 228.

6. 425 is decreased to 319.

In 7–12, find the percent of increase if:

7. 100 is increased to 125.

8. $115 is increased to $147.

9. 135 is increased to 165.

10. 153 is increased to 232.

11. $243 is increased to $298.

12. $327 is increased to $415.

APPLICATION PROBLEMS

1. Tom earned $8,650 last year, and this year he will earn $9,750. What is the percent of increase? ____________

2. The price of peanuts went up from $1.29 a pound to $1.47 a pound. Find the percent of increase. ____________

3. A school had a student enrollment of 1,628 last year. This year the enrollment is 1,865. Find the percent of increase. ____________

4. Patrick weighed 235 pounds. After dieting for six months, his weight was 185 pounds. What percent of his weight did he lose? ____________

5. A retailer sold $23,475 worth of merchandise last month. This month her sales were only $20,560. What is the percent of decrease in sales? ____________

6. A refrigerator selling for $489.50 was reduced to $318.18. What was the percent of reduction? ____________

7. The Fernandos bought a house for $48,000. Three years later, they sold it for $60,000. What was the percent of profit? ____________

8. Lynn's rent was increased from $423 to $456.84. What was the percent of increase? ____________

9. A retailer sold 121 more shirts this month than he did last month. If he sold 484 shirts last month, what was the percent of increase in the sale of shirts? ____________

10. The price of a new car went up from $8,530 to $9,127. What was the percent of increase? ____________

For 11–12, use a calculator. Answer to the nearest *percent.*

11. The Green Lake School District compared the elementary school populations at dates ten years apart. Find the difference in the numbers of students in each school, and find the percent of increase or decrease from the earlier figure. A sample result is shown.

	Population		Increase or Decrease	
School	Earlier Census	Ten Years Later	Amount	Percent
Cordell	347	328	19	5% decrease
Estes	205	152		
Founders	190	218		
Hill	308	345		
Nashua	134	148		
Valley	218	182		

12. Lynn compared prices in two supermarkets. For each item, find what percent higher or lower the price is in Market B compared with the price in Market A.

	Price		Higher or Lower	
Item	A	B	Amount	Percent
Milk	$.89	$1.12	$.23	26% higher
Bread	.85	.79		
Eggs	.89	.75		
Cheese	.79	.85		
Lettuce	.59	.69		

Review of Part XIV (Units 57–64)

In 1–12, change each percent to a decimal.

1. 25% **2.** 5% **3.** 12.5% **4.** 250% **5.** $\frac{1}{4}$% **6.** .5%

7. $\frac{3}{4}$% **8.** $\frac{4}{5}$% **9.** $6\frac{1}{2}$% **10.** $15\frac{3}{4}$% **11.** $12\frac{3}{5}$% **12.** 24.35%

In 13–22, change each decimal to a percent.

13. .35 **14.** .05 **15.** .135 **16.** .075 **17.** .9

18. .005 **19.** 3.75 **20.** .5 **21.** .30 **22.** .105

In 23–25, round off to the nearest *penny*.

23. Find 35% of $245.85. **24.** What is 8% of $353.50? **25.** $5\frac{1}{2}$% of $568.85 =

In 26–32, round off each answer to the nearest *whole number*.

26. Write 28 out of 153 as a fraction. Change the fraction to a percent.

27. 35 is what percent of 125? **28.** Find what percent 47 is of 265.

29. 35% of a number is 75. Find the number. **30.** 15% of what amount is $125?

31. $5\frac{1}{2}$% of the cost is $82. Find the cost. **32.** $\frac{3}{4}$% of what number is 8.35?

In 33–36, find the percent of increase or decrease. Round off to the nearest *whole percent*.

33. The expenses increased from $165 to $210.

34. The enrollment of 235 pupils decreased to 175 pupils.

35. Arnold's bank balance of $568 decreased to $365.

36. The 472 tons of paper increased to 595 tons.

Cumulative Review of Parts *XI–XIV*

In 1–3, write each decimal as a word phrase.

1. 3.04 ____________________

2. .026 ____________________

3. 1.0027 ____________________

In 4–6, write each word phrase as a decimal.

4. six tenths __________ **5.** nineteen thousandths __________

6. four and two hundred fifteen ten-thousandths __________

In 7–10, change each decimal fraction to a common fraction, and reduce to lowest terms.

7. .62 **8.** .465 **9.** .001 **10.** .0208

In 11–12, arrange each group of decimals in ascending order of size (smallest value first).

11. .4 .405 .004 .03 **12.** .567 .5678 .6 .06

In 13–15, change each decimal to a complex fraction, and simplify.

13. $.12\frac{1}{2}$ **14.** $.15\frac{1}{5}$ **15.** $.66\frac{2}{3}$

In 16–19, change each common fraction to a decimal fraction, and round off the answer to the nearest *hundredth.*

16. $\frac{15}{16}$ **17.** $\frac{5}{9}$ **18.** $\frac{7}{12}$ **19.** $\frac{11}{32}$

In 20–27, compute.

20. $245.6 + 15.035 + .71$ **21.** $9.6 \div .24$

22. $38.23 - 6.125$ **23.** $6.87 + 25 + 15.3 + .7654$

24. $70 - 8.775$ **25.** $4.32 \times .05$

26. $1.296 \div .027$ **27.** 698.5×2.4

In 28–31, compute and round off each answer to the nearest penny.

28. $42.25 × .67

29. $18.75 ÷ 12

30. $368.76 ÷ 36

31. $2,035.50 × 12.45

In 32–37, calculate the answer mentally.

32. 16.4 × 100

33. 37 ÷ 1,000

34. .074 × 10,000

35. .2 ÷ 10

36. 4,758 ÷ 100,000

37. .86 × 10

In 38–43, change each mixed number to a mixed decimal, and compute.

38. $285 \times 14\frac{3}{5}$

39. $568 \div 12\frac{1}{2}$

40. $25\frac{1}{2} \div 5$

41. $68\frac{5}{8} \times 102$

42. $5\frac{3}{4} \times 2\frac{1}{2}$

43. $107\frac{16}{25} \div 46\frac{4}{5}$

In 44–47, change each percent to a decimal.

44. 7%

45. 10.8%

46. 150%

47. $5\frac{1}{2}\%$

In 48–51, change each decimal to a percent.

48. 1.02

49. .65

50. .9

51. .001

In 52–57, find each answer to the nearest *whole number*.

52. 45 is what percent of 200?

53. What is 65% of 428?

54. 75% of what number is 288?

55. $\frac{1}{2}\%$ of what number is 45?

56. What is $12\frac{1}{2}\%$ of 104?

57. What percent is 32 of 400?

In 58–60, find the percent of increase or decrease. Round off to the nearest *whole percent*.

58. 18 students passed a test. 24 students passed the next test.

59. A $465 refrigerator was reduced to $348.

60. When made with water, Tammy's tomato soup has 760 mg of sodium per serving. When made with milk, it has 810 mg.

In 61–70, solve each problem.

61. James had the following deductions from his weekly salary check: federal income tax, $127.43; pension fund, $21.38; state income tax, $80.75; union dues, $9.72; health insurance plan, $.87. What is the total of all his deductions? ______________

62. Find the total of all the following weights: 4.71 pounds, .87 pound, 1.50 pounds, 12.97 pounds, .58 pound, and 1.06 pounds. ______________

63. A salesperson earned the following commissions last month: $375.80, $462.65, $392.75, and $483.42. The month before, his total commission earnings were $1,836.75. By how much did his commission earnings drop last month? ______________

64. A salesclerk sold the following lengths of fabric: .84 meter, 3.23 meters, .52 meter, 13.60 meters, and 4.20 meters. How many meters of fabric did the clerk sell? ______________

65. Sloan's Department Store ordered 3 dozen shirts at $165 a dozen, 26 pairs of slacks at $14.25 a pair, and 10 dozen pairs of socks at $16.38 per dozen pairs. Sloan's has a credit with the supplier as the result of the return of an earlier purchase. If the credit amount is $582.65, how much does Sloan's still owe on this order? ______________

66. Of the 540 seniors at Avalon High School, 35% are going on a school trip. If the buses ordered for the trip seat 42, how many buses will be needed so that each student will have a seat? ______________

67. A manufacturer has on hand 423.8 meters of a wool fabric. How many coats can be made if the average amount of fabric per coat is 3.26 meters? ______________

68. For a project on genetics, a biology class did a survey of eye color. Of 285 people, there were 123 with brown eyes, 77 with hazel eyes, 54 with blue eyes, 20 with gray eyes, and 11 with eyes of other colors. Find the percent of the 285 people with each eye color to the nearest *whole percent*.

brown ______________

hazel ______________

blue ______________

gray ______________

other ______________

69. It was determined that 42% of the people in the 40–50 age group in Coral City were overweight. If 2,835 people of that group were overweight, what was the total number of people in the group? ______________

70. At the start of the school year, there were 3,486 textbooks in the Science bookroom. In the first full week of classes, the following numbers of books were distributed to students: Monday, 326; Tuesday, 575; Wednesday, 1,288; Thursday, 693; Friday, 132. What percent of the books were distributed each day?

Monday ______

Tuesday ______

Wednesday ______

Thursday ______

Friday ______

What percent of the books remained in the bookroom at the end of the week? ______

For 71–74, use a calculator.

71. A field representative submitted the following expense statement. Find the expense item totals, the daily totals, and the grand total for the week.

Expense Item	Mon.	Tues.	Wed.	Thurs.	Fri.	TOTALS
Motel	$48.60	42.88	51.16	53.54	63.80	
Breakfast	3.65	2.84	3.42	3.75	2.89	
Lunch	5.80	6.47	6.32	5.73	6.59	
Dinner	13.82	12.67	12.85	13.75	12.49	
Taxi	8.75	9.40	7.75	8.65	9.90	
Car Rental	28.60	32.15	34.80	35.16	42.70	
DAILY TOTALS						

GRAND TOTAL

72. Below is the payroll for the ACE Realty Co. Find:

a. the total deductions and the net pay for each employee.

b. the totals of all the columns.

Check for accuracy. The total of all the deductions added to the total of all the net pay should be the same as the total of all the gross pay.

		DEDUCTIONS					
Employee	Gross Pay	Fed. Tax	FICA Tax	State Tax	City Tax	Total Deductions	Net Pay
A	$423.52	92.45	21.82	11.15	9.23		
B	421.43	96.52	19.63	9.57	8.93		
C	426.59	94.63	22.15	10.64	9.65		
D	421.31	87.28	33.35	12.52	9.27		
E	429.50	95.32	22.43	11.72	8.83		
TOTALS							

73. Find the total wages, the total deductions, and the net pay for each employee. Check for accuracy by finding the grand totals. The total net pay added to the grand total of deductions should equal the grand total of wages.

Card No.	Earnings at Regular Rate	Bonus for Overtime	Total Wages	DEDUCTIONS			Net Pay
				Soc. Sec. (FICA)	Fed. With. Tax	Total Deductions	
01	$386.54	38.47		23.29	81.53		
02	392.65	24.22		25.68	84.68		
03	428.73	36.74		29.78	95.47		
04	419.86	34.65		28.38	93.86		
05	432.44	46.97		34.75	95.67		
TOTALS							

74. Place the decimal point in the correct position in the answer by estimating what the result should be. Then use a calculator to check whether you were right.

a. 3.5 + 10.9 = 1 4 4

b. 20.52 × .4 = 8 2 0 8

c. 35.28 ÷ 7.2 = 4 9

d. 25.32 − 13.6 = 1 1 7 2

e. 12.7 + 5.68 + .943 = 1 9 3 2 3

f. 162.5 − 93.52 = 6 8 9 8

g. 49.8 × 2.05 = 1 0 2 0 9

h. 188.75 ÷ 62.5 = 3 0 2

PART XV. Geometric Measures: Length, Area, and Volume

UNIT 65. Understanding Measurements: Customary and Metric Units

As early humans changed from hunters to farmers, they needed to measure lengths: the lengths of fields, the lengths of building materials, the lengths of cloth, etc. The first units of length were based on the human body. The *cubit* was the length of a man's forearm from his elbow to the tip of his middle finger. The cubit was divided into *palms* (the width of the hand), and the palm was divided into *digits* (the width of a finger). The *mile* was a thousand paces, and the *foot* was based on the actual length of someone's foot.

Of course, since people are of different sizes, each of these units of length differed considerably. To avoid confusion, lengths and other measures were made *standard* (alike and unchanging). Today, the National Bureau of Standards in Washington, D.C., determines the units that you see on rulers and yardsticks. Every time you measure something with a ruler or a tape measure, you are comparing it with these *standard units*.

There are two major systems of measurement, the **Customary system** (also known as the *English system*) and the **metric system**. The system used in most countries is the metric system. (The metric system is formally known as SI, for Système International, the French words for International System.) The United States is also phasing in the metric system as a means of standardizing measurements worldwide.

The metric system is named after the basic unit of length, the meter, whose standard length was established by scientific calculations. It is equal to about 39.37 inches. The metric system is easy to use because the units of measurement are related by powers of 10. That is, to change to different size units, you simply multiply or divide by 10, 100, or 1,000, etc.

For example, in the Customary system, to change from inches to yards, you must divide by 36. To do this, you would need to use pencil and paper or a calculator. In the metric system, to change from meters to kilometers, you would only need to move a decimal point! (As you learned from your work on decimals, moving a decimal point is equivalent to multiplying or dividing by a power of 10.)

Some of the basic metric measures are the ***meter*** (for *length*), the ***liter*** (for *volume*), and the ***gram*** (for *weight*). Prefixes tell how different sizes compare to the basic unit. Thus, for example, a *milli***gram** (mg) is *one thousandth* of a **gram,** and a *kilo***meter** (km) is *one thousand* **meters.**

Table IV: Metric Prefixes

Prefix	Means Multiply by	Prefix Abbreviation
milli	one thousandth (.001)	m
centi	one hundredth (.01)	c
deci	one tenth (.1)	d
deka	ten (10)	da
hecto	one hundred (100)	h
kilo	one thousand (1,000*)	k

*In this book, we continue the customary practice of using commas in writing large numbers. However, work done entirely in metric units uses spaces instead of commas. In metric notation, one million is written 1 000 000.

As you get used to metric measure, it will seem easier all the time. Already, soda pop bottles are marked in terms of liters, packages of cookies give the weight in grams, and track events are measured in meters. A liter is a little more than a quart, a gram is about the weight of a paper clip, and a meter is a little longer than a yard. A table of some of the common equivalent measures appears on page 275.

UNIT 66. Units of Length

The distance between the ends of something, or the distance between two separate objects, is called the *measure of length*. When you measure the length of something, you find the *distance* between its two ends by *measuring in a straight line in one direction only*.

If you know that a certain fence is "fifty feet long," you known only the *distance* from one end to the other. You know nothing about how high it is. When you buy "1 meter of fabric," you know that the fabric is one meter long. This does not tell you how wide it is.

At times the expression "measure of length" is awkward to use because such measurements as "width" and "height" and "altitude" and "breadth" are all really measures of length. Consequently, the expression *linear measure* is used in place of "measure of length" where confusion is possible.

Remember

Measure of length and linear measure both mean straight-line distance.

UNITS OF LINEAR MEASURE

The Customary units of linear measure are the *inch*, the *foot*, the *yard*, and the *mile*. These units are described in the following table:

Table V: Customary Units of Linear Measure

Unit	Abbreviation	Description
inch	in.	There are 12 in. in 1 ft. There are 36 in. in 1 yd.
foot	ft.	One ft. contains 12 in. There are 3 ft. in 1 yd. There are 5,280 ft. in 1 mi.
yard	yd.	One yd. contains 3 ft. One yd. contains 36 in. There are 1,760 yd. in 1 mi.
mile	mi.	One mi. contains 5,280 ft. One mi. contains 1,760 yd.

In the metric system, the prefixes milli-, centi-, etc., tell how the measures are related to the basic unit, the meter.

Both the Customary and the metric systems include measures that are not in everyday use. The Customary system, for example, includes such units as rod and league, which are not commonly used. In the following metric table, the bold (heavy) type shows the measures that are most commonly used.

Table VI: Metric Units of Linear Measure

Unit	Abbreviation	Description
millimeter	**mm**	.001 m, .01 dm, .1 cm
centimeter	**cm**	.01 m, .1 dm, 10 mm
decimeter	dm	.1 m, 10 cm, 100 mm
meter	**m**	10 dm, 100 cm, 1,000 mm .1 dam, .01 hm, .001 km
dekameter	dam	10 m, .1 hm, .01 km
hectometer	hm	100 m, 10 dam, .1 km
kilometer	**km**	1,000 m, 100 dam, 10 hm

CHANGING A SMALLER UNIT TO A LARGER UNIT

RULE

Unit Change Rule 1: To change a smaller unit to a larger unit:

Step 1: Determine how many smaller units are contained in 1 unit of the larger unit.
Step 2: *Divide* the given number of smaller units by this number.

This rule is very important, as it will be used over and over again. You will use *Rule 1* whenever you change any smaller unit to any larger unit. Remember that when you change from smaller units to larger units, you end up with fewer units. (Think of changing dimes to dollars.) As you will learn, this rule is used with measures of area, volume, liquid, time, and counting.

EXAMPLE 1. Change 72 inches to *feet*.

Solution:

Step 1: 12 in. = 1 ft.
Step 2: 72 in. = 72 ÷ 12 = 6 ft.

Answer: 72 in. = 6 ft.

EXAMPLE 2. How many miles are there in a distance of 4,400 yards?

Solution:

Step 1: 1,760 yd. = 1 mi.
Step 2: 4,400 yd. = 4,400 ÷ 1,760 = 2.5 mi.

```
          2.5
1,760 )4,400.0
        3 520
          880 0
          880 0
```

Answer: 4,400 yd. = 2.5 mi. or $2\frac{1}{2}$ mi.

EXAMPLE 3. How many centimeters are there in 138 millimeters?

Solution:

Step 1: 10 mm = 1 cm
Step 2: 138 mm = 138 ÷ 10 = 13.8 cm

138 ÷ 10 = 13.8. = 13.8

Answer: 138 mm = 13.8 cm

CHANGING A LARGER UNIT TO A SMALLER UNIT

RULE

Unit Change Rule 2: To change a larger unit to a smaller unit:

Step 1: Determine how many smaller units are contained in 1 unit of the larger unit.
Step 2: *Multiply* the given number of larger units by this number.

As with *Rule 1*, this rule will be used over and over again. You will use *Rule 2* whenever you change any larger unit to any smaller unit. Remember that when you change from larger units to smaller units, you end up with more units. (Think of changing dollars to dimes.)

EXAMPLE 4. Change 7 yards to (*a*) *feet* (*b*) *inches*.

Solution:

(*a*) 1 yd. = 3 ft.
7 yd. = 7 × 3 = 21 ft.

(*b*) 1 yd. = 36 in.
7 yd. = 7 × 36 = 252 in.

Answer: (*a*) 7 yd. = 21 ft. (*b*) 7 yd. = 252 in.

EXAMPLE 5. Change 2.416 kilometers to (*a*) *millimeters* (*b*) *centimeters* (*c*) *meters*.

Solution:

(*a*) 1 km = 1,000,000 mm
2.416 km = 2.416 × 1,000,000
= 2,416,000 mm

(*b*) 1 km = 100,000 cm
2.416 km = 2.416 × 100,000 = 241,600 cm

(*c*) 1 km = 1,000 m
2.416 km = 2.416 × 1,000 = 2,416 m

Answer:

(*a*) 2.416 km = 2,416,000 mm
(*b*) 2.416 km = 241,600 cm
(*c*) 2.416 km = 2,416 m

Remember

When changing *metric* units:

1. To change a smaller unit to a larger unit, *divide* by moving the decimal point to the *left*.
2. To change a larger unit to a smaller unit, *multiply* by moving the decimal point to the *right*.

You may need two conversions in one problem.

EXAMPLE 6. Change 5 yards 16 inches to *feet.*

Solution:

$$5 \text{ yd.} = 5 \times 3 = 15 \text{ ft.}$$

$$16 \text{ in.} = 16 \div 12 = 1\frac{1}{3} \text{ ft.}$$

$$5 \text{ yd.} + 16 \text{ in.} = 15 \text{ ft.} + 1\frac{1}{3} \text{ ft.} = 16\frac{1}{3} \text{ ft.}$$

Alternate solution:

$$5 \text{ yd. } 16 \text{ in.}$$

$$= 5\frac{16}{36} \text{ yd.} = 5\frac{4}{9} \text{ yd.} = 5\frac{4}{9} \times 3$$

$$= \frac{49}{9} \times \frac{3}{1} = 16\frac{1}{3} \text{ ft.}$$

Answer: 5 yards 16 inches $= 16\frac{1}{3}$ feet

Remember

In both of the Unit Change Rules, find how many smaller units are contained in one unit of the larger unit. Then:

1. To change a smaller unit to a larger unit, *divide* by the number you found. (You end up with *fewer* units.)
2. To change a larger unit to a smaller unit, *multiply* by the number you found. (You end up with *more* units.)

EXERCISES

In 1–8, change each given length to *inches.*

1. 17 ft. **2.** 14 ft. **3.** $4\frac{1}{2}$ ft. **4.** 6.25 ft.

5. 6 ft. 11 in. **6.** 4 yd. **7.** 3 yd. 2 ft. **8.** $4\frac{1}{2}$ yd.

In 9–16, change each given length to *feet.* When necessary, use mixed numbers or mixed decimals to express your answers.

9. 25 yd. **10.** $15\frac{3}{5}$ yd. **11.** 18.7 yd. **12.** 87 in.

13. 102 in. **14.** 8 yd. 30 in. **15.** 12 yd. 34 in. **16.** 137 in.

In 17–24, change each given length to *yards*. When necessary, use mixed numbers or mixed decimals to express your answers.

17. 35 ft. **18.** 247 in. **19.** 3.7 mi. **20.** 147 ft.

21. $75\frac{1}{3}$ ft. **22.** 325 in. **23.** $2\frac{1}{2}$ mi. **24.** 448 in.

In 25–28, change each given length to *millimeters*.

25. 3.7 cm **26.** .035 km **27.** 2 m **28.** 25 cm

In 29–32, change each given length to *centimeters*.

29. 6.24 m **30.** 548 mm **31.** .248 km **32.** .9 m

In 33–36, change each given length to *meters*.

33. 57 cm **34.** 91 mm **35.** 4.3 km **36.** .075 km

In 37–40, change each given length to *kilometers*.

37. 394 m **38.** 204,400 cm **39.** 10,500 m **40.** 5,500,000 mm

APPLICATION PROBLEMS

1. Tom is $5\frac{1}{2}$ feet tall. Express his height in *inches*. __________

2. Jane bought 163 feet of ribbon. How many yards did she buy? (Express your answer as a *mixed number*.) __________

3. To the nearest *tenth*, how many miles high is a plane flying if its altitude is 32,063 feet? __________

4. A manufacturer bought 254 yards of material. If he uses 2 feet of material for one flag, how many flags can he make? ____________

5. How many feet high will a stack of 7-inch boxes be if the stack contains 237 boxes? (Express your answer as a *mixed number* or a *mixed decimal.*) ____________

6. The diameter of a cylinder is 28 millimeters. What is its diameter in centimeters? ____________

7. Janet runs 4.8 kilometers each day. How many meters does she run in a week? ____________

8. How many meters are contained in $\frac{3}{4}$ kilometer? (*Hint:* First change the common fraction to a decimal fraction.) ____________

9. A length of wire measures .6 meter. How many sections measuring 1.5 millimeters can be cut from it? ____________

10. The scale on a map is 5 millimeters = 75 kilometers. How many kilometers are represented by 4.5 centimeters on the map? ____________

UNIT 67. Measuring Length: Perimeter and Circumference

WORDS TO KNOW

The shape or outline of something forms a **figure.** Common figures include the *triangle*, the *quadrilateral*, the *rectangle*, the *square*, and the *circle.*

A **triangle** is a closed figure that has *three straight sides.*

A **quadrilateral** is a closed figure that has *four straight sides.*

A **rectangle** is a special kind of quadrilateral: It has four sides, but the four angles are all 90-degree angles, or *right angles.* Also, the opposite sides are equal in length.

A **square** is a special kind of rectangle. It has four sides and four 90-degree angles, but all four sides are the *same length.*

A **circle** is a closed curved line with all its points equally distant from its center. The **radius** (plural *radii*) is the line segment connecting the center with any point on the circle. The **diameter** is the line segment connecting any two points on the circle while passing through its center. (The diameter is really two radii drawn in a straight line.)

The distance completely around any straight-sided figure is found by adding together the lengths of all the sides. This measure of length is called the **perimeter.** The distance around a circle is called the **circumference.**

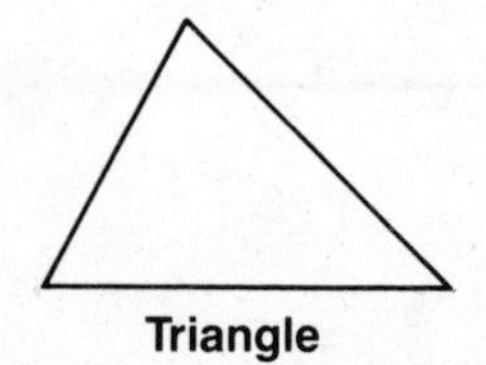
Triangle

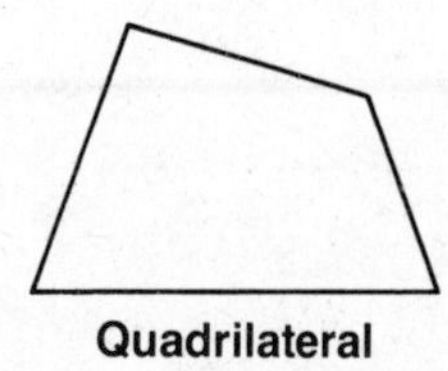
Quadrilateral

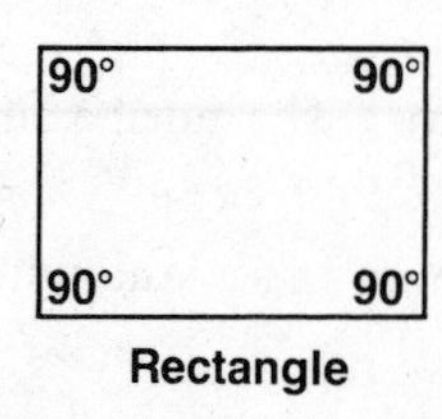

Rectangle

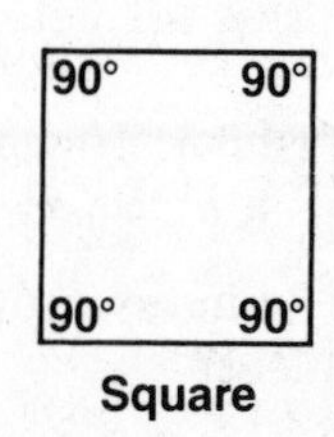

Square

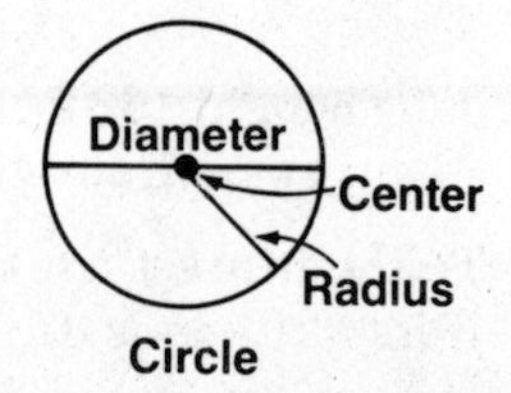

Circle

PERIMETER

Many problems deal with the perimeters of figures such as **triangles, quadrilaterals, rectangles,** and **squares.** To determine the perimeter of such a figure, find the length of each side of the figure and then add all the lengths together. Be sure to add *all* of the lengths.

EXAMPLE 1. Find the perimeter of each of the following figures:

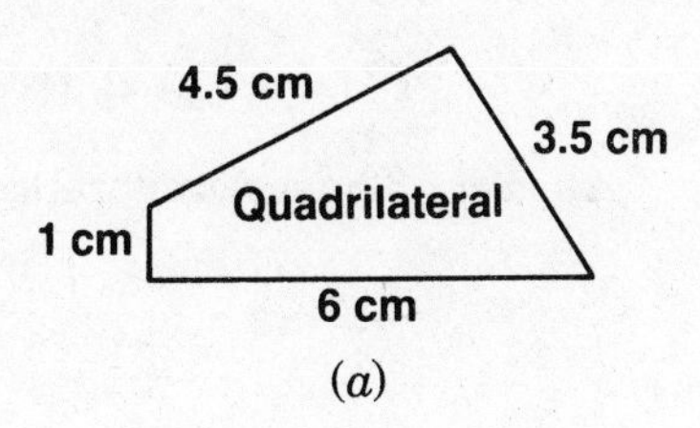

(*a*)

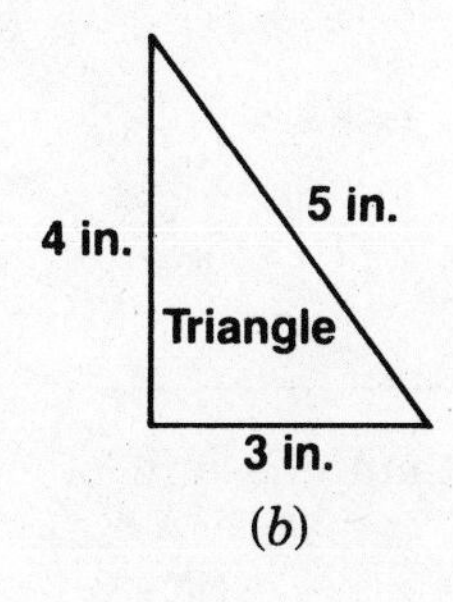

(*b*)

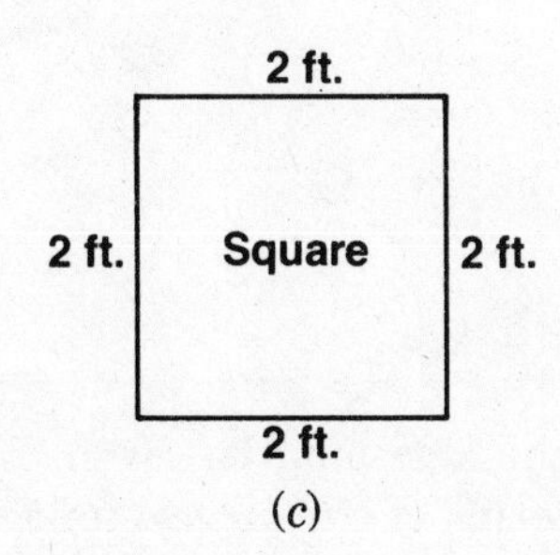

(*c*)

Solution:

(*a*) 1 cm + 4.5 cm + 3.5 cm + 6 cm = 15.0 cm

Answer: The perimeter of the quadrilateral is 15 centimeters.

(*b*) 3 in. + 4 in. + 5 in. = 12 in.

Answer: The perimeter of the triangle is 12 inches.

(*c*) Every square has *four equal sides.* Therefore, you can find the perimeter by multiplying the length of a side by 4:

$$2 \text{ ft.} \times 4 = 8 \text{ ft.}$$

If you prefer, you can add the lengths of the four equal sides:

$$2 \text{ ft.} + 2 \text{ ft.} + 2 \text{ ft.} + 2 \text{ ft.} = 8 \text{ ft.}$$

Answer: The perimeter of the square is 8 feet.

EXAMPLE 2. Find the perimeter of a rectangle whose length is 8 cm and whose width is 6 cm.

Solution: A *rectangle* is a very important geometric figure. Remember the following facts:

1. A rectangle has *four* sides and *four* right angles.
2. The *opposite sides* of a rectangle are always the *same length.*

Using these facts, you can draw a picture of the rectangle whose dimensions are 8 cm × 6 cm. (*Note:* A notation such as "8 cm × 6 cm" is often used to describe rectangles. It is read, "8 centimeters long by 6 centimeters wide." Usually, the longer dimension is called the *length,* and the shorter dimension is called the *width.*)

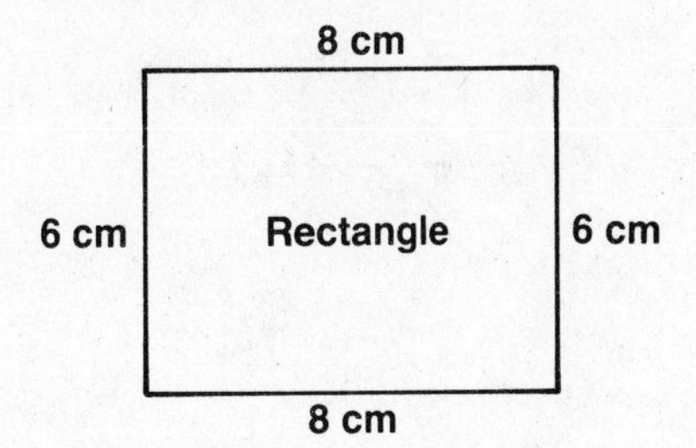

From the figure, the perimeter is:

$$6 \text{ cm} + 8 \text{ cm} + 6 \text{ cm} + 8 \text{ cm} = 28 \text{ cm}$$

Answer: The perimeter of the rectangle is 28 centimeters.

EXAMPLE 3. Find the perimeter of a triangle whose sides are 75 millimeters, 24.3 centimeters, and 26.7 centimeters. Express the answer in *centimeters.*

Solution: Before adding the lengths, change the millimeters to centimeters.

10 mm = 1 cm
75 mm = 75 ÷ 10 = 7.5. = 7.5 cm

Add the three lengths.

$$\begin{array}{r} 7.5 \text{ cm} \\ 24.3 \text{ cm} \\ 26.7 \text{ cm} \\ \hline 58.5 \text{ cm} \end{array}$$

Answer: The perimeter of the triangle is 58.5 centimeters.

EXAMPLE 4. Find the number of feet in the perimeter of the triangle whose sides measure 18 inches, $2\frac{1}{2}$ feet, and 1 yard.

Solution: Change to feet.

$$18 \text{ in.} = \frac{18}{12} \text{ ft.} = 1\frac{1}{2} \text{ ft., and } 1 \text{ yd.} = 3 \text{ ft.}$$

Add the three lengths.

$$1\frac{1}{2} \text{ ft.} + 2\frac{1}{2} \text{ ft.} + 3 \text{ ft.} = 7 \text{ ft.}$$

Answer: The perimeter of the triangle is 7 feet.

CIRCUMFERENCE

To find the circumference C of a circle, multiply the length of the diameter d by π. The symbol π (pronounced "pie") is a Greek letter that represents a number approximately equal to 3.14 or $\frac{22}{7}$.

$$C = \pi \times d$$

Example 5. Find the circumference of a circle whose diameter is 14 cm.

Solution: $C = \pi \times d$

$$C = \pi \times 14 \text{ cm}$$

$$C = \frac{22}{\cancel{7}_1} \times \frac{\cancel{14}^2}{1} = 44 \text{ cm}$$

Answer: The circumference is 44 cm.

Example 6. The radius of a circle is 5 m long. Find the circumference.

Solution: Since the length of a diameter is equal to two radii, a circle with a radius of 5 m has a diameter of 10 m.

$$C = \pi \times d$$

$$C = 3.14 \times 10 \text{ m}$$

$$C = 31.4 \text{ m}$$

Answer: The circumference is 31.4 m.

EXERCISES

In 1–10, find the perimeter of each triangle, using the side lengths given. If more than one unit of measure appears in a problem, give your answer in terms of the *largest* unit.

1. 4 ft., 8 ft., and 6 ft.

2. 13 in., 13 in., and 7 in.

3. $5\frac{3}{4}$ ft., $11\frac{7}{8}$ ft., and $15\frac{1}{3}$ ft.

4. 23.8 in., 15.5 in., and 18.9 in.

5. 7 yd., 9 ft., and 5 yd.

6. 30 in., $3\frac{1}{4}$ ft., and $5\frac{1}{2}$ ft.

7. $3\frac{1}{2}$ m, $2\frac{1}{3}$ m, and $4\frac{3}{4}$ m

8. 442 mm, 32 cm, and 28.3 cm

9. 7.3 km, 6.8 km, and 5.7 km

10. $5\frac{2}{3}$ m, $4\frac{3}{5}$ m, and $6\frac{5}{6}$ m

In 11–16, find the perimeter of each quadrilateral, using the side lengths given. Answers that are not whole numbers should be written as decimal fractions. If more than one unit of measure appears in a problem, give your answer in terms of the *largest* unit.

11. 15 yd., 12 yd., 8 yd., and 27 yd.

12. 12.6 in., 5.3 in., 13.7 in., and 15.9 in.

13. 15 ft., 27 ft., 15 yd., and 18 ft.

14. 25 in., 16 in., 12.7 in., and 21.5 in.

15. 8 m, $12\frac{7}{10}$ m, $7\frac{3}{4}$ m, and $15\frac{3}{5}$ m

16. 15 cm, 124 mm, 160 mm, and 18 cm

In 17–28, find the perimeter of each rectangle, given the length and the width. Answers that are not whole numbers should be written as decimal fractions.

17. 3 ft. by 7 ft.

18. $5\frac{1}{2}$ in. × $12\frac{1}{4}$ in.

19. 7.5 in. × 4.2 in.

20. 25 ft. by 137 ft.

21. 5 in. by $3\frac{1}{2}$ in.

22. 12.6 ft. × 14.8 ft.

23. $5\frac{1}{2}$ ft. × $8\frac{3}{10}$ ft.

24. 9.6 ft. by 3.25 ft.

25. 18 in. × 5 ft.

26. 15.7 m × 34.2 m

27. .65 km × 2.5 km

28. $8\frac{2}{5}$ cm × $14\frac{1}{2}$ cm

In 29–34, find the perimeter of each square, given the length of one side.

29. $3\frac{1}{3}$ in.

30. 225 in.

31. .09 ft.

32. $5\frac{3}{8}$ m

33. .87 km

34. 5.2 cm

In 35–36, find the circumference of each circle. (Use $\pi = \frac{22}{7}$.)

35. A circle whose radius is 63 cm.

36. A circle whose diameter is $4\frac{2}{3}$ in.

In 37–38, find the circumference of each circle. (Use $\pi = 3.14$.)

37. A circle whose diameter is 2.5 m.

38. A circle whose radius is 3.2 ft.

APPLICATION PROBLEMS

1. Mr. Gordon plans to build a rectangular fence around his home. If the property measures 39 feet 6 inches by 105 feet 6 inches, how many feet of fencing will he need? (*Hint:* 6 inches is $\frac{1}{2}$ foot.) __________

2. Mrs. Doheny wants to sew ribbon around her bedspread, which measures $4\frac{5}{8}$ feet by $8\frac{3}{4}$ feet. To the nearest *whole yard*, how many yards of ribbon must she buy? __________

At 30¢ per yard, how much will the ribbon cost? __________

3. In a yacht race, the distance from the starting point to the first marker buoy is 6.8 miles. From the first buoy to the second marker buoy, the distance is 10.3 miles. From the second buoy back to the starting point is 7.4 miles. How long is the triangular race course? __________

4. A baseball "diamond" is really a square each of whose sides is 90 feet in length. If a player hits three home runs, how far will he run around the bases? __________

5. How many times must Ben run around a rectangular schoolyard that measures 367 feet by 293 feet in order to run a mile? (*Hint:* 1 mi. = 5,280 ft.) __________

6. A round tabletop has a diameter of 1.5 meters. How many meters of brass will be needed to make a rim around the edge? __________

7. A circular athletic field has a diameter of $1\frac{3}{4}$ kilometers. If there are five exits all spaced equally apart, how far, to the nearest *kilometer*, are they spaced from each other? __________

8. The length and width of a rectangular swimming pool are $85\frac{3}{4}$ meters and $43\frac{1}{2}$ meters. What is the distance around the pool? __________

9. A painting measures 24 inches by 36 inches. If picture frame molding costs $4.75 per foot, how much will the molding for the frame cost? __________

10. A kitchen floor measures 9 feet by 12 feet. If the door to the kitchen is 3 feet wide, find the number of feet of molding needed to place around the edge of the floor. __________

UNIT 68. Units of Area

Words to Know

The Customary units of *linear* measure are the *inch*, the *foot*, the *yard*, and the *mile*. All linear measures, such as distances or perimeters, are obtained by measuring in *one direction only*. Measures of length are said to be *one-dimensional.*

The number of *square* units in a given figure is the **area** of that figure. As you will learn, measures of area are obtained by measuring in *two directions*. Measures of area are said to be *two-dimensional.*

The Customary units of *square measure* include the *square inch*, the *square foot*, the *square yard*, the *acre*, and the *square mile.*

The metric units of *square measure* include the *square centimeter*, the *square decimeter*, the *square meter*, the *hectare*, and the *square kilometer.*

Measures of **area** are quite common in everyday life. It may be necessary to figure out how many *square feet* of tile are needed to make a bathroom floor. Dress designers must calculate how many *square meters* of material are needed to make a certain dress. Farmers measure the sizes of their farms in *acres* or *hectares.*

Remember

When you find the *measure of area* of a figure, you find out how many times a standard unit of *square measure* is contained in that figure.

Customary Units of Square Measure

The basic Customary units of square measure are described in the following table:

Table VII: Customary Units of Square Measure

Unit	Abbreviation	Description
square inch	sq. in.	There are 144 sq. in. in 1 sq. ft. There are 1,296 sq. in. in 1 sq. yd.
square foot	sq. ft.	One sq. ft. contains 144 sq. in. There are 9 sq. ft. in 1 sq. yd.
square yard	sq. yd.	One sq. yd. contains 9 sq. ft. One sq. yd. contains 1,296 sq. in. There are 4,840 sq. yd. in 1 A.
acre	A.	One A. contains 43,560 sq. ft. One A. contains 4,840 sq. yd. There are 640 A. in 1 sq. mi.
square mile	sq. mi.	One sq. mi. contains 640 A.

When dealing with acres, keep in mind that an acre is already a unit of square measure. There is no such unit as a "square acre."

It is easy to picture a square inch. Just think of a square whose sides are 1 inch in length. (Recall that a square has four equal sides.) Similarly, a square foot may be thought of as a square whose sides are 1 foot in length.

Use the reduced-size diagram below to see that a square with a side measuring 1 foot contains 144 squares that each have a side measuring 1 inch. Thus, 1 square foot contains 144 square inches.

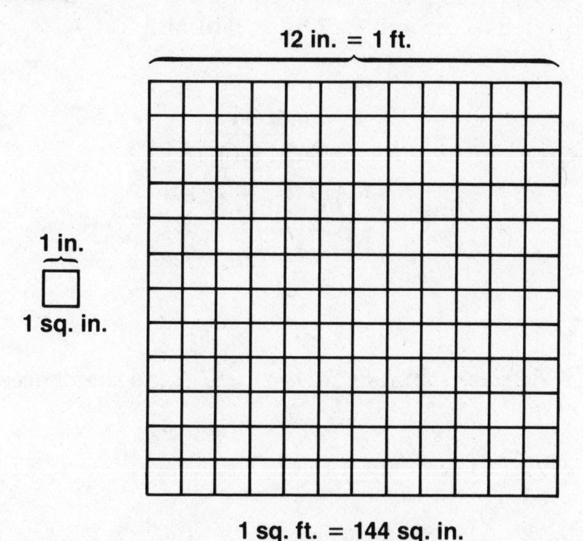

Note in the following reduced-size figure that a square with a side of 1 yard contains 9 squares that each have a side of 1 foot. This is why 1 square yard contains 9 square feet.

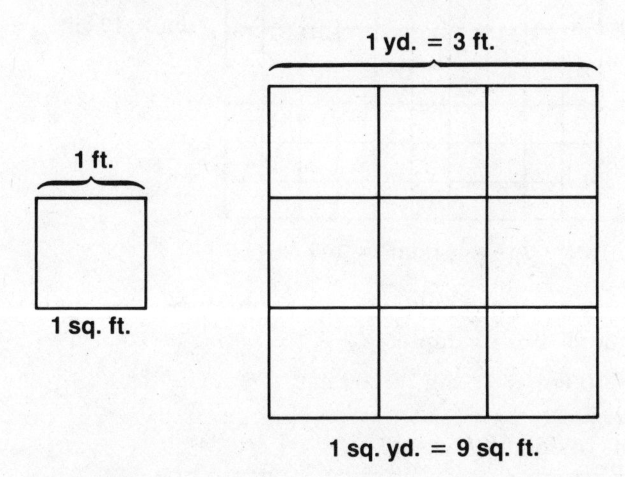

The rules for changing units of square measurement are the same as the rules you learned in Unit 66 for linear measurement.

Remember

In both of the Unit Change Rules, find how many smaller units are contained in one unit of the larger unit. Then:

1. To change a smaller unit to a larger unit, *divide* by the number you found.
2. To change a larger unit to a smaller unit, *multiply* by the number you found.

Example 1. Change 720 square inches to *square feet.*

Solution: To change smaller units to larger units, divide.

144 sq. in. = 1 sq. ft.
720 sq. in. = 720 ÷ 144 = 5 sq. ft.

$$\begin{array}{r} 5 \\ 144\overline{)720} \\ 720 \\ \hline \end{array}$$

Answer: 720 sq. in. = 5 sq. ft.

Example 2. Change 75.6 square feet to *square yards.*

Solution:

9 sq. ft. = 1 sq. yd.
75.6 sq. ft. = 75.6 ÷ 9 = 8.4 sq. yd.

$$\begin{array}{r} 8.4 \\ 9\overline{)75.6} \\ 72 \\ \hline 3\ 6 \\ 3\ 6 \\ \hline \end{array}$$

Answer: 75.6 sq. ft. = 8.4 sq. yd.

Example 3. Change 1,600 acres to *square miles.*

Solution:

640 A. = 1 sq. mi.
1,600 A. = 1,600 ÷ 640 = 2.5 sq. mi.

$$64\not{0}\overline{)\,1{,}60\not{0}}$$

$$\begin{array}{r} 2.5 \\ 64\overline{)160.0} \\ 128 \\ \hline 32\ 0 \\ 32\ 0 \\ \hline \end{array}$$

Answer: 1,600 A. = 2.5 sq. mi. or $2\frac{1}{2}$ sq. mi.

Example 4. Change 100 square feet to *square inches.*

Solution: To change larger units to smaller units, multiply.

1 sq. ft. = 144 sq. in.
100 sq. ft. = 100 × 144 = 14,400 sq. in.

$$100 \times 144 = 144.\underset{\curvearrowright}{00}. = 14{,}400$$

Answer: 100 sq. ft. = 14,400 sq. in.

EXAMPLE 5. Change $7\frac{2}{3}$ square yards to *square feet.*

Solution:

$$1 \text{ sq. yd.} = 9 \text{ sq. ft.}$$

$$7\frac{2}{3} \text{ sq. yd.} = 7\frac{2}{3} \times 9 = 69 \text{ sq. ft.}$$

$$7\frac{2}{3} \times 9 = \frac{23}{\cancel{3}_1} \times \frac{\cancel{9}^3}{1} = 69$$

Answer: $7\frac{2}{3}$ sq. yd. = 69 sq. ft.

EXAMPLE 6. Change 2.37 square miles to *acres.* Round off to the nearest *acre.*

Solution:

1 sq. mi. = 640 A.
2.37 sq. mi. = 2.37 × 640 = 1,517 A.

```
   2.37
  ×640
  94 80
 1422
 1516.(80)
     +1
 1517
```

Answer: 2.37 sq. mi. = 1,517 A., to the nearest acre.

METRIC UNITS OF SQUARE MEASURE

Recall that *linear* measurements in the metric system increase and decrease by *powers of 10.* In **area** measurements, ***square*** *measurements increase and decrease by* ***powers of 100.*** For example, a decimeter contains 10 centimeters, but a square decimeter contains 100 square centimeters, as seen in the reduced-size figure.

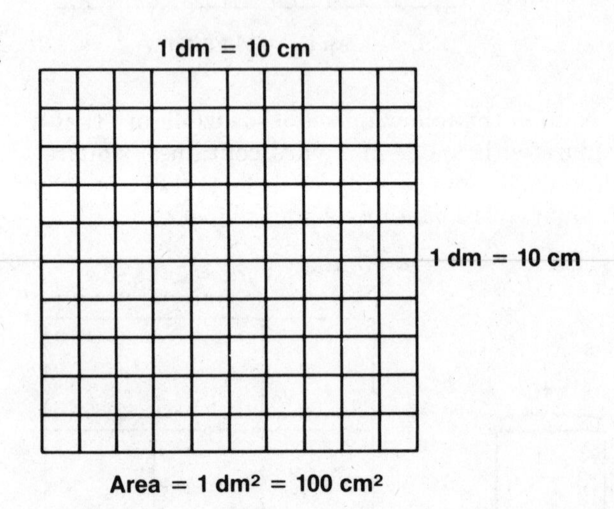

Area = 1 dm² = 100 cm²

Note that in the metric system, square measurements are indicated by the small number 2 (exponent) to the right and slightly above the unit of measurement. For example:

$$1 \text{ dm} \times 1 \text{ dm} = 1 \text{ dm}^2 \qquad 10 \text{ cm} \times 10 \text{ cm} = 100 \text{ cm}^2$$

Table VIII: Metric Units of Square Measure

Unit	Abbreviation	Description
square centimeter	cm^2	.01 dm^2
square decimeter	dm^2	100 cm^2
square meter	m^2	100 dm^2, 10,000 cm^2
hectare	ha	10,000 m^2
square kilometer	km^2	100 ha, 1,000,000 m^2

Note that the hectare, like the acre, is already a unit of square measure. There is no such unit as a square hectare.

EXAMPLE 7. Change $143\frac{1}{2}$ square centimeters to square decimeters.

Solution:

$$100 \text{ cm}^2 = 1 \text{ dm}^2$$

$$143\frac{1}{2} \text{ cm}^2 = 143.5 \text{ cm}^2 = 143.5 \div 100$$

$$= 1.43.5 = 1.435 \text{ dm}^2$$

Answer: $143\frac{1}{2} \text{ cm}^2 = 1.435 \text{ dm}^2$

EXAMPLE 8. Change 6.83 square kilometers to hectares.

Solution:

$$1 \text{ km}^2 = 100 \text{ ha}$$

$$6.83 \text{ km}^2 = 6.83 \times 100 = 6.83. = 683 \text{ ha}$$

Answer: $6.83 \text{ km}^2 = 683$ ha

EXERCISES

In 1–8, change the given area to *square inches*. Round off to the nearest *whole number*.

1. 16 sq. ft. **2.** $4\frac{1}{2}$ sq. ft. **3.** 18.75 sq. ft. **4.** 18 sq. ft.

5. $\frac{4}{5}$ sq. yd. **6.** $6\frac{4}{5}$ sq. ft. **7.** .5 sq. yd. **8.** 60 sq. ft.

In 9–16, change the given area to *square feet*. Round off to the nearest *tenth*.

9. 367 sq. in. **10.** 37 sq. yd. **11.** $12\frac{1}{2}$ sq. yd. **12.** 2,465 sq. in.

13. 37.25 sq. yd. **14.** 3,216 sq. in. **15.** .75 sq. yd. **16.** $\frac{1}{4}$ A.

In 17–24, change the given area to *square yards*. Round off to the nearest *tenth*.

17. 567 sq. ft. **18.** 247.30 sq. ft. **19.** 5,184 sq. in. **20.** 6,804 sq. in.

21. $3\frac{1}{2}$ sq. mi. **22.** .7 sq. mi. **23.** $\frac{3}{4}$ A. **24.** $2\frac{1}{2}$ A.

In 25–28, change the given area to *square centimeters*.

25. .35 m^2 **26.** $3\frac{1}{2}$ dm^2 **27.** 5.25 dm^2 **28.** .035 m^2

In 29–36, change the given area to *square meters*.

29. 53 dm^2 **30.** 86.9 cm^2 **31.** $25\frac{3}{4}$ dm^2 **32.** 15.35 dm^2

33. 3.6 ha **34.** $45\frac{1}{2}$ dm^2 **35.** .625 ha **36.** .082 km^2

APPLICATION PROBLEMS

1. A kitchen has 123 square feet of floor space. Assuming no waste, how many square yards of linoleum are needed to cover the floor? (Figure to the nearest *square yard*.) __________

2. A tabletop contains 3,024 square inches. How many square feet of plywood will be needed to cover it? __________

3. How many square yards are contained in a field that has 5.25 acres? __________

4. How much will 216 square feet of linoleum cost at $4.25 a square yard? __________

5. How much will it cost to shingle a roof with an area of 21,600 square inches? Shingles cost $.82 per square foot. __________

6. How many hectares are contained in a field measuring 3.25 square kilometers? __________

7. How much will 25,000 square centimeters of leather cost at $58 per square meter? __________

8. A farmer averages 8 bushels of corn from 25 square meters of land. How many bushels of corn will he get from a 1.5-hectare farm? __________

9. An artist is designing a mosaic, using tiles that measure 1 square centimeter. How many tiles are needed to cover a surface measuring 1.25 square meters? __________

10. An office measuring 2,000 square yards is to be partitioned into 8 work stations of equal area. How many square feet will be contained in each area? __________

UNIT 69. Measuring Area

WORDS TO KNOW

A **formula** is a rule for figuring out something. In a formula, the rule is written with letters, numbers, and symbols instead of words. In Unit 67, you found the perimeter of a square by multiplying the length of one side by 4. As a formula, this rule is written $P = 4 \times s$. In this formula, P means the *perimeter* of a square and s means the length of one *side*. Also, in Unit 67, you used the formula $C = \pi \times d$ to find the circumference of a circle.

You also learned that *perimeter* is the *linear* distance *around* any figure. **Area** is the surface *inside* the perimeter. When measuring *area*, you are finding how many *square units* are contained *within* the perimeter.

When you find the measure of the area of a figure, you find how many times a standard unit of *square measure* is contained by that figure.

A given unit of square measure is the area contained by a square with a side whose length is the given linear measure. For example, one square inch means a surface that is 1 inch by 1 inch. The following reduced-size figure represents a square whose side is 1 inch:

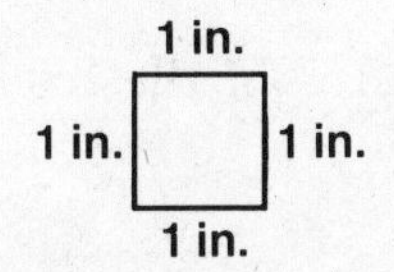

And here is a square whose side represents 5 inches:

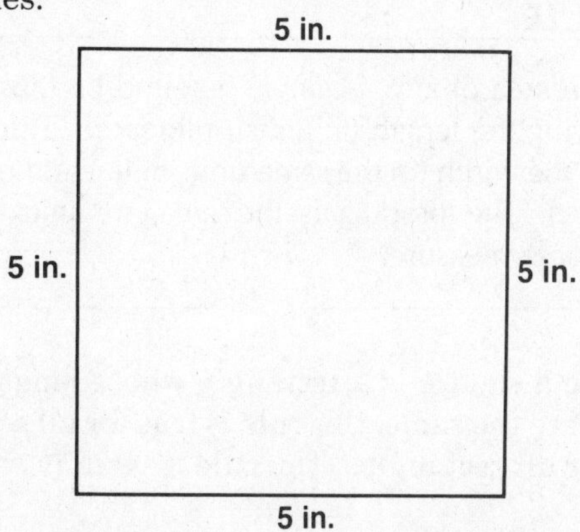

When you divide the 5-inch square into 1-inch squares, you see that the 5-inch square contains 25 1-inch squares:

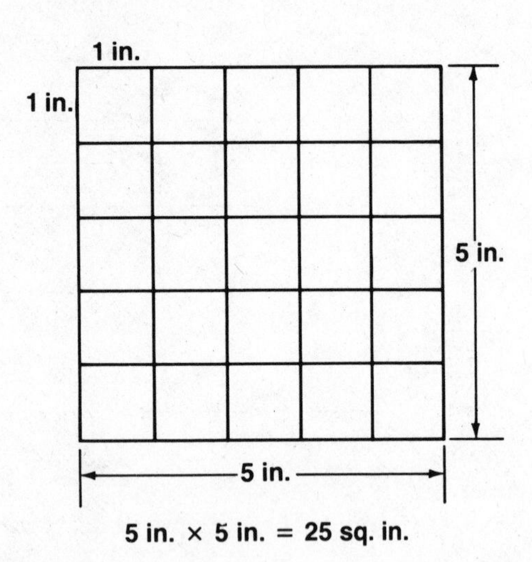

5 in. × 5 in. = 25 sq. in.

If you have a *rectangle* whose length is 6 dm and whose width is 4 dm, you will obtain an area of 24 square dm.

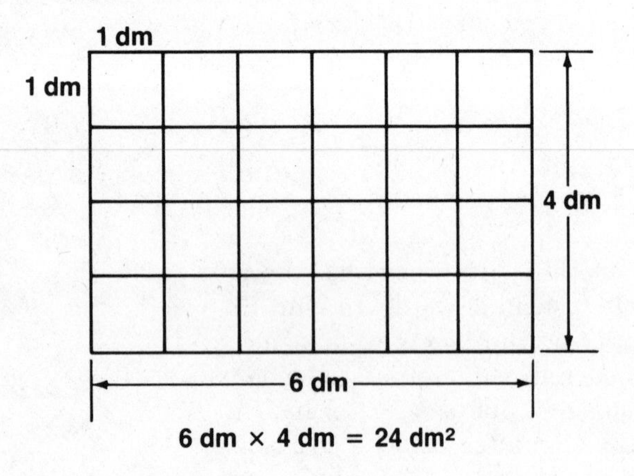

6 dm × 4 dm = 24 dm²

AREA OF A RECTANGLE

The preceding examples illustrate the following rule:

RULE

The area of any *rectangle* is found by multiplying the length (in units of linear measure) by the width (in the same units of linear measure). The product is the area, in units of square measure.

Since a square is a rectangle whose length and width are the same, this rule is true for all squares and for all rectangles. The rule is usually written as a **formula** in which A means *area* of the rectangle, ℓ means *length* of the rectangle, and w means *width* of the rectangle:

$$A = \ell \times w$$

EXAMPLE 1. A rectangular kitchen measures 9 feet by 12 feet. How many square feet of tiles are needed to cover the surface of the floor?

Solution: From the problem, ℓ = 12 feet and w = 9 feet.

$$A = \ell \times w$$
$$A = 12 \text{ ft.} \times 9 \text{ ft.}$$
$$A = 108 \text{ sq. ft.}$$

Answer: 108 square feet of tiles are needed.

EXAMPLE 2. Find the area of a rectangular room that measures 8 feet 4 inches × $10\frac{1}{2}$ feet.

Solution: Whenever you calculate with units of measure, you *must* change all units to the same measure. Therefore, change 8 feet 4 inches to *feet.* Since there are 12 inches in 1 foot, 4 inches are $\frac{4}{12}$ of a foot.

$$8 \text{ ft. } 4 \text{ in.} = 8\frac{4}{12} \text{ ft.} = 8\frac{1}{3} \text{ ft.}$$

Now use the formula to solve for the area. Replace ℓ with $10\frac{1}{2}$ feet and w with $8\frac{1}{3}$ feet.

$$A = \ell \times w$$
$$A = 10\frac{1}{2} \text{ ft.} \times 8\frac{1}{3} \text{ ft.}$$
$$A = \frac{\overset{7}{\cancel{21}}}{2} \times \frac{25}{\underset{1}{\cancel{3}}}$$
$$A = \frac{175}{2} \text{ sq. ft.} = 87\frac{1}{2} \text{ sq. ft.}$$

Answer: The area of the room is $87\frac{1}{2}$ square feet.

Example 3. How many hectares are contained in a rectangular field that is 523 meters by 827 meters?

Solution:

Step 1: Find the area in square meters.

$$A = \ell \times w$$

$$A = 523 \text{ m} \times 827 \text{ m}$$

$$\begin{array}{r} 523 \\ \times 827 \\ \hline 3\,661 \\ 10\,46 \\ 418\,4 \\ \hline 432{,}521 \end{array}$$

$$A = 432{,}521 \text{ m}^2$$

Step 2: Change the square meters to hectares.

$$10{,}000 \text{ m}^2 = 1 \text{ ha}$$

$$432{,}521 \text{ m}^2 = 432{,}521 \div 10{,}000$$

$$= 43.2521. = 43.2521 \text{ ha}$$

Answer: The field contains 43.2521 hectares.

Area of a Triangle

The length and the width of a rectangle are often referred to as the **base** and the **height**. Thus, the formula for the area of a rectangle can also be written as $A = b \times h$.

A triangle having the same base b and the same height h as a rectangle will have half the area of the rectangle. This can be seen in the following diagrams, in which each shaded triangle has half the area of the rectangle.

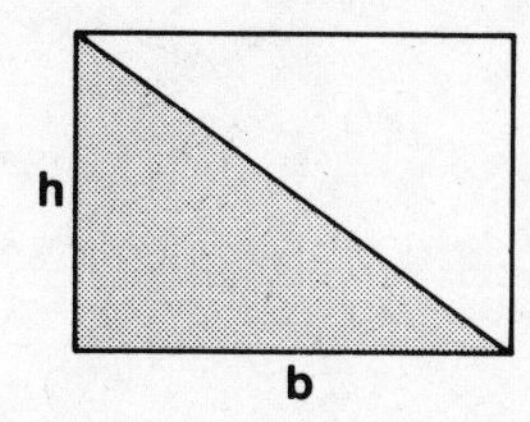

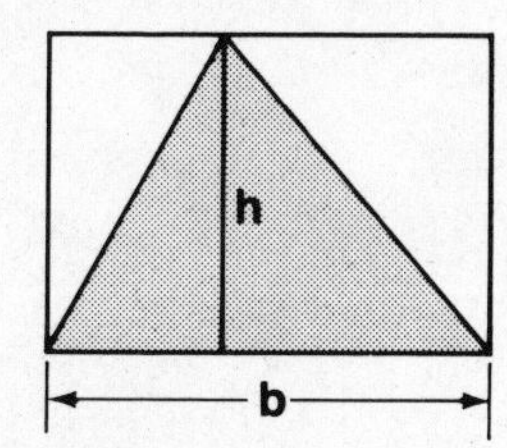

The formula for the area of a triangle is written:

$$A = \frac{1}{2} \times b \times h$$

Example 4. Find the area of a triangle whose base is 6 inches and whose height is 8 inches.

Solution:

$$A = \frac{1}{2} \times b \times h$$

$$A = \frac{1}{2} \times 6 \text{ in.} \times 8 \text{ in.}$$

$$A = \frac{1}{\cancel{2}_1} \times \frac{\cancel{6}^3}{1} \times \frac{8}{1} = 24 \text{ sq. in.}$$

Answer: The area of the triangle is 24 square inches.

Area of a Circle

The formula for the area of a circle is:

$$A = \pi \times r^2$$

This formula means that the area of a circle is equal to the *value of* π multiplied by the *radius squared.* The *exponent 2* to the right of and above the r means r *squared*, or $r \times r$.

Example 5. Find the area of a circle whose radius measures 21 cm. (Use $\pi = \frac{22}{7}$.)

Solution:

$$A = \pi \times r^2$$

$$A = \pi \times (21 \text{ cm})^2$$

$$A = \frac{22}{\cancel{7}_1} \times \frac{\cancel{21}^3}{1} \times \frac{21}{1} = 66 \times 21$$

$$\begin{array}{r} 66 \\ \times 21 \\ \hline 66 \\ 1\,32 \\ \hline 1{,}386 \end{array}$$

$$A = 1{,}386 \text{ cm}^2$$

Answer: The area of the circle is 1,386 cm^2.

EXAMPLE 6. What is the area, to the nearest *tenth*, of a round tabletop if the diameter measures 2.4 feet? (Use $\pi = 3.14$.)

Solution:

A radius is half a diameter.

$$r = \frac{1}{2} \times 2.4 \text{ ft.} = 1.2 \text{ ft.}$$

$$A = \pi \times r^2$$

$$A = 3.14 \times (1.2 \text{ ft.})^2$$

$$A = 3.14 \times 1.44$$

```
   3.14
  ×1.44
   1256
  1 256
  3 14
 4.5216
```

$A = 4.5$ sq. ft. (to the nearest tenth)

Answer: The area of the tabletop is 4.5 sq. ft.

Remember

A linear measure times a linear measure gives a square measure.

Customary Measure	Metric Measure
in. × in. = sq. in.	dm × dm = dm^2
ft. × ft. = sq. ft.	cm × cm = cm^2
yd. × yd. = sq. yd.	m × m = m^2
mi. × mi. = sq. mi.	km × km = km^2

In the preceding unit, we said that measures of area are *two-dimensional.* Now that you know how to calculate areas, you can see why: In order to obtain a measure of area, you must obtain *two* linear measures, or dimensions, the length and the width. These two linear measures are then multiplied to obtain the area measure.

EXERCISES

In 1–9, the given linear measures represent the lengths and widths of rectangles. Find the area of each rectangle. Be sure to include a unit of square measure with each answer. If there are two units in the question, give your answer in terms of the larger unit.

1. 12 mi. × 22 mi.

2. $2\frac{1}{2}$ in. × $8\frac{1}{4}$ in.

3. 23 ft. × $35\frac{1}{5}$ ft.

4. $17\frac{1}{3}$ in. × $17\frac{1}{3}$ in.

5. 15.6 m × 27 m

6. 36.25 km × 68.71 km

7. $12\frac{1}{2}$ mi. by 34.4 mi.

8. 7.3 km by 123.5 m

9. 18 cm by 79 mm

In 10–12, find the area of the triangle with the given base and height.

10. $b = 14$ m; $h = 20$ m

11. $b = 5.2$ in.; $h = 4.5$ in.

12. $b = 4\frac{1}{2}$ ft.; $h = 12$ ft.

In 13–15, find the area of the circle with the given radius or diameter.

13. $r = 4.3$ cm (Use $\pi = 3.14$.)

14. $r = 4\frac{2}{3}$ m (Use $\pi = \frac{22}{7}$.)

15. $d = 5$ in. (Use $\pi = 3.14$.)

APPLICATION PROBLEMS

1. Find the cost of cementing a driveway 8 feet wide by 46 feet long at a cost of $1.25 per square foot. __________
2. A floor measures 18 feet by 26 feet. How much will it cost to carpet the floor if the carpeting costs $7.95 a square yard? __________
3. What is the area in square yards of a room that measures 15 feet by 22 feet? __________
4. How many hectares are contained in a rectangular field measuring 240 meters by 450 meters? __________
5. How much will the material cost to make a bedspread measuring 6 feet by 9 feet if the material sells for $3.85 a square yard? __________
6. Find the value of a rectangular field 648 yards by 897 yards at $875 per acre. (Find the area to the nearest *whole acre.*) __________
7. A round table has a radius of 75 centimeters. How much will the glass cost for a tabletop if plate glass sells at $74.35 per square meter? __________
8. Jenny is making a triangular shawl measuring 60 inches across and 36 inches deep. If the wool fabric for the shawl costs $9.60 a square yard, what will be the cost of the fabric? __________
9. Land sells at $500 per hectare. What is the value of a tract of land measuring 2.25 kilometers by 3.5 kilometers? __________
10. A backyard measures 20 meters by 22 meters. If sod is bought in squares that are 40 centimeters on a side, how many squares of sod are needed to cover the backyard? __________

UNIT 70. Units of Volume

WORDS TO KNOW

The units of *square measure* include the *square inch*, the *square foot*, the *square yard*, the *square centimeter*, the *square decimeter*, and the *square meter*.

Area measures are obtained by making *two linear measures*, length and width, and then multiplying these two measures. Measures of area are said to be *two-dimensional*.

The units of *volume measure* include the *cubic inch*, the *cubic foot*, the *cubic yard*, the *cubic centimeter*, the *cubic decimeter*, and the *cubic meter*.

Measures of **volume** are obtained by making *three linear measures*, length, width, and height, and then multiplying these three measures. Measures of volume are said to be *three-dimensional*.

To measure the **volume** means to measure the capacity or contents of a three-dimensional figure. Such three-dimensional figures as boxes or cartons are called *solids*. The three linear measures of a solid are called the *length*, the *width*, and the *height*. (Words such as *altitude*, *breadth*, and *depth* may also be used.)

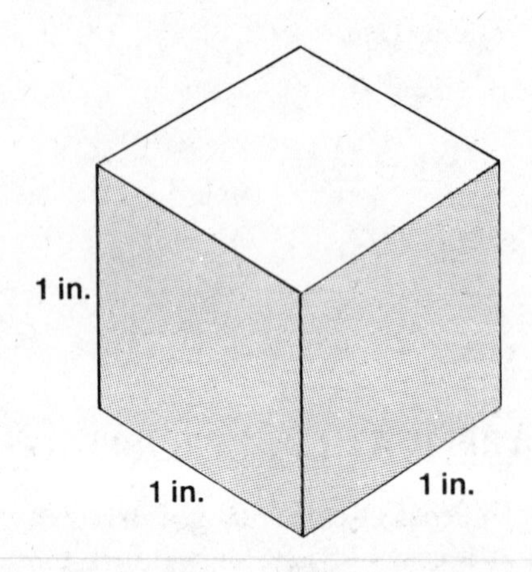

A *cube* is a solid that is made up of six squares, all coming together at right angles. A *cubic inch* is a cube that is 1 inch long, 1 inch wide, and 1 inch deep. The dimensions of such a cube are written "1 inch × 1 inch × 1 inch." A cubic inch is shown in the drawing at the right.

Some standard units of volume are described in the following table:

Table IX: Units of Volume Measure

Unit	Abbreviation	Description
	Customary System	
cubic inch	cu. in.	A cube 1 in. × 1 in. × 1 in.
cubic foot	cu. ft.	A cube 12 in. × 12 in. × 12 in. 1 cu. ft. = 1,728 cu. in.
cubic yard	cu. yd.	A cube 3 ft. × 3 ft. × 3 ft. 1 cu. yd. = 27 cu. ft.
	Metric System	
cubic centimeter	cm^3	A cube 1 cm × 1 cm × 1 cm
cubic decimeter	dm^3	A cube 10 cm × 10 cm × 10 cm 1 dm^3 = 1,000 cm^3
cubic meter	m^3	A cube 10 dm × 10 dm × 10 dm 1 m^3 = 1,000 dm^3 = 1,000,000 cm^3

To change smaller units of volume to larger units, follow *Unit Change Rule 1*, which you learned in Unit 66. To change larger units of volume to smaller units, follow *Unit Change Rule 2*, which you learned in the same Unit.

EXAMPLE 1. Change 2,435 cubic centimeters to *cubic decimeters.*

Solution:

$$1{,}000\text{ cm}^3 = 1\text{ dm}^3$$
$$2{,}435\text{ cm}^3 = 2{,}435 \div 1{,}000 = 2.435.$$
$$= 2.435\text{ dm}^3$$

Answer: $2{,}435\text{ cm}^3 = 2.435\text{ dm}^3$

EXAMPLE 2. Change 10 cubic feet to *cubic inches.*

Solution:

1 cu. ft. = 1,728 cu. in.
10 cu. ft. = 10 × 1,728 = 1728.0.
= 17,280 cu. in.

Answer: 10 cu. ft. = 17,280 cu. in.

EXAMPLE 3. How many cubic feet are contained in $\frac{1}{10}$ of a cubic yard?

Solution:

1 cu. yd. = 27 cu. ft.

$\frac{1}{10}$ cu. yd. = $\frac{1}{10} \times 27$ = 2.7 cu. ft.

$$\frac{1}{10} \times \frac{27}{1} = \frac{27}{10} = 2.7$$

or

.1 cu. yd. = .1 × 27 = 2.7 cu. ft.

Answer: $\frac{1}{10}$ cu. yd. = 2.7 cu. ft.

EXAMPLE 4. Change .643 cubic meter to *cubic decimeters.*

Solution:

$$1\text{ m}^3 = 1{,}000\text{ dm}^3$$
$$.643\text{ m}^3 = .643 \times 1{,}000 = .643.$$
$$= 643\text{ dm}^3$$

Answer: $.643\text{ m}^3 = 643\text{ dm}^3$

Other standard units of volume include measures of *liquid volume,* which you will study in Unit 72, and certain special measures that are used in different professions. For example, one *board foot* is defined as 144 cubic inches. (A board foot is a piece of board that is 1 foot long by 1 foot wide by 1 inch thick.) One *cord* is defined as 128 cubic feet. (A cord is a unit of chopped wood equal to a pile 4 ft. × 4 ft. × 8 ft.) One *bushel* is defined as 2,150.42 cubic inches. (A bushel is equal in volume to a round bucket 8 in. deep, with a diameter of $18\frac{1}{2}$ in.)

EXERCISES

In 1–8, change the given volume to *cubic inches.* When necessary, round off to the nearest *whole number.*

1. 8 cu. ft.

2. 25 cu. ft.

3. 15 cu. yd.

4. $\frac{3}{4}$ cu. yd.

5. 12.7 cu. yd.

6. $3\frac{1}{2}$ cu. ft.

7. 9.25 cu. ft.

8. $7\frac{4}{5}$ cu. yd.

In 9–16, change the given volume to *cubic yards.* When necessary, round off to the nearest *whole number.*

9. 135 cu. ft.

10. 11,664 cu. in.

11. 688.5 cu. ft.

12. $37\frac{1}{2}$ cu. ft.

13. 69,984 cu. in.

14. 2,538 cu. ft.

15. 233,280 cu. in.

16. 472 cu. ft.

In 17–24, change the given volume to *cubic feet.* When necessary, round off to the nearest *whole number.*

17. 5,184 cu. in.

18. 25 cu. yd.

19. 7.5 cu. yd.

20. 3,888 cu. in.

21. 29,376 cu. in.

22. $8\frac{3}{4}$ cu. yd.

23. $\frac{1}{2}$ cu. yd.

24. $4\frac{1}{5}$ cu. yd.

In 25–28, change the given volume to *cubic meters.*

25. 35,635 dm^3

26. 214,000 cm^3

27. 748 dm^3

28. 2,565,000 cm^3

In 29–32, change the given volume to *cubic centimeters*.

29. $1\frac{3}{4}$ dm^3 **30.** .248 m^3 **31.** .63 dm^3 **32.** $\frac{1}{8}$ m^3

In 33–36, change the given volume to *cubic decimeters*.

33. $24\frac{1}{4}$ cm^3 **34.** 2.375 m^3 **35.** .78 m^3 **36.** 5.125 cm^3

APPLICATION PROBLEMS

1. A certain metal weighs .41 pound per cubic inch. Find the weight of 3 cubic feet of this metal. ________
2. A truck holds 4 cubic yards of sand. If a cubic foot of sand costs 35¢, what is the cost of a truckful of sand? ________
3. How much will 5 cubic yards of topsoil cost at $.45 per cubic foot? ________
4. What is the weight of a cubic inch of a metal alloy if a cubic foot weighs 345.6 pounds? ________
5. How much will 405 cubic feet of cement cost at $12.50 a cubic yard? ________
6. A bin holds 2.45 cubic meters of wheat. How many cubic decimeters of wheat does the bin hold? ________
7. Copper sells at $1.25 per cubic decimeter. How much will 585 cubic centimeters cost? ________
8. The volume of a carton is 1,200 cubic decimeters. How many cartons can fit into a truck with a capacity of 393.6 cubic meters? ________
9. If rice sells at $235 per cubic meter, how much will 2,463 cubic decimeters cost? ________
10. A cubic centimeter of chrome weighs 125 grams. How much will 8.35 cubic decimeters weigh? ________

UNIT 71. Measuring Volume

WORDS TO KNOW

A very important three-dimensional figure is the **rectangular solid**. This is a solid made up of rectangles or squares. In a rectangular solid, all the surfaces meet at right angles. Everyday examples of such solids include books, shoe boxes, boards, suitcases, and classrooms.

To find the volume of a **rectangular solid**, you must know all three dimensions: length, width, and height.

RULE

To find the volume of a rectangular solid, multiply the length by the width. Then multiply this product by the height. (All three dimensions must be in the same units of linear measure.) The final product is the volume, in units of cubic measure.

This rule is usually written as a *formula* in which V means *volume*, ℓ means *length*, w means *width*, and h means *height*:

$$V = \ell \times w \times h$$

EXAMPLE. How much water will you need to fill a rectangular fish tank whose measurements are 1 foot × 2 feet × 3 feet? Leave a space of 1 inch at the top so that the water will not overflow.

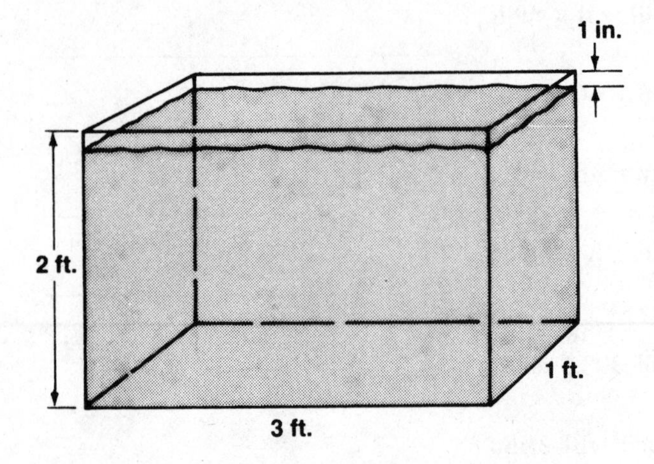

Solution: Since the volume of the water will be 1 inch below the full height of the tank, subtract 1 inch from the 2 feet, the height of the tank, leaving the water height as 1 ft. 11 in., or $1\frac{11}{12}$ ft.

$$V = \ell \times w \times h$$

$$V = 3 \text{ ft.} \times 1 \text{ ft.} \times 1\frac{11}{12} \text{ ft.}$$

$$V = \frac{3}{1} \times \frac{1}{1} \times 1\frac{11}{12}$$

$$V = \frac{3}{1} \times \frac{1}{1} \times \frac{23}{12} = \frac{69}{12} \text{ cu. ft.}$$

$$\begin{array}{r} 5\frac{3}{4} \\ 12\overline{)69} \\ \underline{60} \\ \frac{9}{12} = \frac{3}{4} \end{array}$$

$$V = 5\frac{3}{4} \text{ cu. ft.}$$

Answer: $5\frac{3}{4}$ cubic feet of water are needed.

Remember

A linear measure times a linear measure times a linear measure gives a cubic measure.

Customary Measure	Metric Measure
in. × in. × in. = cu. in.	cm × cm × cm = cm^3
ft. × ft. × ft. = cu. ft.	dm × dm × dm = dm^3
yd. × yd. × yd. = cu. yd.	m × m × m = m^3

EXERCISES

In 1–12, find the volume of each rectangular solid. If there is more than one unit of measure in the question, write your answer in terms of the larger unit.

1. 12 ft. × 3 ft. × 4 ft.

2. 8 in. × 10 in. × 4 in.

3. $6\frac{1}{2}$ ft. × $3\frac{1}{4}$ ft. × $4\frac{5}{8}$ ft.

4. 36 in. × $5\frac{1}{2}$ ft. × 40 in.

5. $2\frac{1}{2}$ ft. × $5\frac{1}{3}$ ft. × 56 in.

6. 3 ft. × 28 in. × 5 ft.

7. 14.7 yd. × $15\frac{1}{5}$ yd. × 8 yd.

8. $12\frac{3}{4}$ ft. × $14\frac{2}{3}$ ft. × $16\frac{4}{5}$ ft.

9. 16 dm × 8 dm × 4 dm

10. 245 cm × 12 dm × 5.3 dm

11. $18\frac{1}{2}$ dm × 6.25 dm × 4 dm

12. .538 m × 24.3 dm × 12 dm

APPLICATION PROBLEMS

1. To prepare for the foundation of a building, a contractor must dig an excavation that is 126 feet by 48 feet by 96 feet. How many cubic yards of dirt will he have to remove? __________

2. What is the capacity of a truck if the inside measurements are 36 feet by 12 feet by 9 feet? __________

3. If an aluminum alloy weighs $\frac{1}{10}$ pound per cubic inch, what will be the weight of an aluminum alloy bar measuring 3 feet by 9 inches by 8 inches? __________

4. How many cubic inches of gold are contained in a bar of gold measuring $11\frac{1}{2}$ inches by $5\frac{1}{5}$ inches by $3\frac{3}{4}$ inches? __________

5. How many cartons, 3 feet by 2 feet by 6 inches, can be stored in a space measuring 15 feet by 9 feet by 24 feet? (Assume that there is no waste space.) __________

6. How many cubic decimeters are there in a box that measures 1.5 meters by 3.6 meters by .75 meter? __________

7. A box measures .35 meter by 1.5 meters by .9 meter. What is the volume of the box in cubic centimeters? __________

8. A metal bar measures 15 centimeters by 6 centimeters by 4 centimeters. If a cubic decimeter of the metal costs $63, what is the total cost of the bar? __________

9. An overseas shipping company charges $3.75 per cubic meter to a certain destination. What is the cost for shipping a crate measuring 2.4 meters by 1.5 meters by .75 meter? __________

10. How many nickel alloy bars measuring 6 centimeters by 3 centimeters by 2 centimeters can be made from a block that measures 1.5 meters by .6 meter by .4 meter? __________

Review of Part XV (Units 65–71)

1. Change:

a. 98 ft. to yd. *b.* 33.5 ft. to in. *c.* 153 in. to ft. *d.* $1\frac{3}{4}$ mi. to yd.

2. Change:

a. 1.3 km to m *b.* $\frac{3}{4}$ m to cm *c.* 763 mm to m *d.* 420 mm to cm

In 3–5, find the perimeter of the geometric figure.

3. A triangle whose sides measure:

a. $4\frac{1}{2}$ ft., $6\frac{2}{3}$ ft., and $5\frac{1}{4}$ ft. *b.* $1\frac{1}{2}$ ft., 30 in., and $\frac{2}{3}$ yd. (Answer in feet.)

4. A rectangle whose dimensions are:

a. 5.3 km by 2.25 km *b.* 6 ft. by 38 in. (Answer in feet.)

5. A square whose sides measure:

a. $5\frac{1}{3}$ yd. *b.* .075 km

6. Find the circumference of a circle with:

a. a radius of $15\frac{3}{4}$ in. $\left(\text{Use } \pi = \frac{22}{7}.\right)$

b. a diameter of 128 cm (Use $\pi = 3.14$.)

7. Change:

a. 23.65 sq. ft. to sq. in.

b. .75 A. to sq. yd.

c. $12\frac{1}{2}$ sq. yd. to sq. ft.

d. $\frac{1}{3}$ sq. yd. to sq. in.

8. Change:

a. $\frac{2}{5}$ m^2 to cm^2

b. .053 km^2 to m^2

c. 4.25 dm^2 to cm^2

d. 12,500 cm^2 to m^2

9. Find the area of a rectangle whose dimensions are:

a. $12\frac{1}{4}$ ft. × 6.5 ft.

b. 18.15 m by 36.7 m

10. Find the area of a triangle with the given dimensions:

a. base = $10\frac{1}{2}$ ft.; height = $9\frac{1}{3}$ ft.

b. base = 2.5 m; height = 1.4 m

11. Using $\pi = 3.14$, find the area of a circle with the given measure. Round off to the nearest *whole number.*

a. a radius of $8\frac{1}{2}$ cm

b. a diameter of 5.5 in.

12. Change:

a. $4\frac{1}{4}$ cu. ft. to cu. in.

b. .53 cu. yd. to cu. ft.

c. 11,232 cu. in. to cu. ft.

d. 306 cu. ft. to cu. yd.

13. Change:

a. .5 m^3 to cm^3

b. $3\frac{1}{4}$ dm^3 to cm^3

c. 2,500 cm^3 to dm^3

d. 738,000 cm^3 to m^3

14. Find the volume of a rectangular solid with the given dimensions. Write your answer in terms of the larger unit.

a. 48 in. × $3\frac{1}{3}$ ft. × 60 in.

b. 12,500 dm by 4,000 dm by .75 m

15. How many acres are there in $3\frac{1}{2}$ square miles? __________

16. Ribbon sells at 98¢ per foot. How much will $5\frac{1}{2}$ yards cost? __________

17. A circular mirror has a diameter of 28 inches. If the frame around it is $3\frac{1}{2}$ inches wide, what is the outside perimeter of the mirror and frame? $\left(\text{Use } \pi = \frac{22}{7}.\right)$ __________

18. Carpeting sells at $19.75 per square yard. How much will it cost to carpet a room that measures 24 feet by 36 feet? __________

19. A cubic centimeter of silver costs $5.85. How much will .05 cubic meter of silver cost? __________

20. How many cubic yards of cement are needed to fill a hole that measures 30 feet by 48 feet by 27 feet? __________

PART XVI. More About Measures

UNIT 72. Measuring Liquid

To measure amounts of liquids, you use *liquid measures*, which are based on cubic measures.

In the Customary system, the basic unit in measuring liquids is the *gallon*, which is defined as 231 cubic inches. The gallon is divided into 4 *quarts*, the quart is divided into 2 *pints*, and the pint is divided into 16 *ounces*.

In the metric system, the basic unit in measuring liquids is the *liter*, which is one cubic decimeter. The units in general use are the *liter* and the *milliliter*.

Table X: Units of Liquid Measure

Customary System
1 gallon (gal.) = 4 quarts (qt.) 1 quart (qt.) = 2 pints (pt.) 1 pint (pt.) = 16 ounces (oz.)
Metric System
1 milliliter (mL) = .001 L 1 liter (L) = 1,000 mL

Using the information in the table, you should be able to change measures of liquids just as you have changed measures of length, measures of area, and measures of volume. Use *Unit Change Rule 1* (from Unit 66) when you change a smaller unit to a larger unit. Use *Unit Change Rule 2* (also from Unit 66) when you change a larger unit to a smaller unit. You have used these same rules in Units 68 and 70.

EXAMPLE 1. (*a*) Change 32 liters to *milliliters*. (*b*) Change 148 milliliters to *liters*.

Solution:

(*a*) 1 L = 1,000 mL
32 L = 32 × 1,000 = 32.000. = 32,000 mL

Answer: 32 liters = 32,000 milliliters

(*b*) 1,000 mL = 1 L
148 mL = 148 ÷ 1,000 = .148. = .148 L

Answer: 148 milliliters = .148 liter

EXAMPLE 2. (*a*) Change $3\frac{1}{2}$ pints to *quarts*.
(*b*) Change $3\frac{1}{2}$ quarts to *pints*.

Solution:

(*a*) 2 pt. = 1 qt.
$3\frac{1}{2}$ pt. = $3\frac{1}{2} \div 2 = 1\frac{3}{4}$ qt.

$$3\frac{1}{2} \div 2 = \frac{7}{2} \div \frac{2}{1} = \frac{7}{2} \times \frac{1}{2} = \frac{7}{4} = 1\frac{3}{4}$$

Answer: $3\frac{1}{2}$ pints = $1\frac{3}{4}$ quarts

(*b*) 1 qt. = 2 pt.
$3\frac{1}{2}$ qt. = $3\frac{1}{2} \times 2 = 7$ pt.

$$3\frac{1}{2} \times 2 = \frac{7}{\cancel{2}_1} \times \frac{\cancel{2}^1}{1} = 7$$

Answer: $3\frac{1}{2}$ quarts = 7 pints

EXAMPLE 3. (*a*) Change 8.2 gallons to *quarts*. (*b*) Change 8.2 quarts to *gallons*.

Solution:

(*a*) 1 gal. = 4 qt.
8.2 gal. = 8.2 × 4 = 32.8 qt.

$$\begin{array}{r} 8.2 \\ \times 4 \\ \hline 32.8 \end{array}$$

Answer: 8.2 gallons = 32.8 quarts

(*b*) 4 qt. = 1 gal.
8.2 qt. = 8.2 ÷ 4 = 2.05 gal.

$$\begin{array}{r} 2.05 \\ 4\overline{)8.20} \\ 8 \\ \hline 20 \\ 20 \\ \hline \end{array}$$

Answer: 8.2 quarts = 2.05 gallons

EXERCISES

In 1–10, change the given liquid measure to *ounces*.

1. 12 pt. **2.** $8\frac{1}{2}$ pt. **3.** 4 qt. **4.** 12.5 pt. **5.** $3\frac{1}{2}$ gal.

6. 4.25 qt. **7.** 28 pt. **8.** 15.8 qt. **9.** 6 gal. **10.** 8.3 gal.

In 11–15, change the given liquid measure to *gallons*.

11. 267 pt. **12.** 78 qt. **13.** 640 oz. **14.** 328 qt. **15.** 75 pt.

In 16–20, change the given liquid measure to *pints*.

16. 15 qt. **17.** 192 oz. **18.** 23 gal. **19.** 15.25 gal. **20.** $29\frac{3}{4}$ qt.

In 21–25, change the given liquid measure to *quarts*.

21. 28 gal. **22.** 27 pt. **23.** $12\frac{3}{4}$ pt. **24.** 13.75 gal. **25.** $14\frac{2}{3}$ gal.

In 26–30, change milliliters to *liters*.

26. 150 mL **27.** 3,275 mL **28.** 42 mL **29.** 25,000 mL **30.** 9 mL

In 31–35, change liters to *milliliters*.

31. 2.5 L **32.** $43\frac{1}{4}$ L **33.** .821 L **34.** $137\frac{1}{2}$ L **35.** .029 L

APPLICATION PROBLEMS

1. How much will a drum containing 63 gallons of oil cost at $.85 per quart? ____________

2. A family uses 3 quarts of milk each day. How many gallons of milk will the family use in 1 year (365 days)? ____________

3. If a can contains 14 ounces of orange juice, how many gallons of juice does a case of 48 cans contain? ____________

4. How much will 175 quarts of paint cost at $6.55 a gallon? ____________

5. How many ounces of water should be added to 6 ounces of frozen lemon juice to make $1\frac{1}{2}$ gallons of lemonade? ____________

6. How many vials each holding 140 milliliters can be filled from a bottle that holds 3.5 liters? ____________

7. A can of paint remover sells for $3. If the can contains $1\frac{1}{2}$ liters, how much will 9,000 milliliters cost? ____________

8. A recipe calls for 6 ounces of apple juice. A 4H club is preparing 8 times the amount in the recipe. How many quarts of apple juice will be needed? ____________

9. A pail contains $2\frac{3}{4}$ gallons of cream. How many pint containers can be filled? ____________

10. A juice drink sells at $.75 per quart. How many gallons of the drink can be bought for $12? ____________

UNIT 73. Measuring Weight

To measure weight, you measure how heavy an object is. The basic unit of weight in the Customary system is the *pound,* which is divided into 16 *ounces.* (The ounce used to measure weight is not the same ounce used to measure liquids.) The *ton* is defined as 2,000 pounds.

In the metric system, the basic unit of weight is the *gram.* Other units are named with the same metric prefixes already used, followed by the word *gram.*

Table XI: Units of Weight Measure

Customary System
1 pound (lb.) = 16 ounces (oz.) 1 ton (T.) = 2,000 pounds (lb.)
Metric System
1 milligram (mg) = .001 g 1 gram (g) = 1,000 mg, .001 kg 1 kilogram (kg) = 1,000 g 1 metric ton (t) = 1,000 kg, 1,000,000 g

Using the information in Table XI, you should be able to change measures of weight just as you have changed measures of length, area, volume, and liquids. Compare the solutions in the following example to the *Unit Change Rules* you first learned in Unit 66.

Example 1. State the rules for changing: (*a*) pounds to ounces, (*b*) ounces to pounds, (*c*) tons to pounds, (*d*) pounds to tons.

Solution:

(*a*) Pounds to ounces: Multiply the number of pounds by 16.

(*b*) Ounces to pounds: Divide the number of ounces by 16.

(*c*) Tons to pounds: Multiply the number of tons by 2,000.

(*d*) Pounds to tons: Divide the number of pounds by 2,000.

Example 2. Change 3.5 kilograms to *grams.*

Solution:

$$1 \text{ kg} = 1{,}000 \text{ g}$$
$$3.5 \text{ kg} = 3.5 \times 1{,}000 = 3.500. = 3{,}500 \text{ g}$$

Answer: 3.5 kilograms = 3,500 grams

EXERCISES

In 1–5, change the given weight to *ounces.*

1. 5 lb. **2.** $8\frac{1}{2}$ lb. **3.** 9.7 lb. **4.** $\frac{3}{4}$ lb. **5.** 37 lb.

In 6–10, change the given weight to *pounds.*

6. 368 oz. **7.** 7.75 T. **8.** $12\frac{1}{2}$ T. **9.** 616 oz. **10.** .8 T.

In 11–15, change the given weight to *tons*. When necessary, round off to the nearest *hundredth*.

11. 25,000 lb. **12.** 17,850 lb. **13.** 5,678 lb. **14.** 56,000 oz. **15.** 23,167 lb.

In 16–25, change the given weight to *grams*.

16. 25,000 mg **17.** .7 kg **18.** 8.2 kg **19.** 100 mg **20.** 46 kg

21. .532 kg **22.** 2,861 mg **23.** 475 mg **24.** .026 t **25.** 95 mg

In 26–30, change the given weight to *kilograms*.

26. 12,500 g **27.** 875,000 mg **28.** 3.6 t **29.** $\frac{1}{2}$ t **30.** 320 g

APPLICATION PROBLEMS

1. How much would 12 ounces of meat cost at $1.89 a pound? ________
2. A farm cooperative sells peanuts at 26¢ per pound. How much will 1.6 tons cost? ________
3. How many 50-pound bags of potatoes can be obtained from a ton of potatoes? ________

4. How many 8-ounce packages of candy can be obtained from 63 pounds of candy? ______
5. How much would 8 ounces of candy cost at $2.37 per pound? ______
6. Candy sells at $1.75 per 250 grams. What will be the cost of .75 kilogram? ______
7. How many milligrams of sodium are there in 2.16 grams? ______
8. If 3 metric tons of ore cost $7,200, what is the cost of 759 kilograms of ore? ______
9. How many boxes of spices, each holding 57 grams, can be filled from a container holding 15.96 kilograms? ______
10. How much will 14 ounces of cheese cost at $3.59 a pound? ______

UNIT 74. Counting Measures

Articles are sometimes sold in counting measures such as the dozen, the gross, and the score.

The basic counting measure is the *unit,* which means 1 article. Standard counting measures include the *dozen,* which is 12 units; the *gross,* which is 12 dozen units; and the *score,* which is 20 units.

Table XII: Counting Measures

1 dozen (doz.)	= 12 units
1 gross (gr.)	= 12 dozen = 144 units
1 score	= 20 units

You change measures of counting by applying the same *Unit Change Rules* you have used for changing measures of length, area, volume, liquid, and weight.

EXAMPLE 1. State the rules for changing: (*a*) dozens to units, (*b*) units to dozens, (*c*) gross to dozens, (*d*) gross to units, and (*e*) units to gross.

Solution:

(*a*) Dozens to units: Multiply the number of dozens by 12.

(*b*) Units to dozens: Divide the number of units by 12.

(*c*) Gross to dozens: Multiply the number of gross by 12.

(*d*) Gross to units: Multiply the number of gross by 144.

(*e*) Units to gross: Divide the number of units by 144.

EXAMPLE 2. Change $6\frac{1}{2}$ gross to *dozens.*

Solution:

$$1 \text{ gr.} = 12 \text{ doz.}$$

$$6\frac{1}{2} \text{ gr.} = 6\frac{1}{2} \times 12 = 78 \text{ doz.}$$

$$\begin{array}{r} 12 \\ \times 6\frac{1}{2} \\ \hline 72 \\ +6 \\ \hline 78 \end{array} \qquad \frac{1}{2} \times \frac{12}{1} = 6$$

Answer: $6\frac{1}{2}$ gross = 78 dozen

EXAMPLE 3. Which is the larger quantity, 10 dozen or 6 score?

Solution:

10 dozen = 10 × 12 = 120 units
6 score = 6 × 20 = 120 units

Answer: The quantities are the same.

EXERCISES

In 1–5, express the given quantity in terms of *units.*

1. 8 doz. **2.** $9\frac{3}{4}$ doz. **3.** 3.5 gr. **4.** $13\frac{5}{6}$ doz. **5.** 5.25 gr.

In 6–10, change the given quantity to *dozens.* Express remainders as *common fractions.*

6. 253 units **7.** $12\frac{1}{4}$ gr. **8.** $8\frac{2}{3}$ gr. **9.** 628 units **10.** 5.75 gr.

In 11–15, express the given quantity in terms of *gross.* Round off to the nearest *tenth.*

11. 865 units **12.** 427 doz. **13.** $629\frac{1}{2}$ doz. **14.** 1,627 units **15.** 329 doz.

In 16–20, change the given times to *scores.* Express the remainders as *units.*

SAMPLE SOLUTION

Change 87 years to scores. 20 years = 1 score

$$\begin{array}{r} 4 \leftarrow \text{score} \\ 20\overline{)87} \\ \underline{80} \\ 7 \leftarrow \text{years} \end{array}$$

4 score and 7 years

16. 50 yr. **17.** 75 yr. **18.** 138 yr. **19.** 15 yr. **20.** 45 yr.

APPLICATION PROBLEMS

1. If shirts sell at $96 a dozen, how much would 4 shirts cost? ______

2. Safety pins are packed 32 pins to a box. How many pins will there be in $3\frac{1}{2}$ gross of boxes? ______

3. Socks sell at $15 a dozen pairs. How much will 8 pairs of socks cost? ______

4. Pencils sell at $.95 a dozen. How much will 7,920 pencils cost? ______

5. Oranges sell at 6 for $.98. How much will $3\frac{1}{2}$ dozen cost? ______

6. If a dozen ballpoint pens are packed in a box, how many boxes are needed for 2,820 pens? ______

7. A chocolate cake recipe calls for 3 eggs. How many dozen eggs are needed to bake 75 cakes? ______

8. Ballpoint pens sell at $2.49 per dozen. How much will $\frac{3}{4}$ gross of pens cost? ______

9. A restaurant had a gross of eggs. If it used $\frac{1}{3}$ of the eggs, how many dozen eggs are left? ______

10. Socks sell at $2.49 for 3 pairs. How much will a dozen pairs cost? ______

UNIT 75. Measuring Time; Adding and Subtracting Times

Measuring time means measuring the duration of time in a given period or between two separate events.

A basic unit for measuring time is the time it takes for the earth to travel once completely around the sun. This measure is the *year,* which is divided into *months* and *weeks.* Another basic time unit is the time it takes for the earth to turn completely around once. This measure is the *day,* which is divided into *hours, minutes,* and *seconds.*

Table XIII: Units of Time Measure

1 minute (min.)	=	60 seconds (sec.)
1 hour (hr.)	=	60 minutes (min.)
1 day (da.)	=	24 hours (hr.)
1 week (wk.)	=	7 days (da.)
1 month (mo.)	=	30 days (da.)
1 year (yr.)	=	365 days (da.)
	=	52 weeks (wk.)
	=	12 months (mo.)

Using the information in the table, you should be able to change measures of time just as you have changed measures of length, area, volume, liquid, and weight. If you have difficulty in understanding the following examples, review the examples in the preceding units.

EXAMPLE 1. State the rules for changing: (*a*) days to years, (*b*) days to hours, and (*c*) days to seconds.

Solution:

(*a*) Days to years: Divide the number of days by 365.

(*b*) Days to hours: Multiply the number of days by 24.

(*c*) Days to seconds: Multiply the number of days by 24 to get hours; multiply the number of hours by 60 to get minutes; then multiply the number of minutes by 60 to get seconds.

Example 2. Change 3 days to *hours.*

Solution:

1 da. = 24 hr.
3 da. = 3 × 24 = 72 hr.

$$\begin{array}{r} 24 \\ \times 3 \\ \hline 72 \end{array}$$

Answer: 3 days = 72 hours

Example 3. What fraction of a day is 144 minutes?

Solution: First, change 1 day to minutes:

$$\begin{array}{r} 24 \text{ hr. per da.} \\ \times 60 \text{ min. per hr.} \\ \hline 1{,}440 \text{ min. per da.} \end{array}$$

Now, (144 minutes is) what fraction (of 1,440) minutes?

$$\frac{\text{IS}}{\text{OF}} = \frac{\overset{1}{\cancel{144}}}{\underset{10}{\cancel{1{,}440}}} = \frac{1}{10}$$

Answer: 144 minutes is $\frac{1}{10}$ of a day.

Other problems relating to the measure of time are those in which you must determine the length of the interval between two given times. If you start work at 8:30 A.M. and work till 5:30 P.M. with one hour off for lunch, how many hours have you worked? If you start a car trip at 7:15 A.M. and drive till 2:45 P.M., how long did the trip take?

The 24 hours that make up a day are divided into two periods, the A.M. (ante meridiem, or before noon) hours from 12 o'clock midnight to 12 o'clock noon, and the P.M. (post meridiem, or after noon) hours from 12 o'clock noon to 12 o'clock midnight. Thus, 2:15 A.M. means 2 hours and 15 minutes after midnight, while 2:15 P.M. means 2 hours and 15 minutes after noon.

To find the length of the interval between two given times that are both A.M. or both P.M., you simply subtract the hours and the minutes of the earlier time from the later time.

Example 4. Find the length of time between 5:15 P.M. and 8:50 P.M.

Solution: Subtract.

$$\begin{array}{r} 8 \text{ hr. } 50 \text{ min.} \\ -5 \text{ hr. } 15 \text{ min.} \\ \hline 3 \text{ hr. } 35 \text{ min.} \end{array}$$

Answer: 3 hours and 35 minutes

Example 5. What is the time interval between 8:30 A.M. and 11:15 A.M.?

Solution: Subtract.

$$\begin{array}{r} 11 \text{ hr. } 15 \text{ min.} \\ -8 \text{ hr. } 30 \text{ min.} \\ \hline \end{array}$$

Since you cannot subtract 30 minutes from 15 minutes, change one of the 11 hours to 60 minutes. Thus, 11 hr. 15 min. = 10 hr. 75 min.

$$\begin{array}{r} 10 \text{ hr. } 75 \text{ min.} \\ -8 \text{ hr. } 30 \text{ min.} \\ \hline 2 \text{ hr. } 45 \text{ min.} \end{array}$$

Answer: 2 hours and 45 minutes

To find the length of an interval between two given times when one is A.M. and the other is P.M., you must work by steps, using the fact that 12:00 o'clock is the time that separates A.M. from P.M. hours. First, find how far each of the given times is from 12:00 o'clock. Then add the two results.

Example 6. Ted's school day starts at 8:10 A.M. and ends at 2:45 P.M. How long is his school day?

Solution:

Step 1: Subtract to find the length of time from 8:10 A.M. to 12:00 noon.

$$\begin{array}{r} 12 \text{ hr. } 00 \text{ min.} \\ -8 \text{ hr. } 10 \text{ min.} \\ \hline \end{array}$$

Change to:

$$\begin{array}{r} 11 \text{ hr. } 60 \text{ min.} \\ -8 \text{ hr. } 10 \text{ min.} \\ \hline 3 \text{ hr. } 50 \text{ min.} \end{array}$$

Step 2: 2:45 P.M. is 2 hours and 45 minutes after 12:00 noon.

Step 3: Add the two intervals.

$$\begin{array}{r} 3\text{ hr. } 50\text{ min.} \\ +2\text{ hr. } 45\text{ min.} \\ \hline 5\text{ hr. } \not{95}\text{ min.} \\ +1\text{ hr. } 35\text{ min.} \\ \hline 6\text{ hr. } 35\text{ min.} \end{array}$$

Change 95 min. to 1 hr. 35 min., and add.

Answer: Ted's school day is 6 hours and 35 minutes long.

Sometimes you may know the length of an interval and you may need to find a time. Given the *starting* time and the length of an interval, *add* to find the time at the end.

EXAMPLE 7. A plane leaves Cincinnati for Atlanta at 6:35 A.M. If the flight time is 1 hour 15 minutes, at what time will it arrive?

Solution: Add.

$$\begin{array}{r} 6\text{ hr. } 35\text{ min.} \\ +1\text{ hr. } 15\text{ min.} \\ \hline 7\text{ hr. } 50\text{ min.} \end{array}$$

Answer: 7:50 A.M.

Given the *ending* time and the length of an interval, *subtract* to find the starting time.

EXAMPLE 8. After traveling 14 hours and 22 minutes, the Bay City Express arrived at its destination at 11:52 P.M. At what time did it start?

Solution: The travel time of 14 hours 22 minutes is 12 hours plus 2 hours 22 minutes. First subtract the 12 hours.

11:52 P.M. − 12 hours = 11:52 A.M.

Now subtract the 2 hours 22 minutes.

$$\begin{array}{r} 11\text{ hr. } 52\text{ min.} \\ -2\text{ hr. } 22\text{ min.} \\ \hline 9\text{ hr. } 30\text{ min.} \end{array}$$

Answer: The Express started at 9:30 A.M.

In some countries, and in the military services of the United States, 24-hour time is commonly used. The hours are counted consecutively starting at midnight, and read as 4-digit numbers. Thus, 1:00 A.M. becomes 0100, and is read "zero one hundred hours." 2:00 P.M. is 14 hours after midnight, and becomes 1400, or "fourteen hundred hours."

EXAMPLE 9. (*a*) Change 1630 hours to regular time. (*b*) Change 3:45 A.M. to military time.

Solution:

(*a*) 1630 hours is 16 hours and 30 minutes past midnight, or 4:30 P.M.

(*b*) 3:45 A.M. is 3 hours and 45 minutes past midnight, or 0345 hours.

EXERCISES

In 1–5, change the given time to *months*.

1. 8 yr. **2.** $\frac{3}{4}$ yr. **3.** $5\frac{1}{4}$ yr. **4.** 7.5 yr. **5.** $6\frac{1}{3}$ yr.

In 6–10, change the given time to *days*. When necessary, round off to the nearest *whole number*.

6. 6 yr. **7.** 26 wk. **8.** $3\frac{4}{5}$ yr. **9.** $13\frac{1}{2}$ wk. **10.** .75 yr.

In 11–15, change the given time to *years.* When necessary, round off to the nearest *hundredth.*

11. 73 mo. **12.** 2,555 da. **13.** 628 wk. **14.** 735 wk. **15.** 115 mo.

In 16–20, change the given time to *hours.*

16. 5 da. **17.** $3\frac{5}{8}$ da. **18.** 4 wk. **19.** 240 min. **20.** $3\frac{3}{7}$ wk.

In 21–30, find the length of the interval between the two given times.

21. 8:30 A.M. to 11:45 A.M.

22. 9:45 P.M. to 10:50 P.M.

23. 7:40 A.M. to 12:00 noon

24. 8:40 A.M. to 10:15 A.M.

25. 11:00 A.M. to 3:00 P.M.

26. 9:00 A.M. to 4:30 P.M.

27. 10:25 A.M. to 2:15 P.M.

28. 10:10 P.M. to 12:35 A.M.

29. 11:42 P.M. to 12:17 A.M.

30. 8:05 A.M. to 2:55 P.M.

In 31–34, the starting time and the length of an interval are given. Find the ending time.

31. 9:45 A.M.; 2 hr. 15 min.

32. 11 A.M.; 5 hr. 10 min.

33. 8:00 A.M.; 12 hr. 10 min.

34. 3:35 P.M.; 1 hr. 45 min.

In 35–36, the ending time and the length of an interval are given. Find the starting time.

35. 3:20 P.M.; 2 hr. 10 min.

36. 2:00 A.M.; 5 hr. 45 min.

37. Change 5:45 P.M. to military time.

38. Change 1800 hours to regular time.

In 39–40, the payroll schedule shows the hours worked by an employee for one week. For convenience, A.M. is used to mean hours before lunch and P.M. means hours after lunch. Find the number of hours worked per day, and find the weekly total.

39.

Day	A.M.		P.M.		A.M. Hours	P.M. Hours	Daily Hours
	In	Out	In	Out			
Mon.	8:30	11:30	12:30	5:00			
Tues.	7:00	12:00	1:00	4:30			
Wed.	8:00	12:00	1:00	5:00			
Thurs.	9:30	11:30	12:00	6:00			
Fri.	10:00	1:00	2:00	7:00			
					Weekly Total		

40.

Day	A.M. In	A.M. Out	P.M. In	P.M. Out	A.M. Hours	P.M. Hours	Daily Hours
Mon.	10:00	12:30	1:00	7:30			
Tues.	9:30	12:00	12:30	7:00			
Wed.	8:30	12:00	12:30	6:00			
Thurs.	7:00	12:30	1:30	5:00			
Fri.	11:00	3:30	4:00	8:30			
					Weekly Total		

APPLICATION PROBLEMS

1. Tom's father works 40 hours a week at $6.75 per hour. What does he earn in a year? (He is paid for 52 weeks each year.) ________
2. The total cost of a car is $9,875, to be paid for over 3 years. If the down payment is $2,600, how much will each of the monthly payments be? ________
3. The Delgados have a 25-year mortgage on their house for $43,500. How much will each of the monthly payments be? ________
4. The Wilsons use 1,875 gallons of fuel oil each year for heating and hot water. If the fuel costs $1.21 per gallon, what is the average cost per month? ________
5. Antonia is paid $685 twice a month. What is her yearly salary? ________
6. A plumber was hired to install bathroom fixtures at a rate of $10.75 per hour. If he started work at 10:30 A.M. and finished at 3:00 P.M., how much was he paid? ________
7. An orchestra charges $75 per hour for its services. How much was the orchestra paid for playing at a wedding reception from 8:00 P.M. to 2:00 A.M.? ________
8. Sandra starts work at 8:30 A.M., leaves for lunch at 1:00 P.M., returns at 1:30 P.M., and works till 6:00 P.M. If she works the same schedule Monday through Friday and is paid $6.25 an hour, what is her salary for a full week? ________
9. Mark works 5 days a week from 8:30 A.M. to 6:00 P.M., with an hour off for lunch. If he is paid $8.45 an hour, how much does he earn a week? ________
10. The Habibs left on a car trip at 10:30 A.M. and arrived at their destination at 5:30 P.M. If they averaged 50 miles an hour, how many miles did they travel? ________

UNIT 76. Adding and Subtracting Measures That Have Mixed Units

WORDS TO KNOW

In everyday life, you often deal with measured quantities that are expressed in "mixed units." For example, "The movie ran for 2 hours and 15 minutes," "The roast weighs 5 pounds 6 ounces," and "He is 6 feet $2\frac{1}{2}$ inches tall." Such measures, which combine two or more different units, are known as **mixed units.**

You have already had some experience with combining mixed units in measuring time (Unit 75). When you must add or subtract mixed units, be sure to arrange the like units one under the other, and then add or subtract each column separately. Whenever possible, simplify the answer.

EXAMPLE 1. Add:
5 ft. 6 in. + 2 ft. 4 in. + 6 ft. 5 in.

Solution: Arrange the columns with feet under feet and inches under inches. Add each column separately.

5 ft.	6 in.
2 ft.	4 in.
+6 ft.	5 in.
13 ft.	15 in.

Since 15 inches is 3 inches more than 1 foot, change 15 inches to 1 foot 3 inches.

13 ft.	~~15 in.~~
+1 ft.	3 in.
14 ft.	3 in.

Answer: 14 ft. 3 in.

EXAMPLE 2. Add:
1 hr. 40 min. 30 sec. + 32 min. 50 sec. + 2 hr.

Solution:

1 hr.	40 min.	30 sec.
	32 min.	50 sec.
+2 hr.		
3 hr.	~~72 min.~~	~~80 sec.~~
1 hr.	12 min.	
+	1 min.	20 sec.
4 hr.	13 min.	20 sec.

To simplify, 72 min. was changed to 1 hr. 12 min. and 80 sec. was changed to 1 min. 20 sec.

Answer: 4 hr. 13 min. 20 sec.

EXAMPLE 3. Subtract 5 pounds 8 ounces from 8 pounds 6 ounces.

Solution: Set up the problem:

8 lb.	6 oz.
−5 lb.	8 oz.

Since 8 ounces is larger than 6 ounces, you cannot subtract. Therefore, rewrite 8 pounds 6 ounces as 7 pounds 22 ounces. This is similar to ordinary borrowing when subtracting dollars and cents. Here, you borrow 1 pound (16 ounces) from the 8 pounds, which leaves 7 pounds; you add the borrowed 16 ounces (1 pound) to the 6 ounces and get 22 ounces. This step can usually be done mentally. Your problem should look like this:

7	22
~~8~~ lb.	~~6~~ oz.
−5 lb.	8 oz.

Now subtract as usual.

7 lb.	22 oz.
−5 lb.	8 oz.
2 lb.	14 oz.

Answer: 2 pounds 14 ounces

EXAMPLE 4. A room measures 9 feet 3 inches long by 6 feet 10 inches wide. By how much does the length exceed the width?

Solution: Find the excess by subtracting the width from the length.

8	15	
~~9~~ ft.	~~3~~ in.	(Borrow 1 ft., which is 12 in.)
−6 ft.	10 in.	
2 ft.	5 in.	

Answer: 2 feet 5 inches

When working with mixed metric units, since the metric system is a decimal system, each smaller unit can be expressed as a decimal fraction of a larger unit, and all measurements can be written as mixed decimals.

In Unit 66, you learned to change smaller metric units to larger units by dividing by the appropriate power of 10, which was easily done by moving the decimal point the correct number of places to the left. This method is used when changing a mixed-unit measurement to a single unit.

EXAMPLE 5. 5 cm 8 mm = __?__ cm

Solution:

Step 1. Change 8 mm to cm. To change millimeters to centimeters, divide by 10 (move the decimal point one place to the left).

8 mm = 8 ÷ 10 = .8. = .8 cm

Step 2. Since 5 cm 8 mm means 5 cm *plus* 8 mm, add the measures.

5 cm + 8 mm = 5 cm + .8 cm
= 5.8 cm

Answer: 5.8 cm

Changing mixed-unit measurements to a single unit simplifies addition or subtraction of metric measures.

EXAMPLE 6. Add: 5 cm 8 mm + 7 cm 9 mm

Solution:

5 cm 8 mm = 5.8 cm (See Example 5.)
7 cm 9 mm = 7.9 cm

```
  5.8 cm
+ 7.9 cm
 13.7 cm
```

Answer: 13.7 cm

In Example 7, remember that to change grams to kilograms, you divide by 1,000 (move the decimal point 3 places to the left).

EXAMPLE 7. Subtract 3 kg 42 g from 5 kg 20 g.

Solution:

Since 42 g = .042 kg, 3 kg 42 g = 3.042 kg.
Since 20 g = .020 kg, 5 kg 20 g = 5.020 kg.

```
  5.020 kg
- 3.042 kg
  1.978 kg
```

Answer: 1.978 kg

EXERCISES

In 1–9, add. Simplify your answer.

1.
```
 6 ft.  8 in.
12 ft. 10 in.
 8 ft.  6 in.
```

2.
```
5 yd. 2 ft.  9 in.
7 yd. 2 ft.  8 in.
8 yd.       10 in.
```

3.
```
12 lb. 12 oz.
 6 lb.  8 oz.
14 lb. 10 oz.
```

4.
```
 8 hr. 42 min.
 9 hr. 50 min.
12 hr. 35 min.
```

5.
```
12 yd. 2 ft. 8 in.
 9 yd. 2 ft. 9 in.
 7 yd. 2 ft. 5 in.
```

6.
```
5 yr. 10 mo. 25 da.
7 yr.  8 mo. 21 da.
3 yr.  6 mo. 15 da.
```

7.
4 kg 235 g
3 kg 648 g
6 kg 865 g
___

8.
15 cm 8 mm
11 cm 7 mm
18 cm 9 mm
___

9.
7 L 275 mL
13 L 593 mL
16 L 87 mL
___

In 10–18, subtract. Simplify your answer.

10.
8 ft. 10 in.
4 ft. 6 in.
___

11.
12 yd. 1 ft.
8 yd. 2 ft.
___

12.
9 gal. 2 qt.
7 gal. 3 qt.
___

13.
8 lb. 5 oz.
3 lb. 12 oz.
___

14.
18 yd. 1 ft. 6 in.
7 yd. 2 ft. 8 in.
___

15.
12 hr. 14 min.
5 hr. 15 min.
___

16.
8 km 15 m
5 km 218 m
___

17.
50 cm 4 mm
35 cm 6 mm
___

18.
50 g 4 mg
35 g 6 mg
___

APPLICATION PROBLEMS

1. John has a part-time job and worked the following hours last week: 5 hours 25 minutes, 4 hours 50 minutes, 6 hours 30 minutes, 3 hours 15 minutes, 5 hours 40 minutes. How many hours and minutes did John work? ______

2. Nancy bought the following amounts of cold cuts: 1 pound 8 ounces of ham, 12 ounces of salami, 1 pound 10 ounces of bologna, and 10 ounces of spiced ham. What was the total weight of all the cold cuts? ______

3. A clerk sold 3 yards 25 inches of material from a bolt that had 28 yards 15 inches. How much material was left on the bolt? ______

4. Three pieces of board were cut from a 23-foot board. The lengths of the cut-off pieces were 3 feet 8 inches, 4 feet 10 inches, and 6 feet 8 inches. What was the length of the remaining board? ______

5. From a 50-pound sack of potatoes, a grocer sold the following weights: 5 pounds 12 ounces, 3 pounds 14 ounces, and 4 pounds 8 ounces. How many pounds of potatoes were left in the sack? ______

6. A train traveled the following distances: 350 kilometers 585 meters, 80 kilometers 92 meters, and 379 kilometers 278 meters. What was the total distance traveled? ______

7. A nut mixture contains 2 kilograms 825 grams of peanuts, 1 kilogram 20 grams of hazelnuts, and 730 grams of cashew nuts. What is the total weight of the nut mixture? ______

8. A drum contained 18 gallons 3 quarts of oil. After using 6 gallons 1 quart, 4 gallons 2 quarts, and 5 gallons 1 quart, how much oil was left in the drum? ______

9. The following lengths of silver wire are cut from a roll measuring 65 centimeters: 12 centimeters 8 millimeters, 14 centimeters 4 millimeters, and 18 centimeters 9 millimeters. How much wire is left on the roll? ______

10. Mrs. Washington had 15 quarts of buttermilk in a milk can, and used most of it in making her famous buttermilk biscuits. In three days, she used 4 quarts 1 pint, 3 quarts $1\frac{1}{2}$ pints, and 5 quarts $\frac{1}{2}$ pint. How much buttermilk was left? ______

UNIT 77. Multiplying and Dividing Measures That Have Mixed Units

MULTIPLYING MIXED UNITS

Customary System

To multiply mixed units, you must treat each different unit as a separate problem. Then combine and simplify the separate products to get your final answer.

Example 1. Multiply: 3 lb. 2 oz. × 4

Solution: Multiply the pounds by 4; multiply the ounces by 4.

$$\begin{array}{r} 3\text{ lb. }\ 2\text{ oz.} \\ \times 4 \\ \hline 12\text{ lb. }\ 8\text{ oz.} \end{array}$$

Answer: 12 lb. 8 oz.

Example 2. Multiply: 4 ft. 8 in. × 7

Solution:

$$\begin{array}{r} 4\text{ ft. }\ 8\text{ in.} \\ \times 7 \\ \hline 28\text{ ft. }\ \cancel{56\text{ in.}} \\ +4\text{ ft. }\ 8\text{ in.} \\ \hline 32\text{ ft. }\ 8\text{ in.} \end{array}$$

(56 in. = 4 ft. 8 in.)

Answer: 32 ft. 8 in.

Metric System

To multiply mixed units, change the mixed units to mixed decimals of the larger units, and multiply as you would mixed decimals.

EXAMPLE 3. Multiply: 6 L 570 mL × 12

Solution: First, change 570 mL to L so that you have a single unit.

1,000 mL = 1 L
570 mL = 570 ÷ 1,000 = .570. = .57 L

Thus, 6 L 570 mL = 6.57 L.

```
  6.57
   ×12
 13 14
 65 7
 78.84
```

Answer: 78.84 L

DIVIDING MIXED UNITS

Customary System

When the divisor goes *evenly* into all units, you simply perform the indicated divisions.

EXAMPLE 4. Divide: 15 ft. 10 in. ÷ 5

Solution:

```
     3 ft.   2 in.
 5 )15 ft.  10 in.
    15      10
```

Answer: 3 ft. 2 in.

When the divisor does not go evenly into the larger units, you must change the remainder to the next smaller units. You then combine the remainder with the given dividend of smaller units.

EXAMPLE 5. Divide: 13 ft. 4 in. ÷ 5

Solution:

```
     2 ft.
 5 )13 ft.  4 in.
    10
     3 ft. ← remainder
```

Change the remainder of 3 feet to 36 inches. Then combine 36 inches with the given dividend of 4 inches (36 inches + 4 inches = 40 inches). Your problem should look like this:

```
     2 ft.
 5 )13 ft.  4 in.
    10
     3 ft. + 4 in. = 40 in.
```

Now divide 5 into 40 in.

```
     2 ft.    8 in. ←──────┐
 5 )13 ft.    4 in.      (8 in.)
    10
     3 ft. + 4 in. =    )40 in.
                         40
```

Answer: 2 ft. 8 in.

EXAMPLE 6. Divide 15 hours 10 minutes by 13.

Solution:

```
      1 hr. 10 min. ←──────────┐
 13 )15 hr. 10 min.            │
     13                   (10 min.)
      2 hr.
         + 10 min. =   )130 min.
                        130
```

Answer: 1 hr. 10 min.

Metric System

To divide mixed units, change the mixed units to mixed decimals of the larger units, and divide as you would mixed decimals.

EXAMPLE 7. Divide 4 cm 8 mm by 5.

Solution: First, change 8 mm to cm so that you have a single unit.

10 mm = 1 cm
8 mm = 8 ÷ 10 = .8. = .8 cm

Thus, 4 cm 8 mm = 4.8 cm.

```
     .96
 5 )4.80
    4 5
      30
      30
```

Answer: .96 cm

EXERCISES

In 1–9, multiply and simplify.

1. 2 ft. 8 in. × 4

2. 5 lb. 12 oz. × 6

3. 6 yd. 2 ft. 8 in. × 7

4. 8 gal. 3 qt. 1 pt. × 6

5. 5 yd. 2 ft. 6 in. × 8

6. 5 da. 7 hr. × 5

7. 5 L 82 mL × 8

8. 14 kg 645 g × 16

9. 8 cm 6 mm × 7

In 10–18, divide and simplify. When necessary, express the smaller quantity of the quotient as a mixed number.

10. 4)8 ft. 12 in.

11. 4)5 yd. 1 ft.

12. 7)28 lb. 14 oz.

13. 5)16 wk. 3 da.

14. 6)14 lb. 4 oz.

15. 6)18 da. 18 hr.

16. 14 m 34 cm ÷ 6   17. 33 kg 600 g ÷ 15   18. 8 L 106 mL ÷ 3

APPLICATION PROBLEMS

1. If a can holds 1 quart 12 ounces, how much do 8 of these cans hold? ________
2. A dressmaker uses 4 yards 2 feet 8 inches of material to make a dress. How much material will she need to make 8 dresses? ________
3. A cake weighs 3 pounds 2 ounces. How much will half of the cake weigh? ________
4. Tom worked 42 hours and 45 minutes last week. What is the average number of hours and minutes per day for the 5-day week? ________
5. A can of juice holds 1 pint 4 ounces. How much juice will 12 of these cans hold? ________
6. Mrs. Grant bought a turkey that weighed 8 kilograms 645 grams. If the price per kilogram was $2.45, what was the cost of the turkey? ________
7. A bottling machine puts 1 liter 575 milliliters of soda pop into each bottle. How much soda pop will be needed to fill 288 bottles? ________
8. A stack of 16 Driver's Education pamphlets is 22 centimeters 4 millimeters thick. What is the thickness of each pamphlet? ________
9. Velvet sells at $9.75 per meter. How much will 12 meters 35 centimeters of the velvet cost? ________
10. A bag of flour weighs 1 kilogram 400 grams. How many bags can be filled with 50 kilograms 400 grams of flour? ________

UNIT 78. Equivalent Measures in the Metric and Customary Systems

An imported ham weighs 6.8 kilograms. How many pounds does it weigh? A foreign car has a gas tank that holds 45 liters. How many gallons does it hold? Which is longer, a 1-mile race or a 1,500-meter race?

As the metric system becomes increasingly popular in the United States, it is often necessary to change measurements in that system to equivalent measurements in the Customary system.

Table XIV shows common equivalents. Note, under Measures of Capacity, that the liquid quart and the dry quart have different metric equivalents. (The dry quart is used for measuring farm products such as berries.)

Table XIV: Equivalent Measures in the Metric and Customary Systems

MEASURES OF LENGTH	
Metric to Customary	Customary to Metric
1 meter = 39.37 inches = 3.28 feet = 1.09 yards 1 centimeter = .39 inch 1 millimeter = .04 inch 1 kilometer = .62 mile	1 inch = 25.4 millimeters = 2.54 centimeters = .03 meter 1 foot = .30 meter 1 yard = .91 meter 1 mile = 1.61 kilometers
MEASURES OF CAPACITY	
Metric to Customary	Customary to Metric
1 liter = 1.06 liquid quarts = .91 dry quart	1 liquid quart = .95 liter 1 dry quart = 1.10 liters
MEASURES OF WEIGHT	
Metric to Customary	Customary to Metric
1 gram = .04 ounce 1 kilogram = 2.20 pounds 1 metric ton = 2,204.62 pounds	1 ounce = 28.35 grams 1 pound = .45 kilogram 1 ton = .91 metric ton

Metric to Customary

To change measurements *from the metric system* to equivalent measurements in the Customary system, find the number of Customary units contained in *one metric unit.* Then multiply the given number of metric units by this value.

To avoid confusion, remember this first step: To change *from the metric,* find the number of Customary units in *one metric unit.*

EXAMPLE 1. How many yards are there in 123 meters? Answer to the nearest *yard.*

Solution:

1 m = 1.09 yd.
123 m = 123 × 1.09 = 134 yd.

```
   123
 ×1.09
  11 07
 123 0
 134.(07)
```

Answer: 123 meters = 134 yards

EXAMPLE 2. Change 6.8 kilograms to *pounds.* Answer to the nearest *tenth* of a pound.

Solution:

1 kg = 2.20 lb.
6.8 kg = 6.8 × 2.20 = 15.0 lb.

```
    6.8
  ×2.20
  1 360
 13 6
 14.9(60) = 15.0
```

Answer: 6.8 kg = 15.0 lb., to the nearest tenth.

EXAMPLE 3. Change 45 liters to *gallons.* Answer to the nearest *tenth* of a gallon.

Solution: Since you do not have a direct conversion from liters to gallons, you can first change liters to quarts.

1 L = 1.06 qt.
45 L = 45 × 1.06 = 47.7 qt.

```
  1.06
   ×45
  5 30
 42 4
 47.70 = 47.7
```

Now change quarts to gallons.

4 qt. = 1 gal.
47.7 qt. = 47.7 ÷ 4 = 11.9 gal.

```
      3
   11.92 = 11.9
4 )47.70
   4
    7
    4
    37
    36
     10
      8
      2 ← remainder equals
          half the divisor
```

Answer: 45 L = 11.9 gal., to the nearest tenth.

EXAMPLE 4. Change 1,500 meters to *miles.* Answer to the nearest *tenth* of a mile.

Solution: Since you do not have a direct conversion from meters to miles, you can first change meters to kilometers.

1,000 m = 1 km
1,500 m = 1,500 ÷ 1,000 = 1.500. = 1.5 km

Now change kilometers to miles.

1 km = .62 mi.
1.5 km = 1.5 × .62 = .9 mi.

```
   1.5
 ×.62
   30
  90
 .9(30)
```

Answer: 1,500 m = .9 mi., to the nearest tenth.

CUSTOMARY TO METRIC

To change *from the Customary system* to the metric system, follow the procedure used in changing metric units to Customary units. However, find the number of metric units in *one Customary unit.* Then multiply the given number of Customary units by this number.

To avoid confusion, remember this first step: To change *from the Customary,* find the number of metric units in *one Customary unit.*

EXAMPLE 5. How many meters are there in 1,000 yards?

Solution:

1 yd. = .91 m
1,000 yd. = 1,000 × .91 = .910. = 910 m

Answer: 1,000 yards = 910 meters

EXAMPLE 6. Change 25 gallons to *liters.*

Solution: Since you do not have a direct conversion from gallons to liters, you can first change gallons to quarts.

1 gal. = 4 qt.
25 gal. = 25 × 4 = 100 qt.

Now change quarts to liters.

1 qt. = .95 L
100 qt. = 100 × .95 = .95. = 95 L

Answer: 25 gallons = 95 liters

Remember

1. To change *from the metric system,* find the equivalent value for *one metric unit.* Then multiply the given number of metric units by this value.
2. To change *from the Customary system,* find the equivalent value for *one Customary unit.* Then multiply the given number of Customary units by this value.

EXERCISES

Refer to Table XIV when solving these exercises. In 1–15, when necessary, round off each answer to the nearest *whole unit.*

Change:

1. 10 m to feet

2. 10 m to yards

3. 100 km to miles

4. 100 L to liquid quarts

5. 500 g to ounces

6. 10 in. to millimeters

7. 100 mi. to kilometers

8. 60 lb. to kilograms

9. 20 dry qt. to liters

10. 1,000 in. to meters

11. $2\frac{7}{10}$ kg to pounds

12. .08 m to inches

13. 10 metric tons to pounds

14. 2.87 mi. to kilometers

15. 1.72 in. to millimeters

In 16–21, the exercises will require two steps. When necessary, round off each answer to the nearest *whole unit.*

Change:

16. 4,000 g to pounds

17. 100 cm to feet

18. 5,000 mm to feet

19. 200 oz. to kilograms

20. .5 mi. to meters

21. $1\frac{3}{4}$ lb. to grams

In 22–25, round off to the required decimal place.

Change:

22. 21.6 cm to inches (nearest *tenth*)

23. 16.4 in. to centimeters (nearest *tenth*)

24. .076 kg to pounds (nearest *hundredth*)

25. .001 metric ton to pounds (nearest *tenth*)

APPLICATION PROBLEMS

1. What is the width in inches of an 8-mm film? ______
2. An imported can of coffee weighs 750 grams. How many ounces does the can of coffee weigh? ______
3. How many liters are there in 3 quarts of milk? ______
4. John jumped 2.5 meters in a high jump competition. How high was his jump in feet? ______
5. A picture measures 8 cm × 12 cm. What are the measurements in inches? ______
6. Nancy is 5 feet 7 inches tall. What is her height in meters? ______
7. A state speed limit is 55 miles per hour. What is the speed limit in kilometers? ______
8. Charles bought 43 liters of gasoline. How many gallons did he buy? ______
9. A package of cheese weighs 325 grams. What is the weight of the cheese in ounces? ______
10. A liter of milk costs 79 cents. Find the price per gallon. ______

Review of Part XVI (Units 72–78)

1. **Change:**

a. $6\frac{1}{4}$ pt. to oz.

b. 464 oz. to qt.

c. 27.75 qt. to pt.

d. 270 pt. to gal.

e. 432 qt. to gal.

f. 13.75 qt. to oz.

2. **Change:**

a. 275 mL to L

b. .008 L to mL

c. $8\frac{1}{8}$ L to mL

3. **Change:**

a. $11\frac{1}{8}$ lb. to oz.

b. 8.25 lb. to oz.

c. 208 oz. to lb.

d. 36,400 lb. to T.

e. .72 T. to lb.

f. 104,000 oz. to T.

4. **Change:**

a. 3.08 kg to g

b. $\frac{1}{5}$ t to kg

c. 4,750 mg to g

5. **Change:**

a. 156 units to doz.

b. $15\frac{1}{2}$ gr. to doz.

c. 1,404 units to gr.

6. **Change:**

a. $3\frac{1}{4}$ yr. to mo.

b. 42 wk. to da.

c. $4\frac{2}{3}$ da. to hr.

7. **Find the length of the given interval.**

a. 2:30 P.M. to 8:45 P.M.

b. 6:30 P.M. to 4:15 A.M.

c. 9:15 A.M. to 6:30 P.M.

In 8–11, simplify your answers.

8. **Add.**

a.
3 yd. 1 ft. 10 in.
5 yd. 2 ft.
6 yd. 2 ft. 7 in.

b.
7 hr. 45 min.
9 hr. 35 min.
6 hr. 55 min.

c.
3 kg 472 g
5 kg 837 g
4 kg 917 g

d.
32 cm 6 mm
15 cm 9 mm
27 cm 7 mm

9. Subtract.

a. 12 lb. 8 oz.
5 lb. 15 oz.

b. 23 yd. 1 ft. 5 in.
9 yd. 2 ft. 7 in.

c. 11 km 145 m
3 km 673 m

d. 14 kg 728 g
5 kg 865 g

10. Multiply.

a. 12 lb. 14 oz.
×8

b. 5 yd. 2 ft. 8 in.
×7

c. 14 cm 8 mm
×5

d. 18 kg 735 g
×9

11. Divide.

a. 4)6 yd. 2 ft.

b. 7)22 lb. 5 oz.

c. 8)12 m 8 cm

d. 14)35 kg 56 g

12. Convert. Answer to the nearest *whole unit.*

a. 14 meters to yards

b. 342 kilometers to miles

c. 475 grams to ounces

d. 425 miles to kilometers

e. 48 pounds to kilograms

f. 525 inches to centimeters

13. On Thursday, Jason worked from 8:15 A.M. to 7:30 P.M. If he had an hour off for lunch, how many hours did Jason actually work? __________

14. A four-ounce can of juice concentrate costs 84¢. How much will $2\frac{1}{2}$ pints of the juice concentrate cost? __________

15. A can contains 475 grams of coffee. If it sells for $3.85, what is the price per kilogram? __________

16. A package of 6 tomatoes costs 98¢. How much will $5\frac{1}{2}$ dozen tomatoes cost? __________

17. Yael left on a car trip at 6:30 A.M. and drove till 4:45 P.M. How many hours did she drive? __________

18. The length of a bolt of fabric was 27 yards 2 feet. A clerk sold the following lengths of the fabric: 5 yards 2 feet, 6 yards 1 foot, and 3 yards. How much fabric was left? __________

19. A diner uses an average of 4 gallons 3 quarts 1 pint of milk per day. How much milk is used in 7 days? __________

20. To the nearest whole number, how many 5-ounce blocks of cheese are contained in 1 kilogram of cheese? __________

Cumulative Review of Parts XV and XVI

In 1–2, answers that are not whole numbers should be rounded off to the nearest *tenth.*

1. Change to *feet:*

a. 136 in. *b.* $8\frac{1}{2}$ yd. *c.* 12 yd. 28 in.

2. Change to *yards:*

a. 43 ft. *b.* 463 in. *c.* $3\frac{1}{4}$ mi.

3. Change to *centimeters:*

a. 23 mm *b.* $2\frac{1}{4}$ m *c.* .85 m

4. Change to *meters:*

a. 850 mm *b.* 62 cm *c.* .057 km

5. Change to *kilometers:*

a. 4,600 m *b.* 62,000 cm *c.* 5,000,000 mm

In 6–7, round off answers that are not whole numbers to the nearest *tenth.*

6. Change to *square feet:*

a. 872 sq. in. *b.* $34\frac{1}{2}$ sq. yd. *c.* half an acre

7. Change to *square yards:*

a. 1,460 sq. ft. *b.* 3,625 sq. in. *c.* .5 sq. mi.

8. Change to *square meters:*

a. .048 km^2 *b.* 2.5 ha *c.* 3,526 cm^2

9. Find the perimeter of each of the following figures:

a. A square whose side measures $1\frac{7}{8}$ inches.

b. A rectangle whose length is 5.3 cm and whose width is 2.5 cm.

c. A triangle with sides that measure $2\frac{1}{2}$ ft., 20 in., and $\frac{1}{2}$ yd. Answer in inches.

d. A rectangle whose dimensions are $8\frac{1}{2}$ in. × $3\frac{3}{4}$ in.

10. Find the circumference of a circle:

a. whose radius is $3\frac{1}{2}$ in. (Use $\pi = \frac{22}{7}$.)

b. whose diameter is 3.6 cm. (Use $\pi = 3.14$.)

11. Find the area of each of the following figures. Round off answers to the nearest *whole unit.*

a. A rectangle 24 in. × 35 in.

b. A triangle whose base is $\frac{4}{5}$ m and whose height is 88 cm. Answer in square centimeters.

c. A square whose side is 5.4 mm.

d. A rectangular field $\frac{1}{2}$ mile wide by $1\frac{1}{2}$ miles long. Answer in acres.

e. A triangle that has a base of 28 in. and a height of .75 ft. Answer in square inches.

12. Find the area of each circle. Round off answers to the nearest *whole unit.*

a. A circle with a radius of $4\frac{2}{3}$ ft. (Use $\pi = \frac{22}{7}$.)

b. A circle with a diameter of 2.5 cm. (Use $\pi = 3.14$.)

13. Find the volume of each of the following solids. Round off your answers to the nearest *whole unit.*

a. A rectangular solid 18 in. × 15 in. × 6 in.

b. A cube whose side measures $5\frac{1}{8}$ cm. (*Hint:* In a cube, the length, width, and height are all the same number of units.)

c. A rectangular solid 9 in. × 3 ft. × 8 ft. Answer in cubic feet.

d. A wooden crate .5 m × $2\frac{1}{4}$ m × 150 cm. Answer in cubic meters.

In 14–17, round off your answers to the nearest *whole unit.*

14. Change:

a. $15\frac{3}{4}$ gal. to qt. *b.* 57 pt. to qt. *c.* 100 oz. to qt.

d. half a gallon to ounces *e.* 1,837 mL to L *f.* 3.45 L to mL

15. Change:

a. 640 oz. to lb. *b.* 18 lb. 5 oz. to oz. *c.* 5,600 lb. to tons

d. $\frac{1}{100}$ ton to ounces *e.* 3.42 g to mg *f.* 26 kg to g

16. Change:

a. 1 week to hours *b.* 6,000 seconds to hours *c.* $\frac{1}{10}$ day to minutes

17. Change:

a. 10 gross to dozens *b.* 240 units to dozens *c.* 240 units to scores

18. Add and simplify.

a.
12 yd. 2 ft. 10 in.
6 yd. 1 ft. 8 in.
16 yd. 2 ft. 9 in.

b.
8 yr. 9 mo. 23 da.
12 yr. 11 mo. 16 da.
19 yr. 7 mo. 28 da.

c.
15 lb. 12 oz.
19 lb. 15 oz.
23 lb. 7 oz.

19. Subtract and simplify.

a.
23 ft. 7 in.
13 ft. 11 in.

b.
18 gal. 2 qt.
5 gal. 3 qt.

c.
21 yd. 1 ft. 7 in.
12 yd. 2 ft. 9 in.

20. Multiply and simplify.

a.
21 lb. 13 oz.
× 5

b.
12 gal. 3 qt. 1 pt.
× 9

c.
15 yd. 2 ft. 8 in.
× 7

21. Divide and simplify.

a. 5) 18 gal. 3 qt. *b.* 3) 16 lb. 5 oz. *c.* 6) 28 yd. 2 ft. 6 in.

22. Find what time it will be:

a. 2 hours 15 minutes after 9:25 P.M. *b.* 4 hours 45 minutes after 10:20 A.M.

23. Find the number of hours and minutes between:

a. 8:15 A.M. and 10:45 A.M.

b. 7:30 A.M. and 3:10 P.M.

24. Change:

a. 1550 hours to regular time

b. 10:30 P.M. to military time

25. Change as specified. Round off your answers to the nearest *tenth* of a unit when necessary.

a. 200 cm to in.

b. 50 mi. to km

c. 100 m to yd.

d. 5 kg to lb.

e. 200 g to oz.

f. 60 liquid quarts to liters

26. Find, to the nearest *whole unit:*

a. the number of grams in 26 pounds

b. the number of feet in half a kilometer

PART XVII. Ratio and Its Applications

UNIT 79. Ratios

WORDS TO KNOW

You often compare two numbers by subtraction. Suppose the State University football team beats Tech by a score of 21 to 7. You could say, "State beat Tech by 14 points," since $21 - 7 = 14$.

You can also compare two numbers by division. Of the football game, you can say, "State scored 3 times as much as Tech," since $7 \times 3 = 21$, or $21 \div 7 = 3$.

When two numbers are compared by *division*, the indicated division is called a **ratio.**

When a newspaper article tells you, "In River City, there are 3 registered Democrats for every 2 registered Republicans," you do not know how many Democrats or Republicans there are. The article tells you how the number of Democrats is *related* to the number of Republicans. It does not tell you the actual numbers. Such a relationship between two numbers is called a *ratio.*

A **ratio** compares two numbers by division.

In a class of 36 students, there are 12 girls and 24 boys. To compare the number of girls to the number of boys, you say, "The ratio of girls to boys is 12 to 24." The word "to" is often represented by a colon (:). Thus, "12 to 24" may be written as "12 : 24."

In fact, the colon (:) is simply another form of the division symbol (÷). Any ratio can also be written as a fraction, since the fraction bar itself is a symbol for division. Thus, "12 : 24" and "$\frac{12}{24}$" can both be read as "12 to 24."

You know that when a fraction is reduced to lowest terms, the value of the fraction does not change. Since 12 to 24 can be written as $\frac{12}{24}$, you can reduce the ratio to lower terms *without changing the ratio of girls to boys.* Note that $\frac{12 \text{ girls}}{24 \text{ boys}}$ is the same ratio as $\frac{6 \text{ girls}}{12 \text{ boys}}$, which is the same ratio as $\frac{3 \text{ girls}}{6 \text{ boys}}$, which is the same ratio as $\frac{1 \text{ girl}}{2 \text{ boys}}$.

Remember

Reducing a ratio to lower terms does not change the value of the ratio.

Similarly, the ratio of boys to girls is 24 to 12, which is written as 24 : 12 or as $\frac{24}{12}$. The ratio $\frac{24}{12}$ can be reduced to $\frac{12}{6}$ or $\frac{6}{3}$ or $\frac{2}{1}$. A ratio *must* have a denominator. Always write the "1" in the denominator when necessary.

The ratio of girls to the total number of students is 12 to 36, which is written as 12 : 36 or as $\frac{12}{36}$. The ratio of boys to the total number of students is 24 to 36, which is written as 24 : 36 or as $\frac{24}{36}$.

In everyday language, the words "out of" are often used when a part of something is compared to the whole. Thus, "12 *out of* 36 students are girls" or "24 *out of* 36 students are boys."

Although the above examples dealt with only *two* groups of students (24 boys and 12 girls), there were many different ratios:

girls to boys	12 : 24 or $\frac{12}{24}$
boys to girls	24 : 12 or $\frac{24}{12}$
girls to total students	12 : 36 or $\frac{12}{36}$
boys to total students	24 : 36 or $\frac{24}{36}$

You could also write the ratio of total students to girls (36 : 12) or the ratio of total students to boys (36 : 24).

In word problems involving ratios, be sure you read the problem carefully in order to write the correct ratio. It is usually easier to solve ratio problems when you write the ratios as fractions. As with all word problems, look for *key words* to help you. Such phrases as "7 out of 10 voters" and "3 girls to 4 boys" tell you that a *ratio* is involved.

EXAMPLE 1. A school bus carries 95 students. If the ratio of girls to boys is 8 : 12, how many girls are on the bus?

Solution: Since the ratio of girls to boys is 8 : 12, you know that there are 8 girls for every 12 boys. In other words, *for every 20 students there are 8 girls and 12 boys.* Write these facts as ratios:

$$\frac{\text{number of girls}}{\text{total number of students}} = \frac{8}{20} = \frac{2}{5}$$

$$\frac{\text{number of boys}}{\text{total number of students}} = \frac{12}{20} = \frac{3}{5}$$

Your problem asks for the number of *girls* on the bus. Therefore, choose the ratio of *girls* to the total number of students, $\frac{2}{5}$. Your problem is now simply, "Find $\frac{2}{5}$ of 95."

$$\frac{2}{\cancel{5}_{1}} \times \frac{\cancel{95}^{19}}{1} = 38$$

Check: If "38 girls" is the correct answer, then there should be 57 boys on the bus (95 − 38 = 57). Does this agree with the given girl-to-boy ratio of $\frac{8}{12}$ or $\frac{2}{3}$?

$$\frac{38 \div 19}{57 \div 19} = \frac{2}{3} \checkmark$$

Answer: 38 girls

EXAMPLE 2. The Centreville Drum and Bugle Corps has 60 members. A member plays either a drum or a bugle. If 3 out of 5 members play bugles, how many members play drums?

Solution: Of every 5 members, 3 play bugles and 2 play drums. From this information, write the ratio of members that play *drums* to the total number of members:

$$\frac{\text{number of members that play drums}}{\text{total number of members}} = \frac{2}{5}$$

Find $\frac{2}{5}$ of the total number of members.

$$\frac{2}{\cancel{5}_{1}} \times \frac{\cancel{60}^{12}}{1} = 24$$

Check: If 24 members play drums, 36 members play bugles (60 − 24 = 36). Does a ratio of "36 out of 60" agree with the given ratio of "3 out of 5"?

$$\frac{36 \div 12}{60 \div 12} = \frac{3}{5} \checkmark$$

Answer: 24 members play drums.

EXAMPLE 3. Change each of the following expressions to a ratio in fraction form. Reduce the ratio to lowest terms.
(*a*) 12 voters out of 60 voters
(*b*) 18 inches to 1 yard
(*c*) 3 quarts to 2 gallons

Solutions:

(*a*) $\frac{12}{60} = \frac{1}{5}$

(*b*) Since you must compare inches and yards, change both numbers to inches.

$$\frac{18 \text{ inches}}{1 \text{ yard}} = \frac{18 \text{ inches}}{36 \text{ inches}} = \frac{18}{36} = \frac{1}{2}$$

(*c*) There are 4 quarts in a gallon. Thus,

$$\frac{3 \text{ quarts}}{2 \text{ gallons}} = \frac{3 \text{ quarts}}{8 \text{ quarts}} = \frac{3}{8}$$

Answers: (*a*) $\frac{1}{5}$ (*b*) $\frac{1}{2}$ (*c*) $\frac{3}{8}$

Note that when working with ratios, you must be careful to express both quantities of the ratio in terms of the same unit. The ratio of 5 *inches* to 1 *foot,* for example, is *not* 5 : 1. You convert the 1 foot to 12 inches. Then, with the same unit for both quantities, the ratio of 5 inches to 12 inches is correctly written as 5 : 12.

RATES

Ratios are also used to compare quantities of *different* types, such as miles per hour or income per year. These comparisons are called **rates.** The key word *per* indicates a rate. Rates may also be indicated in other ways; for example, 3 pens for a dollar, 99 miles in 3 hours, or 18 miles on a gallon of gas.

EXAMPLE 4. Sally uses 7 tablespoons of sugar per 2 quarts of lemonade. How much sugar will she need for 6 quarts of lemonade?

Solution: 7 tablespoons per 2 quarts can be written as the ratio $\frac{7 \text{ tablespoons}}{2 \text{ quarts}}$. You want an equivalent ratio for 6 quarts.

$$\frac{7 \text{ tbs.}}{2 \text{ qts.}} = \frac{? \text{ tbs.}}{6 \text{ qts.}}$$

Recall how you changed fractions to higher equivalents (Unit 24).

$$\frac{7 \times 3}{2} = \frac{21}{6}$$

2 into 6 is 3; 7 × 3 is 21

Answer: 21 tablespoons

EXAMPLE 5. On a trip through the Southwest, Mr. Olson is driving his camper at an average rate of 42 miles per hour. How long should it take him to drive the 147 miles to the next campsite?

Solution: His rate of 42 miles per hour can be written as the ratio $\frac{42 \text{ miles}}{1 \text{ hour}}$. You want an equivalent ratio for 147 miles.

$$\frac{42 \text{ miles}}{1 \text{ hour}} = \frac{147 \text{ miles}}{? \text{ hours}}$$

You again use the method for obtaining an equivalent fraction. This time the unknown quantity is in the denominator.

$$\frac{42}{1 \times 3\frac{1}{2}} = \frac{147}{3\frac{1}{2}}$$

42 into 147 is $3\frac{1}{2}$; 1 × $3\frac{1}{2}$ is $3\frac{1}{2}$

$$\begin{array}{r} 3\frac{21}{42} = 3\frac{1}{2} \\ 42\overline{)147} \\ \underline{126} \\ 21 \end{array}$$

(21 over 42)

Answer: $3\frac{1}{2}$ hours

A special kind of rate that is used in baseball is the "batting average." It describes the number of hits a batter gets, compared to the number of times at bat, and is expressed in decimal form to the nearest thousandth.

$$\text{batting average} = \frac{\text{number of hits}}{\text{number of times at bat}}$$

EXAMPLE 6. Joe Slugger is the best batter on the Grover School baseball team. In 109 times at bat, he got 34 hits. What is his batting average?

Solution: Batting average $= \frac{34}{109}$

Divide:

$$\begin{array}{r} .311 = .312 \\ 109\overline{)34.000} \\ \underline{32\ 7} \\ 1\ 30 \\ \underline{1\ 09} \\ 210 \\ \underline{109} \\ 101 \end{array}$$

half of the divisor → $54\frac{1}{2}$; 101 ← remainder

Recall: Since the remainder is larger than half the divisor, you add one unit to the quotient. Since the unit here is thousandths, you add .001 to .311, getting .312.

Answer: .312

EXERCISES

In 1–21, change each expression to a ratio in fraction form. Reduce each ratio to lowest terms.

1. 6 to 8 **2.** 9 to 12 **3.** 5 to 5

4. 15 to 9 **5.** 10 in. to 1 ft. **6.** 25 : 40

7. 50 : 30

8. 24 : 3

9. 4 oz. to 1 lb.

10. 2 pt. : 1 gal.

11. 7 out of 28

12. 15 out of 45

13. 6 out of 48

14. 4 out of 10

15. 12 out of 50

16. 24 to 24

17. 28 in. : 4 ft.

18. 9 out of 45

19. 6 in. to $1\frac{1}{2}$ ft.

20. 20 : 60

21. $2\frac{1}{2}$ ft. to 1 yd.

In 22–27, express each rate as a fraction in lowest terms.

22. 5 oranges for 98 cents

23. 3 lollipops for a quarter

24. growing 2 meters in 6 years

25. 30 pizzas for 45 people

26. 96 miles in 2 hours

27. 6 pounds of cabbage per 8 quarts of coleslaw

APPLICATION PROBLEMS

Sample Solution

On a political issue, the voting ran 5 in favor to 2 against. If the total number of votes was 28,000, how many voted in favor of the issue?

Of every 7 voters, 5 voted in favor.

$$\frac{\text{number in favor}}{\text{total number}} = \frac{5}{7}$$

$$\frac{5}{7} \text{ of } 28{,}000 = \frac{5}{\cancel{7}_1} \times \frac{\cancel{28{,}000}^{4000}}{1} = 20{,}000$$

20,000 voted in favor.

1. The ratio of boys to girls in a school is 3 : 2. If there are 1,850 students, how many are boys and how many are girls? ____________ boys

   ____________ girls

2. A team won 150 games and lost 65 games. What is the ratio of games won to games played? ____________

3. If 30 students in a class are present and 5 students are absent, what is the ratio of students absent to students present? ____________

   What is the ratio of students absent to the total number of students? ____________

4. Two partners in a business share profits in the ratio of 5 : 4. How much money would each partner get if the profits for the year were $18,000? ____________

5. A retailer had the following sizes in men's sport shirts: 35 small, 65 medium, and 45 large. What is the ratio of:

   a. small size to large size? ____________

   b. medium size to large size? ____________

   c. large size to total? ____________

6. For every 5 customers that pay cash for their purchases in a local jewelry store, 4 customers charge their purchases on credit cards. How many customers out of 972 charge their purchases? ____________

7. In a taste test, 8 people preferred Corky's Cookies to every 3 people who chose Brand X. Of the 231 people who participated, how many chose Corky's Cookies? ____________

8. Steve's car averages 720 kilometers on a full tank (40 liters) of gasoline. At that rate, how many liters of gasoline will he need to drive from his home to Vancouver, a distance of 2,160 kilometers? ____________

9. The freshman class ordered T-shirts at a rate of a dozen shirts for $65. What will be the cost of 132 of these shirts? ____________

10. At the start of the baseball season, Hank Stone has 13 hits for 57 times at bat. What is his batting average? ____________

UNIT 80. Proportions

Words to Know

When two ratios are equal, they form a **proportion.**

You learned that reducing a ratio to lowest terms does not change the value, or relationship, of the ratio.

When you write $5 : 25 = 1 : 5$ or $\frac{5}{25} = \frac{1}{5}$, you have in fact written a proportion. You are stating that the relationship of the numbers 5 to 25 is the same as the relationship of the numbers 1 to 5, or that 5 is to 25 as 1 is to 5.

You know that the ratio 8 : 12 has the same value as the ratio 2 : 3, because reducing a ratio to lower terms does not change its value. Thus, you may write the following **proportion:**

$$8:12 = 2:3 \quad \text{or} \quad \frac{8}{12} = \frac{2}{3}$$

This proportion tells you that the two ratios are equal.

The two numbers at the ends of the proportion are called the **extremes:**

$$8:12 = 2:3 \quad \text{or} \quad \frac{8}{12} = \frac{2}{3}$$

(8 and 3 marked: extremes)

In the above proportion, the extremes are the numbers 8 and 3.

The two numbers inside the proportion are called the **means:**

$$8:12 = 2:3 \quad \text{or} \quad \frac{8}{12} = \frac{2}{3}$$

(12 and 2 marked: means)

In the above proportion, the means are 12 and 2. In this proportion, the product of the extremes (8 × 3) is 24; the product of the means (12 × 2) is also 24. Thus, in a true proportion, the product of the means equals the product of the extremes.

RULE

In a proportion, the product of the means equals the product of the extremes.

EXAMPLE 1. From the information that is given, write a proportion. Then tell whether the proportion is *true* or *false*.

(*a*) On Saturday, Cal drove 400 miles in 10 hours. On Sunday, he drove 200 miles in 5 hours.

Solution: 400 : 10 = 200 : 5

means: 10 × 200 = 2,000 } products
extremes: 400 × 5 = 2,000 } are equal

The proportion is *true*.

(*b*) Patricia feeds 12 cats with 4 quarts of milk. She can feed 25 cats with 8 quarts of milk.

Solution: 12 : 4 = 25 : 8

means: 4 × 25 = 100 } products are
extremes: 12 × 8 = 96 } not equal

The proportion is *false*.

(*c*) George has an LP record that plays music for 15 minutes. With 5 such records, he can listen to 75 minutes of music.

Solution: 1 : 15 = 5 : 75

means: 15 × 5 = 75 } products
extremes: 1 × 75 = 75 } are equal

The proportion is *true*.

Recalling how you reduced fractions to lowest terms (Unit 23), and how you raised fractions to higher equivalents (Unit 24), you see that the ratios of a proportion are really two equivalent fractions.

To solve word problems involving proportions, you can use the method of finding equivalent fractions or the preceding rule that the product of the means equals the product of the extremes.

EXAMPLE 2. A coat regularly selling for $90 is on sale at $\frac{1}{3}$ off. What is the sale price of the coat?

Solution: A $\frac{1}{3}$-off sale means that the regular price is reduced by $\frac{1}{3}$ ($1 out of $3). Therefore, the sale price is the remaining $\frac{2}{3}$ ($2 out of $3).

To find the sale price, write the proportion "$2 is to $3 as SP (Sale Price) is to $90."

Equivalent Ratio Method:

$$\frac{\$2}{\$3} = \frac{SP}{\$90}$$

$$\frac{2}{3} = \frac{?}{90}$$

Recall:

$$\frac{2 \times 30 \text{ is} \rightarrow 60}{3 \text{ into} \rightarrow 90 \text{ is} \rightarrow 30}$$

$$\frac{2}{3} = \frac{60}{90}$$

$$SP = \$60$$

Proportion Method:

$$\frac{\$2 \text{ (extreme)}}{\$3 \text{ (mean)}} = \frac{SP \text{ (mean)}}{\$90 \text{ (extreme)}}$$

$$\frac{2}{3} = \frac{SP}{90}$$ Multiply the means. Multiply the extremes.

$$3 \times SP = 2 \times 90$$

$$3 \times SP = 180$$

$$SP = \frac{180}{3}$$ Use division, the opposite of multiplication, to undo the multiplication.

$$SP = \frac{180}{3} = 60$$

$$SP = \$60$$

Check:

$$\frac{\$2}{\$3} = \frac{SP}{\$90}$$

$$\frac{2}{3} \stackrel{?}{=} \frac{\not{60}^{\,2}}{\not{90}_{\,3}}$$

$$\frac{2}{3} = \frac{2}{3} \ \checkmark$$

Answer: The sale price is $60.

EXERCISES

In 1–5, write a proportion from the information that is given. Tell whether the proportion is *true* or *false*.

1. A sales representative earns a commission of $100 on sales of $2,000. She will earn a commission of $150 on sales of $3,000. ____________
2. Last year, Mr. Davis paid $6,000 in income tax on earnings of $24,000. This year, he paid $7,200 in income tax on earnings of $25,600. ____________
3. Jane earned $7.15 for $5\frac{1}{2}$ hours of babysitting. She will earn $3.90 for 3 hours of babysitting. ____________
4. Dewey read 6 books in $25\frac{1}{2}$ hours. He can read 5 books in $21\frac{1}{4}$ hours. ____________
5. Pierre's car traveled $98\frac{3}{4}$ miles on $6\frac{1}{4}$ gallons of gasoline. Pierre's car can travel $173\frac{4}{5}$ miles on 11 gallons. ____________

In 6–17, find the missing number that will make each expression a true proportion. Use the equivalent ratio method.

6. $\frac{5}{7} = \frac{?}{56}$ **7.** $\frac{35}{14} = \frac{?}{2}$ **8.** $\frac{4}{5} = \frac{36}{?}$

9. $\frac{35}{42} = \frac{5}{?}$

10. ? : 48 = 5 : 8

11. 2 : 3 = ? : 21

12. 5 : 8 = 45 : ?

13. ? : 6 = 40 : 48

14. 5 : ? = 45 : 63

15. 24 : ? = 3 : 4

16. 56 : 72 = 7 : ?

17. 2 : 13 = ? : 1,105

In 18–26, find the missing number. Use the proportion method: the product of the means equals the product of the extremes.

18. $\frac{4}{5} = \frac{?}{\$575}$

19. $\frac{8}{12} = \frac{520 \text{ ft.}}{?}$

20. $\frac{4}{16} = \frac{?}{592 \text{ lb.}}$

21. $\frac{2}{3} = \frac{1{,}268}{?}$

22. 7 : 9 = ? : 1,287

23. 5 : 11 = ? : 693

24. 3 : ? = 156 : 364

25. ? : 8 = 360 : 576

26. 11 : 12 = 242 : ?

APPLICATION PROBLEMS

SAMPLE SOLUTION

A nut mixture is made up of 3 pounds of cashews and 5 pounds of peanuts. How many pounds of each are needed to make 40 pounds of the mixture?

Of every 8 lb., 3 are cashews and 5 are peanuts.

$\frac{\text{cashews}}{\text{mixture}} = \frac{3}{8} = \frac{?}{40}$ $\frac{3}{8} = \frac{15}{40}$ *15 lb.* cashews

$\frac{\text{peanuts}}{\text{mixture}} = \frac{5}{8} = \frac{?}{40}$ $\frac{5}{8} = \frac{25}{40}$ *25 lb.* peanuts

1. At the E-Z Cut-Rate Store, the following items are on sale:

   Erasers: 2 for 35¢, 6 for $1.00.
   Pencils: 3 for 25¢, 12 for $1.00.
   Ballpoint Pens: 2 for 39¢, 6 for $1.10.

   Write each price as a proportion. Which proportion is *true*? ______
2. If 3 pairs of socks sell for $4, how much would 15 pairs cost? ______
3. A plane averages 1,200 miles in 2 hours. How many hours will it take to travel 3,300 miles? ______
4. A color TV sells for $875. If $\frac{2}{5}$ of the selling price is the retailer's profit, find the amount of profit. ______
5. A survey concluded that for every 5 people who preferred television, 3 people preferred reading. If 950 people in the survey preferred television, how many preferred reading? ______
6. For every 3 color TV sets sold, 5 black and white TV sets are sold. If there are 1,585 black and white sets sold, how many color TV sets are sold? ______
7. For every 5 reserved seats sold for a concert, 4 general admission seats were sold. If the total number of seats sold was 27,360, how many general admission seats were sold? ______
8. The ratio of the width of a store to the length is 3 to 5. If the width of the store is 45 feet, what is the length? ______
9. If 1 inch on a map represents 85 kilometers, how many kilometers are represented by 8 inches? ______
10. A baker mixes 5 cups of flour to 2 cups of sugar to make a tray of doughnuts. How much sugar is needed to make 125 trays of doughnuts? ______

UNIT 81. Probability

Words to Know

Probability is a mathematical theory used to predict the likelihood that an event, or **outcome**, will happen. Probability is expressed as a ratio that compares the number of ways an event can occur successfully to the total number of ways the event can possibly happen.

$$\textbf{Probability of event} = \frac{\textbf{Number of \textit{successful} outcomes}}{\textbf{Number of \textit{possible} outcomes}}$$

Probabilities may be expressed as decimals or percents as well as fractions.

Example 1. What is the probability of getting heads when tossing a coin?

Solution: There is just *one successful* outcome: heads. There are *two possible* outcomes: heads or tails.

$$\frac{\text{Number of successful outcomes}}{\text{Number of possible outcomes}} = \frac{1}{2}$$

Answer: $\frac{1}{2}$ or .5 or 50%

In a toss of a coin, there is a 50% probability of getting heads (and a matching 50% probability of getting tails). A common expression is that there is a fifty-fifty chance of getting heads.

Example 2. What is the probability that in the spinner shown in the diagram, the pointer will land on an even number?

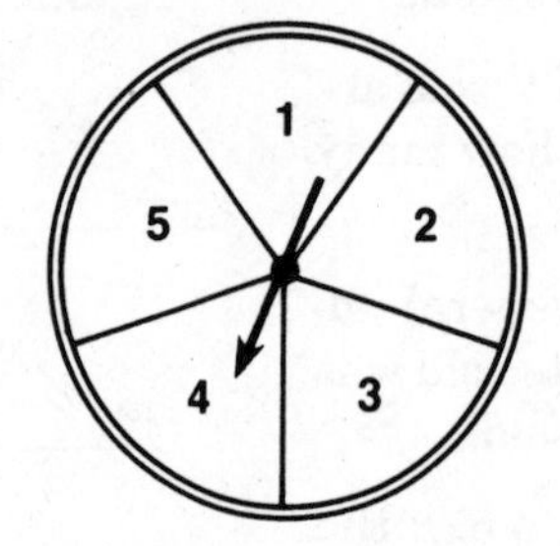

Solution:
There are *two successful* outcomes: the even numbers 2 and 4. There are *five possible* outcomes: 1, 2, 3, 4, and 5.

$$\frac{\text{Number of successful outcomes}}{\text{Number of possible outcomes}} = \frac{2}{5}$$

Answer: $\frac{2}{5}$ or .4 or 40%

Sometimes you must think carefully to determine the number of possible outcomes. Suppose you were to look at the sum of the dots facing up when you roll a pair of dice.

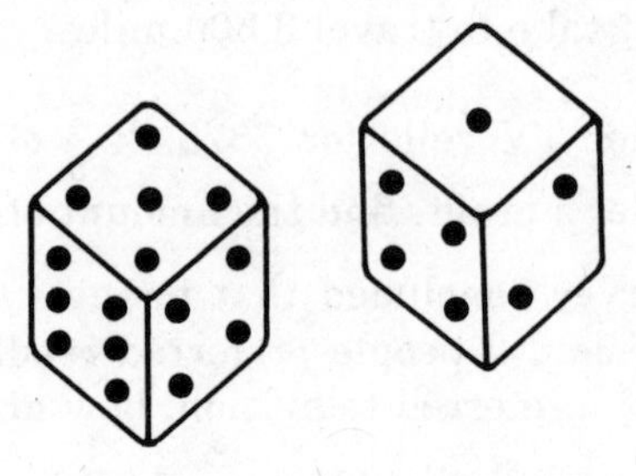

Each die (singular of dice) has six sides, and there are 1 to 6 dots on a side. Since you can get any number from 1 to 6 on the first die, and any number from 1 to 6 on the second die, you can get a sum of 2 or 3 or 4, or any sum up to 12. That makes eleven possible sums, but that is not the total number of possible outcomes.

From the following chart, you can see that there are 36 possible outcomes.

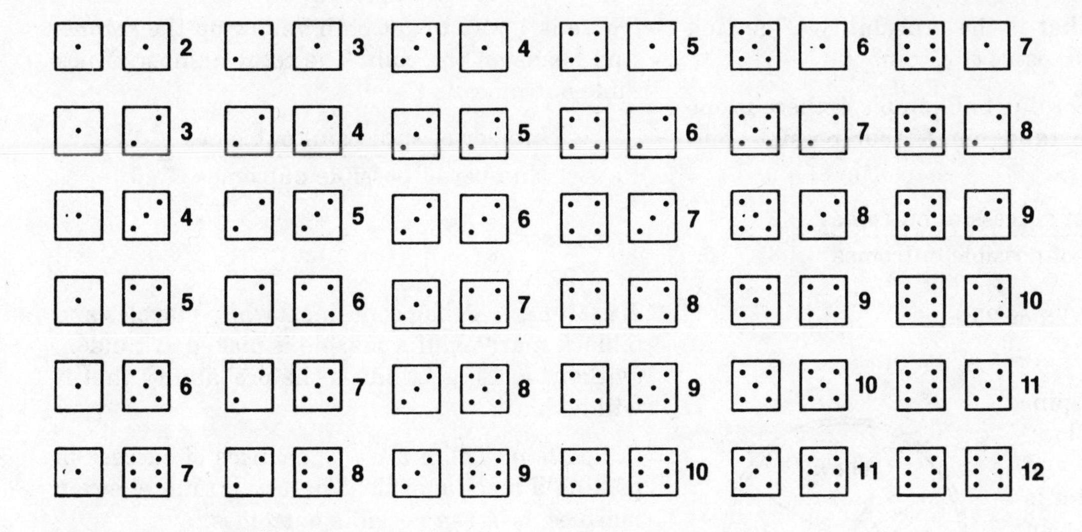

Example 3. What is the probability of rolling a sum of 5 with a pair of dice?

Solution: There are *four* ways to get a sum of 5: 1 + 4, 2 + 3, 3 + 2, and 4 + 1. The chart shows a total of 36 possible outcomes.

$$\frac{\text{Number of successful outcomes}}{\text{Number of possible outcomes}} = \frac{4}{36} = \frac{1}{9}$$

Answer: $\frac{1}{9}$

Three ways to be sure of getting a complete set of possible outcomes are by using a chart, a list, or a tree diagram.

Example 4. Using H for heads and T for tails, show all the possible outcomes of tossing a coin twice:

(*a*) by making a chart.

(*b*) by writing a list.

(*c*) by drawing a tree diagram.

Solution:

(*a*) Make a chart. Pair outcomes as indicated by the arrows.

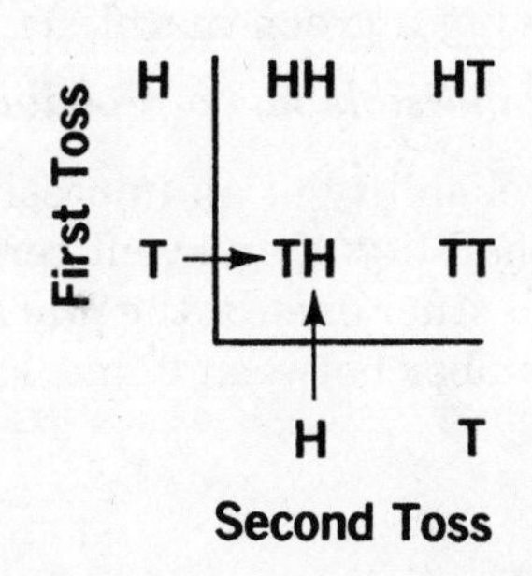

(*b*) Write a list.

HH

HT

TH

TT

(*c*) Draw a tree diagram.

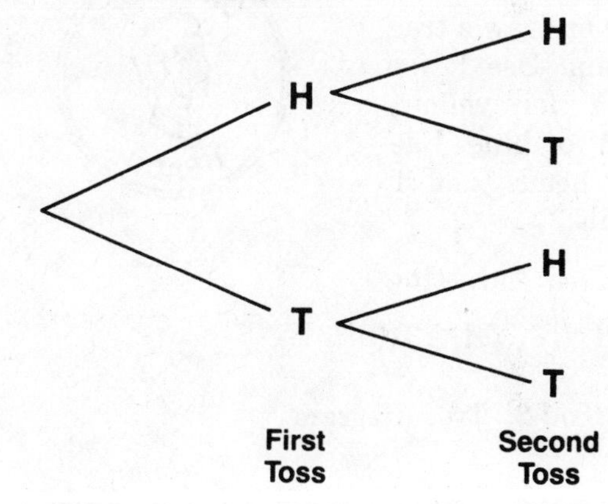

Read the tree diagram by following each branch from left to right.

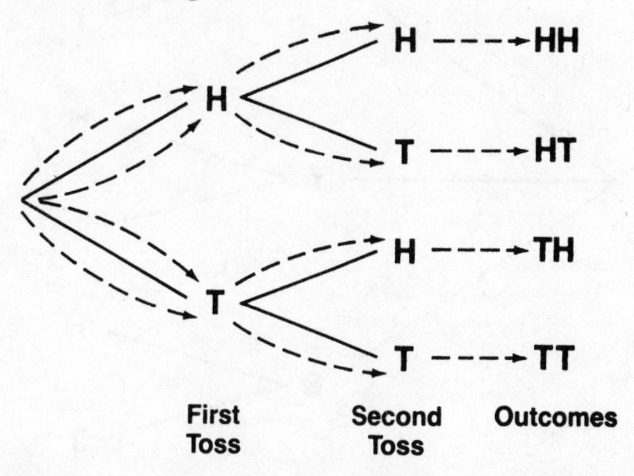

Answer: The possible outcomes are HH, HT, TH, and TT.

Example 5. What is the probability of getting both tails on two tosses of a coin?

Solution: According to Example 4, there is one way to get both tails, out of four possible outcomes.

$$\frac{\text{Number of successful outcomes}}{\text{Number of possible outcomes}} = \frac{1}{4}$$

Answer: $\frac{1}{4}$ or .25 or 25%

Example 6. A spinner has three colors, red, yellow, and blue. If the pointer is spun, and then a coin is tossed, what is the probability of getting yellow on the spinner and heads on the coin?

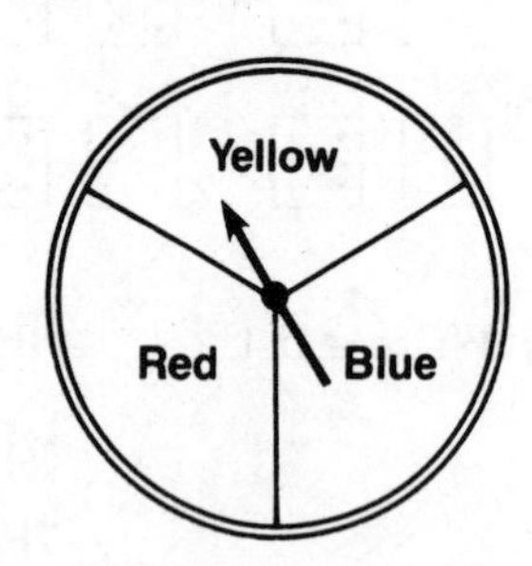

Solution: List all the possible outcomes or draw a tree diagram. Use R for red, Y for yellow, and B for blue. Use H for heads, and T for tails.

Method 1: Listing

RH, RT, YH, YT, BH, BT

Method 2: Tree diagram

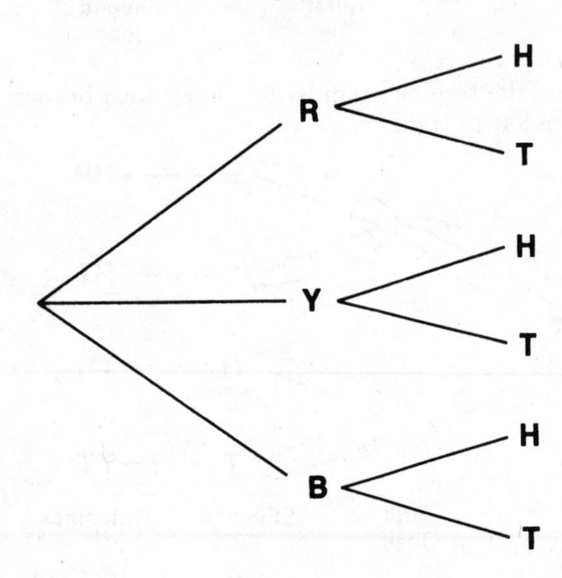

There is 1 way to get both yellow on the spinner and heads on the coin. The total number of possible outcomes is 6.

$$\frac{\text{Number of successful outcomes}}{\text{Number of possible outcomes}} = \frac{1}{6}$$

Answer: $\frac{1}{6}$

Example 7. A bag contains 2 white marbles and 5 black marbles. If a marble is picked at random (without looking), what is the probability that it will be black?

Solution: There are *five* chances of success in picking a black marble. There is a total of seven marbles, or *seven* possible outcomes.

$$\frac{\text{Number of successful outcomes}}{\text{Number of possible outcomes}} = \frac{5}{7}$$

Answer: $\frac{5}{7}$

Sometimes it is convenient to take a probability expressed as a percent and give it as an equivalent ratio.

Example 8. If there is a 25% chance of rain, what is the probability ratio?

Solution: Write 25% as a fraction.

$$25\% = \frac{25}{100} = \frac{1}{4}$$

Answer: The probability is $\frac{1}{4}$. There is 1 chance out of 4 that it will rain.

Some events are either certain or impossible. If a bag contains only 8 red marbles, the probability of drawing a red marble is $\frac{8}{8}$, or 1. The event of drawing a red marble is certain to occur because no other result is possible. *Any event that is certain to occur has a probability of 1.*

From the same bag of marbles, the event of drawing a green marble is impossible. The probability of drawing a green marble is $\frac{0}{8}$, or 0. *Any event that is impossible has a probability of 0.*

Thus, the probability of an impossible event is 0, and the probability of an event certain to happen is 1. For all other events, the probability ratio is always a number between 0 and 1.

APPLICATION PROBLEMS

In 1–5, write answers in fraction form.

1. A die numbered from 1 to 6 is rolled. Find the probability that the side facing up will show the following number of dots:

a. 4 dots ________ *b.* an odd number ________

c. an even number ________ *d.* a number less than 5 ________

e. an even number greater than 3 ________ *f.* an odd number less than 5 ________

g. an odd number less than 1 ________ *h.* a number between 0 and 7 ________

2. A spinner is divided into sections numbered from 1 to 8, not in order. Half the sections are shaded, as shown. The pointer is spun. Find the probability that it will land on:

a. an odd number ________

b. a shaded section ________

c. an unshaded even number ________

d. a shaded prime number ________

e. a number greater than 10 ________

f. a number less than 10 ________

g. an unshaded number less than 10 ________

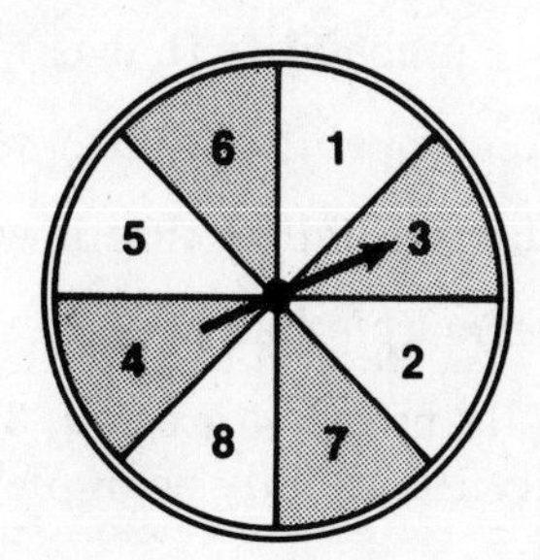

3. A coin is tossed 3 times.

a. Show all the possible outcomes by drawing a tree diagram.

b. Write a listing of all the possible outcomes.

____

c. Determine the probability of getting:

(1) all heads ________

(2) 2 heads and then 1 tail, in that order ________

(3) 2 heads and 1 tail, in any order ________

(4) no heads ________

4. A pair of dice is rolled. Find the probability that the sum of the dots showing on top will be:

a. 12 ________ *b.* 7 ________ *c.* an even number greater than 7 ________

d. an odd number less than 4 ________ *e.* less than 20 ________

5. A pair of dice is rolled once.

a. What sum is most likely to occur? __________

b. What two sums are least likely to occur? __________

c. The probability that the dice will show the sum in part *a* is __________ times as great as the probability of getting either of the sums in part *b*. (Fill in.)

In 6–8, write answers in both decimal and percent form.

6. A question on a multiple-choice test has a choice of 4 answers.

a. What is the probability of guessing the right answer? __________

b. What is the probability of not guessing the right answer? __________

7. A bag contains 3 red marbles, 5 green marbles, and 2 black marbles. One marble is picked at random.

a. What is the probability that it will be black? __________

b. What is the probability that it will be red? __________

c. What is the probability that it will be green? __________

d. What color marble has the best chance of being picked? __________

8. In a bag of jelly beans, 4 are red, 3 are black, 6 are yellow, 2 are purple, and 5 are green. Find the probability that a jelly bean picked at random:

a. will be purple __________ *b.* will be yellow __________ *c.* will be green __________

d. will not be red __________ *e.* will be white __________

In 9–10, write the percent prediction as a fraction in lowest terms, and fill in the blanks.

9. On an early-morning weather report, the forecaster says there is a 60% chance of rain. There are __________ chances out of __________ that it will rain.

10. Ms. Jones believes that Alex has an 80% chance of passing the math test. There are __________ chances out of __________ that Alex will pass.

UNIT 82. Scale Drawings

When architects draw the floor plans of houses they have designed, it is impossible to make the drawing the same size as the actual house. Instead, they make a **scale drawing** that shows the exact shape of the house, but with greatly reduced dimensions. They may let $\frac{1}{4}$ inch on the plan represent 1 foot of the actual size of the house. They then reduce all actual dimensions *proportionally,* according to the same **scale.** A scale is usually shown with an equals sign: $\frac{1}{4}$ in. = 1 ft.

The first term in the scale always represents a measurement on the drawing. The second term represents the actual distance represented by the first term. The second term is often called the "equivalent measurement."

When the first term is smaller than the second term, you know that the drawing is smaller than the actual object being represented.

There are times when the actual object is extremely small, as with certain machine parts. In this case, the scale drawing will be larger than the object being represented. The engineer who is making the scale drawing may let 1 inch on the plans represent $\frac{1}{10}$ of an inch of the actual size of the machine part. All the dimensions are then enlarged by this ratio. The scale would appear on the plans as: 1 in. = $\frac{1}{10}$ in.

Since the first term (which always represents a measurement on the drawing) is larger than the second term, you know that the drawing is larger than the actual object.

Example 1. On a floor plan, the length of a room is 3 inches. If the scale is $\frac{1}{4}$ in. = 1 ft., what is the actual length of the room?

Solution: Write a proportion relating the quantities: $\frac{1}{4}$ inch is to 1 foot as 3 inches is to ? feet, or

$$\frac{\frac{1}{4}\text{ in.}}{1\text{ ft.}} = \frac{3\text{ in.}}{?\text{ ft.}}$$

To solve, use the method of multiplying the means and multiplying the extremes.

$$1 \times 3 = \frac{1}{4} \times ?$$

$$3 = \frac{1}{4} \times ?$$

Use division to undo the multiplication.

$$? = 3 \div \frac{1}{4} = \frac{3}{1} \times \frac{4}{1} = 12$$

Answer: The actual length of the room is 12 feet.

Example 2. A schoolyard measures 96 feet by 168 feet. Find the scale-drawing size of the yard on a plan whose scale appears as: $\frac{1}{4}$ in. = 3 ft.

Solution: Write and solve a proportion for each measurement. First, find the scale-drawing measurement corresponding to 96 feet.

$$\frac{\frac{1}{4}\text{ in.}}{3\text{ ft.}} = \frac{?\text{ in.}}{96\text{ ft.}}$$

$$3 \times ? = \frac{1}{4} \times 96$$

$$3 \times ? = 24$$

$$? = 24 \div 3 = 8$$

Now find the scale-drawing measurement corresponding to 168 feet.

$$\frac{\frac{1}{4}\text{ in.}}{3\text{ ft.}} = \frac{?\text{ in.}}{168\text{ ft.}}$$

$$3 \times ? = \frac{1}{4} \times 168$$

$$3 \times ? = 42$$

$$? = 42 \div 3 = 14$$

Answer: On the plan, the scale-drawing size of the yard is 8 inches × 14 inches.

Example 3. On the blueprint of a machine, the diameter of a certain bolt measures $1\frac{1}{2}$ in. If the scale is $\frac{1}{2}$ in. = 3 mm, what is the actual diameter of the bolt?

Solution:

$$\frac{\frac{1}{2}\text{ in.}}{3\text{ mm}} = \frac{1\frac{1}{2}\text{ in.}}{?\text{ mm}}$$

$$3 \times 1\frac{1}{2} = \frac{1}{2} \times ?$$

$$\frac{1}{2} \times ? = 3 \times 1\frac{1}{2} = \frac{3}{1} \times \frac{3}{2} = \frac{9}{2}$$

$$? = \frac{9}{2} \div \frac{1}{2} = \frac{9}{\cancel{2}_1} \times \frac{\cancel{2}^1}{1} = 9$$

Answer: The diameter is 9 mm.

EXAMPLE 4. The scale on a map shows that $\frac{1}{2}$ centimeter equals 35 kilometers. What measurement on the map will represent 175 kilometers?

Solution:

$$\frac{\frac{1}{2}\text{ cm}}{35\text{ km}} = \frac{?\text{ cm}}{175\text{ km}}$$

$$35 \times ? = 175 \times \frac{1}{2} = \frac{175}{2}$$

$$? = \frac{175}{2} \div 35 = \frac{\overset{5}{\cancel{175}}}{2} \times \frac{1}{\underset{1}{\cancel{35}}}$$

$$= \frac{5}{2} = 2\frac{1}{2}$$

Answer: $2\frac{1}{2}$ centimeters on the map will represent 175 kilometers.

EXERCISES

In 1–10, find the actual measurements or distances.

	Scale	Measurement on Scale Drawing	Actual Measurement or Distance
1.	$\frac{1}{4}$ in. = 25 mi.	$3\frac{1}{2}$ in.	
2.	$\frac{1}{8}$ in. = 4 ft.	$1\frac{3}{4}$ in.	
3.	1 in. = 20 mi.	$4\frac{5}{8}$ in.	
4.	$\frac{1}{8}$ in. = 1 ft.	$6\frac{1}{2}$ in.	
5.	1 in. = $\frac{1}{8}$ in.	6 in.	
6.	1 in. = 40 mi.	$6\frac{3}{4}$ in.	
7.	1 in. = 1 in.	5 in.	
8.	$\frac{1}{16}$ in. = 5 ft.	$3\frac{1}{4}$ in.	
9.	$\frac{3}{4}$ cm = 150 km	$5\frac{3}{4}$ cm	
10.	$\frac{1}{2}$ cm = 5 mm	$3\frac{1}{2}$ cm	

In 11–20, find the measurement on the scale drawing.

	Scale	Measurement on Scale Drawing	Actual Measurement or Distance
11.	$\frac{1}{8}$ in. = 60 mi.		960 mi.
12.	$\frac{1}{2}$ cm = 3 mm		45 mm
13.	$\frac{1}{4}$ in. = 6 ft.		30 ft.
14.	$\frac{1}{8}$ cm = 85 km		1,360 km
15.	1 in. = $\frac{1}{2}$ in.		$12\frac{1}{2}$ in.
16.	$\frac{1}{16}$ in. = 4 yd.		96 yd.
17.	1 in. = 60 mi.		345 mi.
18.	$\frac{1}{4}$ in. = 18 in.		270 in.
19.	$\frac{1}{16}$ in. = $\frac{3}{4}$ in.		18 in.
20.	$\frac{1}{8}$ in. = 1 ft.		$16\frac{1}{2}$ ft.

APPLICATION PROBLEMS

1. What are the distances represented by line segments *a*, *b*, and *c* if the scale is $\frac{1}{4}$ in. = 15 mi.?

a. ______ *b.* ______

c. ______

a. ______
b. ______
c. ______

2. A garage measures 22 ft. × 18 ft. Give the dimensions of the scale drawing of the garage, using the scale: $\frac{1}{4}$ in. = 2 ft. ______

3. The scale on a map is $\frac{3}{4}$ in. = 350 mi. How many miles are represented by $4\frac{1}{2}$ inches? ______

4. The distance between two cities is 1,125 kilometers. If the scale on a map is $\frac{1}{4}$ cm = 75 km, what will be the distance on the map between the two cities? ______

5. The blueprint of a small gear has a scale of $\frac{1}{2}$ in. = $\frac{3}{32}$ in. If the width of the gear is $\frac{9}{32}$ in., what will the width on the blueprint be? ____________

6. A floor plan has a scale of $\frac{1}{4}$ in. = 2 ft. What are the floor-plan measurements of a room measuring 10 feet by 28 feet? ____________

7. A plan of a house uses a scale of $\frac{1}{2}$ cm = 2 m. How many meters are in the length and width of a backyard that measures 3 cm by 2 cm on the drawing? ____________

8. The scale on a blueprint is 1 in. = 3 ft. What length on the blueprint would represent 27 feet? ____________

9. In a photograph, Jack's height measures $4\frac{1}{2}$ inches, and his sister's height measures 3 inches. If Jack is really 6 feet tall, how tall is his sister? ____________

10. If a scale is given as the ratio 1 : 5, what would be the scale size of a room measuring 15 feet by 25 feet? ____________

Review of Part XVII (Units 79–82)

In 1–6, find the missing number that will make each expression a true proportion. Use the equivalent ratio method.

1. $\frac{4}{5} = \frac{?}{45}$ **2.** $\frac{?}{60} = \frac{3}{12}$ **3.** $\frac{12}{13} = \frac{24}{?}$

4. 3 : 13 = ? : 39 **5.** ? : 48 = 5 : 12 **6.** 8 : 9 = 56 : ?

In 7–12, find the missing number. Use the proportion method.

7. $\frac{3}{11} = \frac{?}{2{,}695}$ **8.** $\frac{5}{9} = \frac{1{,}315}{?}$ **9.** $\frac{6}{?} = \frac{1{,}326}{1{,}547}$

10. 4 : ? = 920 : 2,070 **11.** 7 : 15 = ? : 1,650 **12.** 7 : 12 = 2,275 : ?

In 13–17, find the missing measurements.

	Scale	Measurement on Scale Drawing	Actual Measurement or Distance
13.	$\frac{1}{4}$ in. = 12 ft.	$2\frac{1}{4}$ in.	
14.	$\frac{1}{2}$ cm = 75 km		525 km
15.	$\frac{1}{2}$ in. = $\frac{1}{16}$ in.		$1\frac{3}{16}$ in.
16.	$\frac{1}{8}$ in. = 3 ft.	$5\frac{3}{4}$ in.	
17.	1 in. = 150 mi.		675 mi.

18. George Foster has two pairs of jeans, one blue and one green, and four shirts, a checked red, a checked yellow, a solid white, and a solid purple.

a. Draw a tree diagram to show all the outcomes of choosing a pair of jeans and a shirt.

b. If a shirt and a pair of jeans are picked at random, find the probability of getting:

(1) blue jeans and a white shirt

(2) green jeans and a solid color shirt

(3) gray jeans and a pink shirt

(4) blue jeans and a checked shirt

(5) green jeans and any shirt

(6) either pair of jeans and a purple shirt

PART XVIII. Statistical Averages

UNIT 83. Finding the Mean

WORDS TO KNOW

The word **average** is used in everyday language to mean "in between." An "average student" in a certain school is somewhere between the best student and the worst student. The "average income" of an office worker is neither the highest income nor the lowest income of a group of office workers.

The words **arithmetic average,** or **mean,** have a special mathematical meaning. Suppose you say, "My math tests average 85%." You mean that you added together all of your mathematics test grades, divided this sum by the number of grades, and got a result of 85%.

To find the mean of a group of numbers:

Step 1: Add all the numbers in the group.

Step 2: Divide the sum obtained in step 1 by the number of items in the group.

EXAMPLE. Find the mean of the following group of test marks:

75% 90% 80% 65% 70%

Step 1: Add all the numbers in the group.

$75 + 90 + 80 + 65 + 70 = 380$

Step 2: Since there are 5 marks in the group, divide the sum, 380, by 5.

$$\begin{array}{r} 76 \\ 5\overline{)380} \\ \underline{35} \\ 30 \\ \underline{30} \end{array}$$

Answer: The mean of all the test marks is 76%.

EXERCISES

In 1–6, find the mean of each group of numbers. Answers that are not whole numbers may be given as mixed numbers.

1. 35 46 27 40

2. 87 93 76 80 74

3. 158 210 163 185

4. 84 75 60 90 80

5. 330 365 390 380

6. 12 28 15 14 19 17 19

In 7–8, find the mean in lowest terms.

7. $3\frac{1}{4}$ $5\frac{3}{4}$ $2\frac{1}{2}$ $1\frac{5}{6}$

8. $1\frac{3}{10}$ $2\frac{1}{3}$ $\frac{4}{5}$ $3\frac{2}{5}$ $1\frac{2}{3}$

In 9–10, find the mean, to the nearest *tenth*.

9. 23.25 17.9 20.55 16.35

10. 437.2 293.841 51.38

APPLICATION PROBLEMS

SAMPLE SOLUTION

In a football game, halfback Terry Reid made the following gains: 5 yards, 1 yard, 8 yards, 2 yards, 5 yards, 21 yards. What was his mean gain per carry?

$5 + 1 + 8 + 2 + 5 + 21 = 42$

$42 \div 6 = 7$

7 yd.

1. A salesperson traveled the following numbers of miles: Monday, 87; Tuesday, 115; Wednesday, 95; Thursday, 130; Friday, 110. What is the mean number of miles he traveled on any one day? ________

2. During the 21 school days last March, the total attendance in a school was 23,688. What was the mean daily attendance? ________

3. In preparing for her vacation, Mary bought five dresses at the following prices: \$43, \$29, \$52, \$26, \$35. What was the mean price? ________

4. John's car traveled 810 miles using 45 gallons of gasoline. What was the mean number of miles per gallon? ________

5. The weights in pounds of the seven linemen of the Centerville Chiefs football team are as follows: 198, 179, 210, 220, 189, 205, 213. What is the mean weight of a Centerville lineman? ______

6. The six art classes in a school have the following numbers of students: 32, 26, 34, 22, 33, and 27. What is the mean class size? ______

7. A salesperson earned the following commissions in the last four weeks: $168 $194 $216 $186
What is her mean weekly commission? ______

8. Last week, the following temperatures were recorded in Dade City: 79°, 81°, 83°, 85°, 84°, 85°, and 82°. What was the mean temperature? ______

9. The mean weekly salary of five employees is $382. If four of the employees earn salaries of $350, $420, $385, and $365, what is the salary of the fifth employee? (*Hint:* If the mean salary is $382, then the total weekly salary for the five employees is $382 × 5.) ______

10. Joel worked five days last week and earned a mean daily salary of $62. If on four of the days he earned $45, $74, $66, and $52, how much did he earn on the fifth day? ______

UNIT 84. Finding the Median

WORDS TO KNOW

A numerical average is a number that is representative of a set of facts, or **data.** It is a *measure of central tendency,* a value in the center of a set of related data.

Often the *mean* best represents the data. At other times, a better value is the **median.** The median is the *middle number* when the data are arranged in order, from the smallest to the largest. Half the values are below it, and half the values are above. The median is often useful when the data relate to rank or position.

To find the median of a group of numbers:

Step 1: Arrange all the numbers in the group in numerical order.

Step 2: Locate the number(s) midway in the group.

Step 3: In an even-numbered group, find the average (mean) of the two middle numbers.

EXAMPLE 1. Find the median of the following set of numbers:

$65, $80, $45, $60, $55

Solution:

Step 1: Arrange the numbers in order:

$45 $55 $60 $65 $80

Step 2: Find the middle number.

$45 $55 ($60) $65 $80

$60 is located midway in the odd-numbered set.

Answer: The median is $60.

Example 2. Find the median of the following set of numbers:

$65, $80, $45, $60, $55, $75

Solution:

Step 1: Arrange the numbers in order:

$45 $55 $60 $65 $75 $80

Step 2: Find the middle numbers.

$45 $55 ($60 $65) $75 $80

Step 3: Find the mean of the two middle numbers.

$$\frac{\$60 + \$65}{2} = \$62.50$$

Answer: The median is $62.50.

EXERCISES

In 1–12, find the median for each set of data. Give non-integer results to the nearest *tenth.*

1. 165 135 150 140 165 170 135 165

2. 60 90 50 80 70 60 90 60 80

3. 250 285 265 255 275 265 270 280

4. 20 15 25 35 25 30 20 25 15

5. 300 700 400 600 500 800 500 900

6. $25 $50 $35 $45 $55 $35 $40 $30

7. 420 455 440 435 450 425 450 440 450

8. 15 30 25 45 20 50 35 40 25 30 25

9. 245 300 255 285 295 280 265 280

10. 500 1,100 800 500 1,000 700 900 600

11. 253.7 194.8 158.4 132.9

12. 36.27 45.9 29.23 20.345

In 13–14, find the median. Give answers in lowest terms.

13. $17\frac{1}{8}$ $14\frac{2}{8}$ $15\frac{5}{8}$ $16\frac{4}{8}$ $19\frac{3}{8}$

14. $24\frac{1}{2}$ $23\frac{1}{3}$ $25\frac{1}{4}$ $22\frac{3}{5}$

In 15–16, you are given the median and all but one of the values in a set of data. Choose the value, A, B, or C, that could be the missing value.

15. The median is 15. The values are 5, 8, 15, 24, __?__ . The missing value is
(A) 10 (B) 14 (C) 20

16. The median is 87. The values are 88, 83, 86, __?__ . The missing value is
(A) 90 (B) 85 (C) 80

APPLICATION PROBLEMS

1. The ages of the children in the Adams family are 1, 2, 4, 7, 8, 11, 12, and 13. What is the median age? ____

2. The heights of the members of a basketball team are 189 cm, 213 cm, 221 cm, 178 cm, and 201 cm. What is the median height? ____

3. In one week, Stan jogged the following distances, in kilometers. What was the median distance? ____

Sun.	Mon.	Tues.	Wed.	Thurs.	Fri.	Sat.
2.5	3.7	4.2	5.4	3.2	4.5	3.8

In 4–5, choose A, B, or C.

4. Sue's median grade for five courses was 81. Four of her grades were 70, 80, 81, and 94. The fifth grade could have been
(A) 75 (B) 80 (C) 85 ____

5. The median hourly wage of six assistant cooks is $8.35. The wages of five of the cooks are $6.75, $8.00, $8.45, $9.50, and $10.05. The hourly rate of the sixth cook must be
(A) less than $8.35 (B) $8.35 (C) more than $8.35 ____

UNIT 85. Finding the Mode

WORDS TO KNOW

Another measure of central tendency is the **mode.** For a set of data, the mode is simply the value that occurs *most often.* It is useful when items are important because of their frequency of occurrence.

To find the mode of a group of numbers:

Step 1: Tally the numbers in the given set.

Step 2: Find the number that appears most frequently.

EXAMPLE. Find the mode of the following set of numbers: 15, 35, 10, 20, 15, 25, 15, 20

Solution:

Step 1:

15, 35, 10, 20, 15, 25, 15, 20 (8 items)
||| | | || | *(8 tally marks)*

Step 2: 15 occurs most frequently.

Answer: The mode is 15.

It is possible to have more than one mode for a set of data. It is also possible to have no mode:

The set of numbers 3, 4, 3, 3, 4, 4 has two modes, 3 and 4.

The set of numbers 5, 6, 7, 8, 9 has no mode.

EXERCISES

In 1–6, find the mode(s), if any.

1. 1 2 3 4 5 6 7

2. 1 5 2 5 3 5 4 5

3. 15 6 7 8 7 11 15 9

4. 19 18 17 18 17 16 17 16 15

5. 1 2 1 2 1 2 1 2

6. 25.3 32.1 15.24 25.3 22.2 19.65

APPLICATION PROBLEMS

1. One Saturday morning, Shelley's Shoe Shop sold shoes of the following sizes: 9, $8\frac{1}{2}$, $6\frac{1}{2}$, $7\frac{1}{2}$, 10, 8, $7\frac{1}{2}$, 9, $7\frac{1}{2}$, and 7. What size is the mode? __________

2. Six employees of the Apex Machine Shop commute the following distances to work: 35.1 km, 22.4 km, 27.3 km, 31.9 km, 28.5 km, and 22.4 km. Which distance is the mode? __________

3. Agustin picked five ripe tomatoes off one plant. Their weights were .19 kg, .12 kg, .09 kg, .23 kg, and .15 kg. Find the mode weight(s), if any. __________

4. Lucy kept a record of all her test grades in one report card period. They were 74%, 82%, 91%, 68%, 82%, 93%, 60%, and 74%. Find the mode grade(s), if any. __________

5. The table below shows an age breakdown of the students in Chestnut Street School. What age is the mode? __________

Age	12	13	14	15	16	17	18	19
Number of Students	3	18	26	28	25	22	17	2

UNIT 86. Comparing the Measures of Central Tendency

In the preceding units, you have learned about three different kinds of measure, the *mean*, the *median*, and the *mode*. A good measure is one that is "typical" of a set of data. Deciding which measure to use will depend on the given data, and which measure will give the most useful information.

The mean is the measure that is most familiar. Students use the mean when averaging their grades. Sometimes, however, when the data are not evenly distributed, the mean may be misleading, and the median or the mode will be more typical of the data.

Example 1. The annual salaries of all the people working at the Palmer Company are $86,780; $15,475; $14,500; $12,225; $13,850; $14,825; and $12,900. Which measure—the mean, the median, or the mode—would best represent the salary of an average employee?

Solution: One salary, $86,780, is much greater than any of the others. (It is the salary of the president of the company, Mr. Palmer.) It raises the mean and makes the value of the mean too high to be a representative average. The *mean* of $24,365 is not typical of the salaries in the group.

No salary appears more often than the others. There is no *mode*.

Let us examine the median.

$12,225 $12,900 $13,850 ($14,500) $14,825 $15,475 $86,780

The *median*, $14,500, is a good representative of most of the salaries.

Answer: The median best represents the salary of an average employee.

Example 2. Marvin's Men's Mart ran a one-day sale on a particular model of jacket. The sales made were sizes 36, 44, 46, 36, 40, 36, 42, 38, 44, and 36. Before ordering more jackets, the buyer checked with a salesperson on the average size sold. What should the salesperson tell him?

Solution: The *mean* value is 39.8, which is not a jacket size.

There is an even number of values. After arranging the sizes in order, we get the median by finding the mean average of the two middle numbers, 38 and 40. The *median* value is 39, which is not one of the jacket sizes.

The *mode*, the size sold most often, was size 36. In this case, the mode is the best average.

Answer: The average size sold was size 36.

For a given set of data, it can happen that two, or even three, of the measures have the same value.

Example 3. Find the mean, the median, and the mode of the following data:

41 45 37 44 38 41

Solution:

Find the mean:

$$\frac{41 + 45 + 37 + 44 + 38 + 41}{6} = \frac{246}{6} = 41$$

Find the median:

37 38 (41 41) 44 45

$$\frac{41 + 41}{2} = \frac{82}{2} = 41$$

Find the mode:

41	45	37	44	38	41
//	/	/	/	/	

The value 41 occurs most often.

Answer: The mean, the median, and the mode all have the value 41.

EXERCISES

In 1–6, tell whether the data are best represented by the mean, the median, or the mode, and find the value of the measure chosen.

1. 3 1 59 5 2

2. 104 100 2 98

3. 5 5 5 9 5 4

4. 72 70 74 82 80 78

5. 45 15 45 45

6. 120 124 123 130 127 126

In 7–15, find the mean, the median, and the mode. (If no value occurs more frequently than the others, then there is no mode.) Give non-integer answers to the nearest *tenth*.

		Mean	Median	Mode
7.	Test grades: 85% 90% 70% 80%			
8.	Ages: 18 15 19 16 17 18			
9.	Weights: 165 kg 180 kg 195 kg 170 kg 150 kg			
10.	Temperatures: 87 93 89 82 80			
11.	84 63 95 78 86 65 75 72 65			
12.	300 800 500 700 400 800 500 600 500			
13.	65 95 75 80 70 85 70 90			
14.	1,500 2,500 1,800 3,000 1,600 2,300 1,800			
15.	20 70 40 60 30 20 50 80 70 50 70 50			

APPLICATION PROBLEMS

In 1–5, find non-integer answers to the nearest tenth.

1. Liza did 27 sit-ups, Clarissa did 19, Sissy 15, Tina 29, and Jill 22.

a. Find the mean number of sit-ups. ____________

b. Find the median. ____________

2. The local Boy Scout troops competed to see which could collect the most newspapers for recycling. Troop 1 collected 153 pounds of newspapers; Troop 2, 98 pounds; Troop 3, 129 pounds; and Troop 4, 216 pounds.

a. Find the mean number of pounds. ____________

b. Find the median number of pounds. ____________

3. Jennifer's report card grades are 82 in English, 75 in Social Studies, 85 in Italian, 78 in Mathematics, and 65 in Biology.

a. What is the mean grade? ____________

b. What is the median grade? ____________

4. In February, the Boonetown Library received reserve requests as follows:

Title	Number of Requests
Tom, Dick, and Harry	43
Computers Made Easy	27
Haunted Nights	55
A Summer Romance	30
Cooking Natural Foods	21
Murder in Red	27

a. Find the mean number of requests. ____________

b. Find the median. ____________

c. Find the mode. ____________

5. A mail-order house placed an ad with an order coupon for tool sets in five newspapers, resulting in the following numbers of orders:

Newspaper	Number of Orders
The Daily Bugle	153
The River City Record	171
The Valley Voice	94
The Tribune	207
The News Clarion	235

a. What is the mean? ____________

b. What is the median? ____________

Review of Part XVIII (Units 83–86)

In 1–5, find the *mean*, the *median*, and the *mode*. Give non-integer answers to the nearest *tenth*.

		Mean	Median	Mode
1.	130 190 140 180 150 170 140			
2.	60 95 75 80 65 100 90 70 80 85			
3.	200 800 500 400 700 900 200 400 600 400			
4.	185 230 195 235 275 240 235 245 270			
5.	453 515 465 525 476 483 465 518 468			

In 6–7, find the *mean*, the *median*, and the *mode*. Give answers in lowest terms.

		Mean	Median	Mode
6.	$36 \quad 24\frac{1}{2} \quad 20\frac{1}{2} \quad 24\frac{1}{2}$			
7.	$12\frac{1}{2} \quad 10\frac{1}{4} \quad 11\frac{1}{2} \quad 9\frac{3}{4} \quad 10\frac{1}{4}$			

In 8–10, choose A, B, or C.

8. For which set of data do the mean, the median, and the mode all have the same value?
(A) 82 80 84 85 84 (B) 83 80 83 86 (C) 84 83 77 85 86 83 ________

9. For which set of data is the median greater than the mean?
(A) 110 114 119 119 113 (B) 96 11 95 98 (C) 128 121 130 120 126 ________

10. For which set of data is the mode greater than the median?
(A) 3 6 7 4 7 (B) 3 7 2 5 3 6 (C) 5 5 5 5 5 ________

11. The school nurse recorded the following heights of a group of second-grade children: 120.3 cm, 132.8 cm, 127.0 cm, 124.4 cm, 121.2 cm, and 118.5 cm. Find the mean height, to the nearest *tenth*. ________

12. In the first five months after moving to a new job in a different city, Janice had the following telephone charges: March, $44.83; April, $12.65; May, $13.12; June, $15.74; and July, $14.36.

a. Which measure—the mean, the median, or the mode—would best represent the monthly telephone charges? ________

b. Find the value of the measure you chose in part *a*. ________

13. Scott had to lose weight to get in shape for the track team. He kept a chart of his monthly weight loss.

Sept.	Oct.	Nov.	Dec.	Jan.
18.3 kg	4.1 kg	3.2 kg	1.9 kg	3.8 kg

a. Which measure—the mean, the median, or the mode—would best represent the monthly weight loss? ________

b. Find the value of the measure you chose in part *a*. ________

14. Rick's bowling scores were 123, 115, 134, 142, and 155. Pat's scores were 147, 122, 131, 160, 135, and 110.

a. Find the mean score for each bowler, to the nearest *whole number.*

Rick: ____________

Pat: ____________

b. Who has the better average? ____________

15. The Lake Plains School District conducted a survey to determine the reading level of 6th-grade pupils, and tabulated the results.

School	Class	Reading Level (Class Mean)
High Street	6-1	5.8
	6-2	6.3
	6-3	6.0
Main Street	6-1	4.9
	6-2	5.2
	6-3	5.8
Elm Street	6-1	6.2
	6-2	5.9

Find, to the nearest *tenth:*

a. the mean reading level for each school ____________

b. the median reading level for each school ____________

c. the mean of all three schools ____________

d. the mode for all three schools ____________

16. Use a calculator for this problem.

A set of U.S. population census figures, by region, is shown in the table.

Region	Population
New England	12,348,493
Middle Atlantic	36,786,790
East North Central	41,682,217
West North Central	17,183,453
South Atlantic	36,959,123
East South Central	14,666,423
West South Central	23,746,816
Mountain	11,372,785
Pacific	31,799,705

a. For the nine regions, what is the mean population, to the nearest *whole number*? ____________

b. Which region has a population closest to the mean? ____________

c. What is the median? ____________

d. Which region has the median population? ____________

PART XIX. Statistical Graphs

Graphs are used to present numerical facts visually. When readers can see a picture that represents numbers or amounts, it is easy for them to understand these facts. Instead of reading paragraphs of information or puzzling over columns of figures, readers can see the facts in the graph and can make comparisons between them, almost at a glance.

UNIT 87. The Picture Graph

A **picture graph** (also called a **pictograph**) uses pictures to represent numerical facts. To make a picture graph, do the following:

Step 1: Gather the data.

Step 2: Round off the figures.

Step 3: Decide on an appropriate picture to represent the items in the graph. For example, use a picture of a pile of money for dollars. Also, decide how many units each picture will represent.

Step 4: Find out how many pictures are needed to represent each rounded-off number in Step 2.

Step 5: Draw the required number of pictures.

Step 6: Label the graph, including a description of what each picture represents.

Example. Construct a picture graph showing the schooling completed in the United States in 19-- by persons from 18 to 24 years old.

Step 1: Gather the data. (The figures in Table XV are from a recent census.)

Step 2: Round off the numbers to the nearest million.

Table XV: Schooling Completed by 18- to 24-Year-Olds in the United States in 19--

Schooling Completed	Number of Persons (nearest thousand)	Rounded to the nearest million
Less than 4 years of High School	6,344,000	6,000,000
High School (4 years)	13,158,000	13,000,000
College (1–3 years)	7,528,000	8,000,000
College (4 or more years)	2,018,000	2,000,000

Step 3: Decide on a symbol to represent 2,000,000 persons. 

Step 4: To find how many pictures are needed to represent each rounded-off number in Step 2, divide the rounded-off number by 2 million (the number of persons that each picture represents).

Schooling Completed	Number of Persons (nearest million)	Number of Pictures
Less than 4 years of High School	6,000,000	$\frac{6{,}000{,}000}{2{,}000{,}000} = 3$
High School (4 years)	13,000,000	$\frac{13{,}000{,}000}{2{,}000{,}000} = 6\frac{1}{2}$
College (1–3 years)	8,000,000	$\frac{8{,}000{,}000}{2{,}000{,}000} = 4$
College (4 or more years)	2,000,000	$\frac{2{,}000{,}000}{2{,}000{,}000} = 1$

Steps 5 and 6: Draw the graph and label it. The completed graph may look like this:

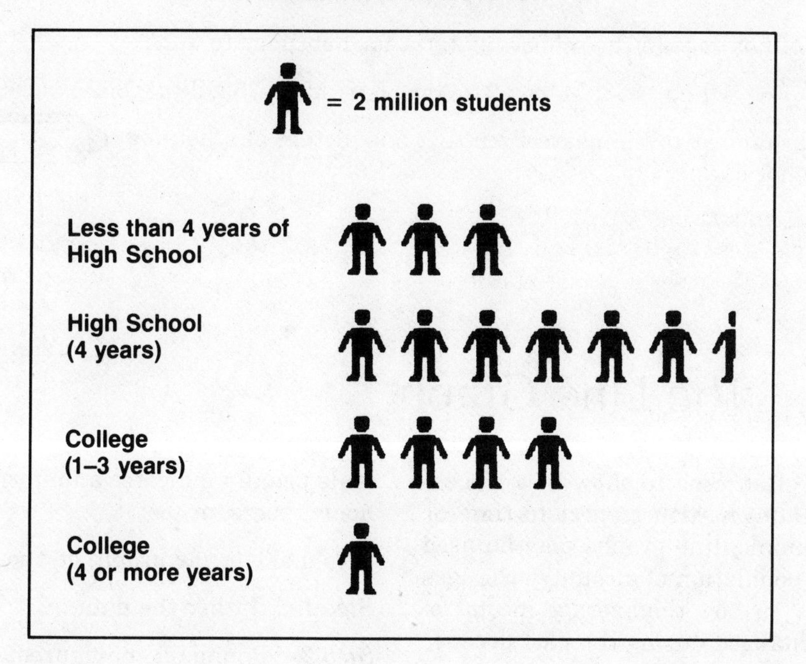

EXERCISES

Use the preceding graph to answer these exercises. Do *not* use the data in Table XV (page 316).

1. What level of education was completed by the greatest number of people? ________

2. What level of education was completed by the smallest number of people? ________

3. To the nearest million, how many people completed 4 years of high school? ______

4. Compare the number of people who completed 4 years of high school with those who had less than 4 years of high school. To the nearest million, how many more people completed 4 years of high school? ______

5. In which category are there about four times as many people as in the group of those completing 4 or more years of college? ______

APPLICATION PROBLEMS

In 1–2, draw and label a picture graph for the given data.

1. Apple harvesting at Lyman's Orchard over a period of 5 years:

   1st year: 1,840 bushels
   2nd year: 2,510 bushels
   3rd year: 3,020 bushels
   4th year: 2,480 bushels
   5th year: 3,230 bushels

   Round off each number to the nearest *hundred* bushels. Let one picture of an apple represent 500 bushels of apples.

2. Consumption of hot dogs in the school cafeteria for February to June:

   Feb., 353; Mar., 497; Apr., 548; May, 753; June, 802

   Round off each number to the nearest *ten.* Let one picture of a hot dog represent 100 hot dogs.

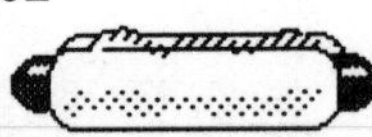

UNIT 88. The Line Graph

A **line graph** is often used to show how a measured quantity *changes* with respect to time or distance. For example, line graphs may be used to show how the population of a country changes from year to year, or how the average income of a certain group changed during the past decade.

A line graph will show at a glance the changes in the quantity being considered. When the line in the graph *rises,* the quantity *increases.* When the line *falls,* the quantity *decreases.* When the line is *horizontal,* the quantity is *not changing.*

A line graph uses two **scales,** which are lines drawn from the same point at right angles to each other. The vertical scale usually indicates the amount of the measured quantity, such as temperature, dollars, or population. The horizontal scale usually indicates a uniform change, such as hours, years, or feet.

To make a line graph, do the following:

Step 1: Gather the data.

Step 2: Round off the figures.

Step 3: Decide on appropriate lengths for both the vertical scale and the horizontal scale. Label the scales in uniform units. It is convenient to use graph paper.

Step 4: Using the two scales, draw a dot on the graph for each set of rounded-off figures in Step 2.

Step 5: Connect the dots, using straight lines.

Step 6: Label the graph.

EXAMPLE. Construct a line graph showing the population of the United States from 1900 to 1980, for every 10-year period.

Steps 1 and 2: The following information was obtained from the United States Census Bureau:

Table XVI: Population of the U.S.

Year	Population	Population to Nearest Ten Million
1900	75,994,575	80,000,000
1910	91,972,266	90,000,000
1920	105,710,620	110,000,000
1930	122,775,046	120,000,000
1940	131,669,275	130,000,000
1950	150,697,361	150,000,000
1960	179,323,175	180,000,000
1970	203,302,031	200,000,000
1980	226,545,805	230,000,000

Steps 3 to 6 are shown in the completed line graph.

In Step 4, to draw a dot on the graph for each set of figures in Table XVI, proceed as follows:

In 1900, the population was 80 million. Find the vertical line labeled "1900" and move up it until you reach the horizontal line labeled "80." Make a dot where the "1900" line and the "80" line meet.

In 1910, the population was 90 million. Find the vertical line labeled "1910" and move up it. Note that there is no "90" on the graph's scale. Since the number 90 is halfway between 80 and 100, move halfway between the horizontal line labeled "80" and the line labeled "100." Make a dot on the "1910" line, where "90" should be.

Continue this process for the rest of the years.

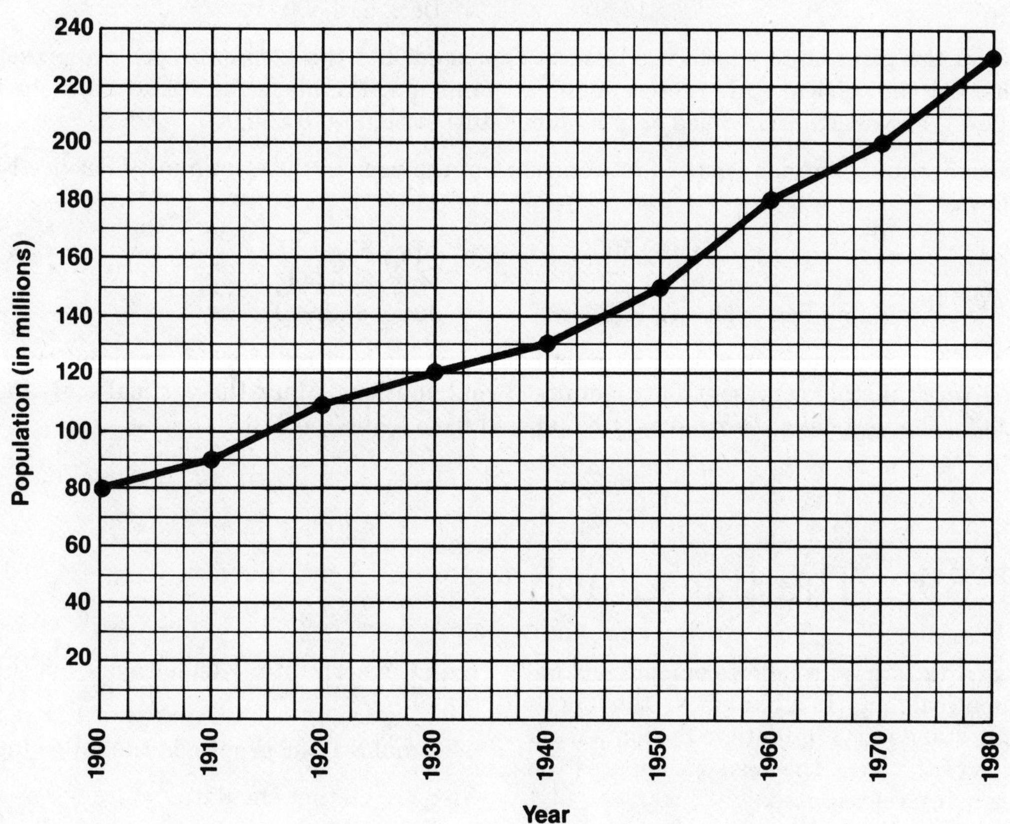

EXERCISES

Use the preceding graph to answer these exercises. Do *not* use the data in Table XVI (page 319).

1. In which three 10-year intervals was the population increase the smallest? ____________
2. In which two 10-year intervals was the population increase the greatest? ____________
3. Since 1900, the U.S. population has ____________
(Answer *increased 50%, doubled,* or *more than doubled.*)
4. What do you think the population of the United States will be in the year 2000? (Answer to the nearest *ten million.*) ____________

APPLICATION PROBLEMS

In 1–2, use graph paper to draw a line graph for the given data.

1. The monthly sales of John Wilson last year were as follows:

Jan., $8,570	July, $14,300
Feb., $11,690	Aug., $13,500
Mar., $10,400	Sept., $15,900
Apr., $13,600	Oct., $17,500
May, $15,800	Nov., $18,300
June, $16,500	Dec., $16,500

Round off the given amounts to the nearest thousand. Let the horizontal scale represent the 12 months. Let the vertical scale represent the amounts of sales. Mark the vertical scale in intervals of $2,000. The vertical scale need only be labeled from $8,000 to $20,000.

2. The monthly contributions for the first 6 months of the year to the Community Service Fund were as follows:

Jan., $9,500	Apr., $13,100
Feb., $11,700	May, $15,500
Mar., $10,500	June, $12,900

Round off the given amounts to the nearest $500. Let the horizontal scale represent the 6 months. Let the vertical scale represent the amounts of contributions. Mark the vertical scale in intervals of $1,000. The vertical scale need only be labeled from $9,000 to $16,000.

UNIT 89. The Bar Graph

The **bar graph** is used to show comparisons between numerical facts. In a bar graph, the length of each bar stands for a quantity. By comparing the lengths of the bars, the reader compares the quantities being represented.

A bar graph is constructed in the same way as a line graph. However, instead of connecting the points as in a line graph, bars are drawn from one of the scales to each point. The bars may be drawn either vertically (straight up from the horizontal scale) or horizontally (sideways from the vertical scale).

To make a bar graph, do the following:

Step 1: Gather the data.

Step 2: Round off the figures.

Step 3: Decide on the horizontal scale and the vertical scale.

Step 4: Decide on a vertical bar graph or a horizontal bar graph.

Step 5: Draw dots that represent the rounded-off figures in Step 2.

Step 6: Draw bars of equal width from the 0 of the vertical or horizontal scale to the dots. Leave the same amount of space between bars.

Step 7: Label the graph.

Note: By using graph paper, which is ruled into squares, it is easy to draw the bars accurately and to leave equal spaces between the bars.

EXAMPLE. Construct a vertical bar graph showing the population of the United States from 1900 to 1980. Use the data from Table XVI (page 319). The completed graph may look like this:

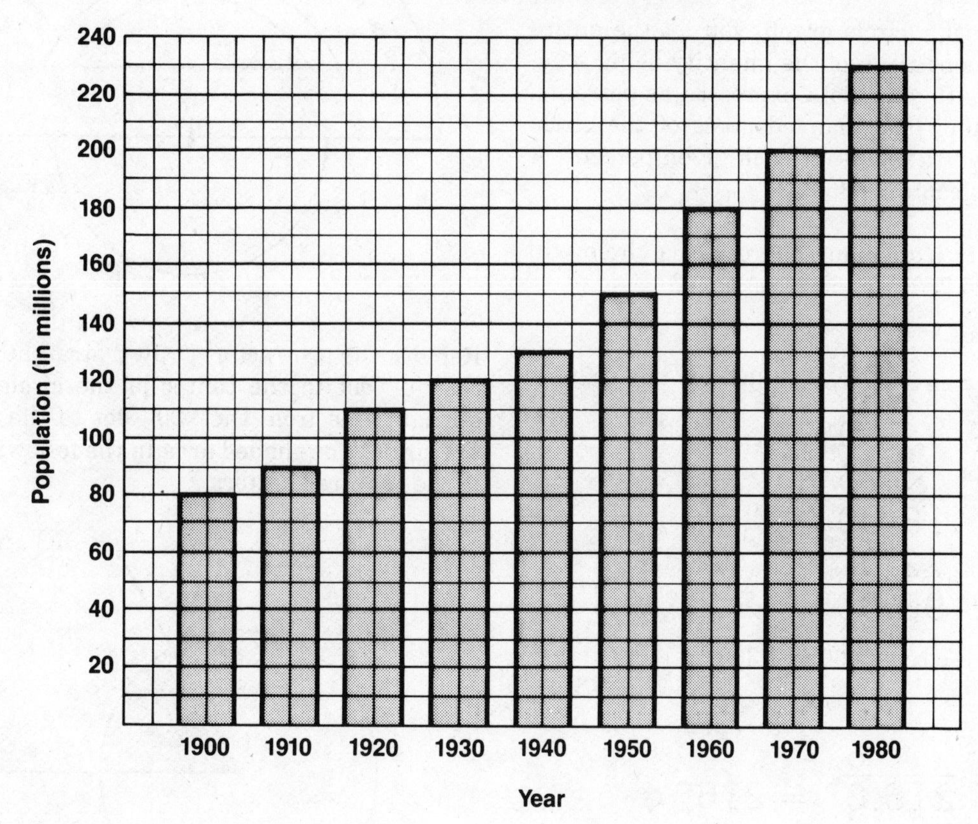

APPLICATION PROBLEMS

In 1–2, use graph paper to draw a bar graph for the given data. (*Note:* The bars in a bar graph can be drawn vertically or horizontally.)

1. The sales of the Kim Appliance Co. for one month, by department, were as follows:

TV sets and Radios, $3,500
Dryers, $1,500
Ranges, $2,250
Freezers, $1,250
Washers, $3,750
Other, $1,800

On the scale that represents sales, let each square of the graph paper represent $250. Mark this scale in intervals of $1,000.

2. Quarterly expenses of the Animal Shelter last year were as follows:

1st quarter, $9,532
2nd quarter, $11,290
3rd quarter, $10,485
4th quarter, $12,621

Arrange the four quarters on either the horizontal scale or the vertical scale. Round off the given amounts to the nearest $100. Let each square of the graph paper represent $500, and mark the scale that represents these amounts in intervals of $1,000. This scale need only be labeled from $9,000 to $13,000.

UNIT 90. The Circle Graph

The **circle graph** (also called a **pie graph**) is used to show how something is divided into parts and how those parts compare with one another. For example, a circle graph labeled "How Your Tax Dollar Is Spent" would show what percent of your tax dollar was spent for defense, what percent was spent on education, etc.

To construct a circle graph, you let the entire circle represent 100% of the quantity being studied. Then, with your data expressed as percents, you determine how many degrees of the circle must be used to represent each given percent. A whole circle contains 360°.

Example 1. How many degrees of a circle will represent (*a*) 25% (*b*) 60% (*c*) 1.1%?

Solutions:

(*a*) 25% of a circle is 25% of 360°.

$$\begin{array}{r} 360° \\ \times .25 \\ \hline 18\,00 \\ 72\,0 \\ \hline 90.00° \end{array} = 90°$$

(*b*) 60% of a circle is 60% of 360°.

$$\begin{array}{r} 360° \\ \times .6 \\ \hline 216.0° \end{array} = 216°$$

(*c*) 1.1% of a circle is 1.1% of 360°.

$$\begin{array}{r} 360° \\ \times .011 \\ \hline 360 \\ 3\,60 \\ \hline 3.960° \end{array} = 4.0° = 4°$$, to the nearest degree

Answers: (*a*) 25% = 90° (*b*) 60% = 216° (*c*) 1.1% = 4°

To draw a circle graph, you must be able to use a **protractor**.

Example 2. Using a protractor, mark off a 60-degree **sector** ("slice of pie") in a circle.

Solution: Place the center of the protractor on the center of the circle. Then make a dot at the "0" mark and make another dot at the "60" mark, as shown in the diagram:

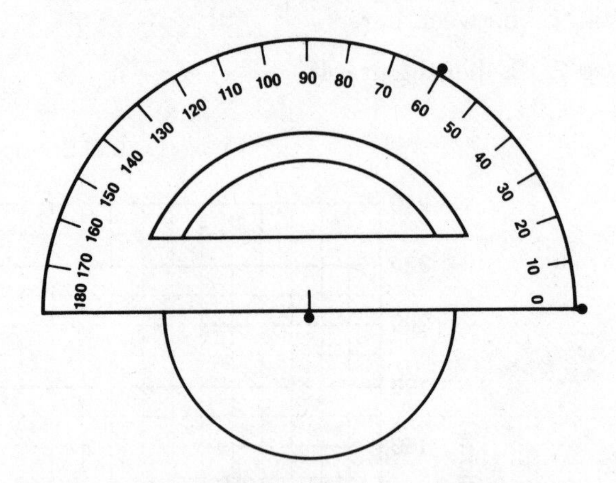

Remove the protractor. Draw a straight line from the "0" dot to the center of the circle; draw a straight line from the "60" dot to the center of the circle. The shaded area in the following figure is the required sector:

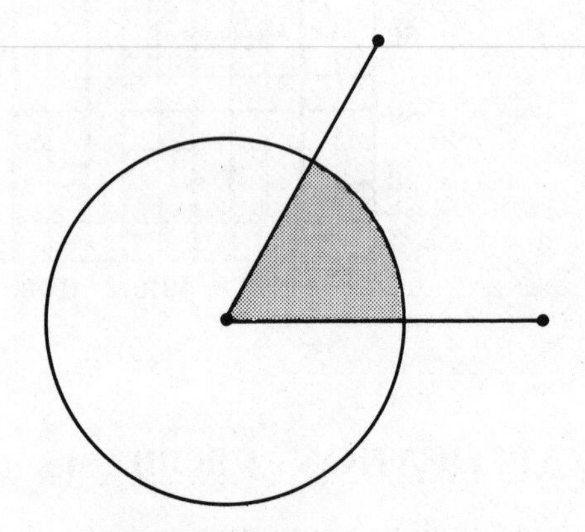

To make a circle graph, do the following:

Step 1: Gather the data.

Step 2: If necessary, change the data to percents.

Step 3: Determine how many degrees of a circle each percent represents.

Step 4: Draw a circle to represent 100% of the quantity being graphed. Then, using the degrees in Step 3, divide the circle into sectors, as shown in Example 2.

Step 5: Label the graph, including the percent in each sector.

Example 3. Mr. Allen's take-home pay last year was \$18,200. During that year, the Allen family used their income as follows:

Savings: \$1,820	Shelter: \$3,640
Clothing: \$2,185	Food: \$4,550
Car: \$1,820	Other: \$4,185

Make a circle graph showing the Allen family budget.

Solution:

Steps 1 and 2: Change each expenditure to a percent of Mr. Allen's pay, \$18,200.

$$\text{Savings} = \frac{\$1{,}820}{\$18{,}200} = .1 = 10\%$$

$$\text{Clothing} = \frac{\$2{,}185}{\$18{,}200} = .12 = 12\%$$

$$\text{Car} = \frac{\$1{,}820}{\$18{,}200} = .1 = 10\%$$

$$\text{Shelter} = \frac{\$3{,}640}{\$18{,}200} = .2 = 20\%$$

$$\text{Food} = \frac{\$4{,}550}{\$18{,}200} = .25 = 25\%$$

$$\text{Other} = \frac{\$4{,}185}{\$18{,}200} = .23 = 23\%$$

Step 3: Change each percent to degrees of a circle.

$$\text{Savings} = 360° \times .10 = 36°$$

$$\text{Clothing} = 360° \times .12 = 43.2° = 43°$$

$$\text{Car} = 360° \times .10 = 36°$$

$$\text{Shelter} = 360° \times .20 = 72°$$

$$\text{Food} = 360° \times .25 = 90°$$

$$\text{Other} = 360° \times .23 = 82.8° = 83°$$

Step 4: With a protractor, mark off the sectors, using the numbers of degrees in Step 3.

Step 5: Label the graph.

The completed graph may look like this:

Note that the percents in the circle graph must total 100%:

$$25 + 23 + 20 + 10 + 12 + 10 = 100$$

EXERCISES

Next year, the Allen family's total income will be \$19,850. Find the amounts spent for each of the items in the preceding graph. (Assume that the percents remain the same.)

a. Savings __________ *b.* Clothing __________

c. Car __________ *d.* Shelter __________

e. Food __________ *f.* Other __________

APPLICATION PROBLEMS

1. Draw a circle graph to represent the Alvarez family's vacation expenses for last year. Use the following data:

Transportation, 25%; Lodging, 30%; Food, 35%; Amusement, 10%

2. Using the information in the graph for Problem 1, find the amounts of money that the Alvarez family spent on each item. The total amount spent was $1,875.

Transportation ____________

Lodging ____________

Food ____________

Amusement ____________

Review of Part XIX (Units 87–90)

1. Draw a picture graph for the numbers of books borrowed in one week from the Denton School library.

Monday, 47
Tuesday, 83
Wednesday, 69
Thursday, 94
Friday, 116

Round off each number to the nearest ten. Let one picture of a book  represent 20 books.

2. Using graph paper, make a line graph showing the quarterly sales of the Franklin Manufacturing Company for last year.

1st Quarter: $21,568
2nd Quarter: $21,320
3rd Quarter: $20,390
4th Quarter: $24,685

Round off each amount to the nearest $1,000. On the vertical scale, choose a convenient number of squares to represent $1,000. The vertical scale need only be labeled from $20,000 to $25,000.

3. Using graph paper, make a vertical bar graph showing the sales by department of the Ajax Department Store for last year.

Clothing: $38,895
Furniture: $36,380
Appliances: $26,595
Toys: $8,419
Shoes: $13,485

Round off each amount to the nearest $1,000. On the vertical scale, let each square represent $2,000 and mark this scale in intervals of $10,000.

4. In a certain office, 8% of the employees are classified as executives, 28% are in the sales department, 22% are in the billing department, 10% are file clerks, and 32% are typists. Find, to the nearest *whole degree*, how many degrees of a circle should be used to represent each of these jobs on a circle graph. Then draw the circle graph.

executives: ____________°
file clerks: ____________°
sales dept.: ____________°
typists: ____________°
billing dept.: ____________°

PART XX. Earning Money

UNIT 91. Wages

WORDS TO KNOW

In many businesses and industries, workers are paid **wages** on an hourly basis for a 35- or 40-hour week. The pay for time worked beyond the regular weekly hours is usually figured at one and one-half times the regular hourly rate. The regular weekly pay, based on the 35- or 40-hour week, is known as **straight-time pay.** The extra pay, based on extra hours at "time-and-a-half," is called **overtime pay.**

In some industries, workers are not paid by the hour. Instead, they receive a certain amount of money for each article or unit that they complete or produce. These articles or units are called **pieces,** and the employee is said to be working on a **piece-work basis.** The amount he or she receives for each piece is called the **rate per piece.**

HOURLY WAGES

EXAMPLE 1. Harry Smith is paid $4.85 an hour for a 40-hour week plus time-and-a-half for overtime. Last week he worked 45 hours. What were his wages for the week?

Solution: First compute the straight-time pay at $4.85 per hour for a regular 40-hour week.

$$\begin{array}{r} 4.85 \\ \times 40 \\ \hline 194.00 \end{array}$$ ← straight-time pay

Next compute how much Smith earns per hour at time-and-a-half.

$$4.85 \times 1\frac{1}{2} = 4.85 \times 1.5$$

$$= 7.275$$ ← overtime pay per hour

Since Smith worked 45 hours, his overtime is 45 − 40 = 5 hours. Compute the overtime pay at $7.275 per hour for 5 hours.

$$\begin{array}{r} 7.275 \\ \times 5 \\ \hline 36\ 37(5) \\ +.01 \\ \hline 36.38 \end{array}$$

+.01 ← Round off the fraction of a cent to 1 cent and add.

36.38 ← overtime pay

(*Note:* The symbol @ means "at." Thus, "40 hr. @ 4.85" means 40 hours at $4.85 for each hour.)

40 hr. @ 4.85	=	194.00
5 hr. @ 7.275	=	36.38
Total earnings	=	$230.38

Answer: $230.38

PIECE-WORK WAGES

To find the wages earned by a worker on a **piece-work basis,** multiply the total number of **pieces** by the **rate per piece.**

EXAMPLE 2. For the week ending July 5, Julia Ward completed the following numbers of pieces:

Mon. 47, Tue. 51, Wed. 50, Thu. 49, Fri. 48

The rate per piece is $1.10. Find her total wages.

Solution: Add the units completed daily: 47 + 51 + 50 + 49 + 48 = 245, the total number of units completed. Multiply this total by the rate per piece, $1.10.

$$\begin{array}{r} 245 \\ \times 1.10 \\ \hline 24\ 50 \\ 245 \\ \hline 269.50 \end{array}$$

Answer: Total wages for the week are $269.50.

APPLICATION PROBLEMS

1. Alan Carter is paid \$6.10 an hour for a 35-hour week, plus time-and-a-half for overtime. Find his total earnings for last week, when he worked 43 hours. ______

2. Dorothy McIntyre earns \$5.25 an hour for a 40-hour week, plus time-and-a-half for overtime. The hours she worked last week are shown in the table.

Mon.	Tue.	Wed.	Thu.	Fri.
$8\frac{1}{2}$	$9\frac{1}{2}$	8	9	9

Find her total earnings for the week. ______

3. Sara is paid \$1.25 for every skirt she completes. Last week, her daily record of skirts completed was 45, 53, 47, 52, and 46. What were her earnings for the week? ______

4. Ernesto assembles portable radios and is paid \$2.85 for each radio assembled. He assembled the following numbers of radios: Monday, 18; Tuesday, 21; Wednesday, 17; Thursday, 22; Friday, 19. How much did he earn for the week? ______

Use a calculator for problems 5 and 6.

5. The payroll record of the Acme Window Washing Company is shown below. Based on a 40-hour week, with time-and-a-half for overtime, enter the following information for each employee: total regular hours, total overtime hours, and total earnings. As a sample solution, this information has been computed for James Bell.

PAYROLL RECORD
Acme Window Washing Company

		Time Record					Hourly Rate	Total Hours		Total Earnings
	Name	M	T	W	Th	F		Regular	Overtime	
	James Bell	8	7	9	10	8	\$4.18	40	2	\$179.74
a.	Joseph Davis	8	6	9	9	8	5.20			
b.	Louis DeMato	8	$9\frac{1}{2}$	9	9	9	6.95			
c.	Juan Diaz	7	9	8	10	10	5.05			
d.	Manuel Ortiz	$8\frac{1}{2}$	9	$7\frac{1}{2}$	9	8	6.15			
e.	Jerry Schultz	9	8	7	$8\frac{1}{2}$	$9\frac{1}{2}$	7.40			
f.	Sam Siegel	$8\frac{1}{2}$	$9\frac{1}{2}$	$8\frac{1}{2}$	8	9	6.35			
g.	Donald Westler	9	8	10	$8\frac{1}{2}$	$9\frac{1}{2}$	7.25			

Bell: $8+7+9+10+8=42=40$ *reg.* $+ 2$ *overtime*

reg. pay $= 4.18 \times 40 = 167.20$

overtime $= 1.5 \times 4.18 \times 2 = 12.54$

$$\begin{array}{r} 167.20 \\ +\ 12.54 \\ \hline 179.74 \end{array}$$

6. From the following piece-work record, find the total number of dozens completed and the total earnings for each employee. As a sample solution, this information has been computed for M. Gold. (*Hint:* $\frac{1}{4} = .25$, $\frac{1}{2} = .5$, and $\frac{3}{4} = .75$)

	Name	Number of Dozens					Total Dozens	Rate per Dozen	Total Earnings
		M	T	W	Th	F			
	Gold, M.	10	$11\frac{1}{2}$	12	$13\frac{1}{4}$	$12\frac{1}{2}$	59.25	$4.85	$287.36
a.	Gomez, D.	11	$13\frac{1}{4}$	$10\frac{3}{4}$	12	$12\frac{1}{2}$		4.84	
b.	Gonzales, M.	12	$11\frac{3}{4}$	$12\frac{1}{2}$	13	$10\frac{1}{4}$		4.76	
c.	Goodman, L.	10	$12\frac{1}{2}$	$13\frac{1}{4}$	$11\frac{3}{4}$	14		4.98	
d.	Gould, I.	$13\frac{1}{2}$	$12\frac{1}{4}$	$11\frac{3}{4}$	13	$10\frac{1}{2}$		4.97	
e.	Green, S.	12	$13\frac{3}{4}$	$11\frac{3}{4}$	$12\frac{1}{2}$	13		4.85	
f.	Hart, D.	$11\frac{3}{4}$	$12\frac{1}{2}$	14	$13\frac{3}{4}$	12		4.74	
g.	Hunt, E.	12	$13\frac{3}{4}$	$14\frac{1}{4}$	$12\frac{3}{4}$	$13\frac{1}{4}$		4.93	
h.	Huntley, B.	$13\frac{1}{2}$	$12\frac{3}{4}$	$11\frac{1}{4}$	12	14		4.87	
i.	Jones, D.	12	$13\frac{3}{4}$	12	$11\frac{3}{4}$	13		4.88	

Gold: $10 + 11.5 + 12 + 13.25 + 12.5 = 59.25$ doz.

$59.25 \times 4.85 = 287.36(25) = 287.36$

UNIT 92. Salary and Commission

Words to Know

In some businesses, a salesperson is paid a **commission** instead of or in addition to a fixed **salary.** The commission usually is a percent of the dollar value of his or her sales. This percent is called the **rate of commission.**

A salesperson who is paid a commission only is said to work on a **straight commission** basis.

In many instances, a salesperson receives a weekly salary, called a **base salary,** in addition to the commission. At times, the commission may be a percent of sales above a fixed amount, known as a **quota.**

To find the amount of **commission** a salesperson has earned, multiply the **rate of commission** by the total sales. Since the rate of commission is a percent, change the given percent to a decimal before multiplying.

EXAMPLE 1. Charles Wilkins receives a commission of 10% on his sales. In July his sales came to $9,350. Find the amount of his commission.

Solution: Multiply the rate of commission, 10%, by the total sales, $9,350.

$$9{,}350 \times 10\% = 9{,}350 \times .10$$
$$= 935.0. = 935$$

Answer: His commission for July was $935.

EXAMPLE 2. Mrs. Lee is paid $145 a week plus a commission of 5% on sales above a quota of $2,500. Last week her sales were $8,100. What were her total earnings for the week?

Solution: Subtract the quota, $2,500, from her total sales, $8,100, to find the part of her sales on which she gets a commission. This part is known as her *net sales* or *commission sales.*

$$\begin{array}{r} 8{,}100 \\ -2{,}500 \\ \hline 5{,}600 \end{array}$$

To find the commission, multiply the net sales, $5,600, by the rate of commission, 5%.

$$\begin{array}{r} 5{,}600 \\ \times .05 \\ \hline 280.00 \end{array}$$

Add the base salary, $145.00, to the commission, $280.00, to compute the total earnings.

$$\begin{array}{r} 280.00 \\ +145.00 \\ \hline 425.00 \end{array}$$

Answer: Her total earnings were $425.

APPLICATION PROBLEMS

1. A real estate salesperson earns a commission of 12%. If she sold a house for $78,600 and another house for $84,750, how much commission did she earn on the two houses? ____________

2. A salesperson earns $125 a week plus a commission of 8% on sales over $4,000. Last week, his sales were $7,882.50. Find his total earnings for the week. ____________

3. From the following sales record, find the amount of commission earned by each salesperson. As a sample solution, the commission for L. Baker has been computed.

	Name	Weekly Sales	Rate of Commission	Amount of Commission
	Baker, L.	$1,872	15%	$280.80
a.	Bates, D.	2,320	12%	
b.	Beam, C.	1,968	15%	
c.	Bean, A.	2,475	13%	
d.	Calb, W.	2,298	12%	
e.	Deeds, M.	2,463	12%	
f.	Donald, R.	2,578	13%	
g.	Dubler, F.	2,648	15%	
h.	Dubrow, S.	2,329	14%	
i.	Engle, R.	2,468	13%	

Baker: 15% = .15

1,872 × .15 = 280.80

4. **Use a calculator for this problem. From the given record, find the net sales and the total earnings for each salesperson. The quota for each salesperson is $2,500. As a sample solution, this information has been computed for G. Evans.**

	Name	Weekly Salary	Weekly Sales	Net Sales	Commission	Total Earnings
	Evans, G.	$164	$5,491.50	$2991.50	8%	$403.32
a.	Farber, G.	163	7,247.25		5%	
b.	Farbstein, D.	160	5,956.80		7%	
c.	Farley, B.	165	6,952.44		6%	
d.	Farraro, V.	162	6,746.25		7%	
e.	Fenway, B.	164	6,535.35		8%	
f.	Ferguson, M.	163	6,805.50		5%	
g.	Garden, D.	165	7,126.05		6%	
h.	Gary, E.	164	7,042.86		7%	
i.	Gutman, B.	165	6,715.35		6%	

Evans:

$$\begin{array}{r} 5{,}491.50 \\ -2{,}500.00 \\ \hline 2{,}991.50 \end{array} \qquad \begin{array}{r} 2{,}991.50 \\ \times\ .08 \\ \hline 239.3200 \end{array} \qquad \begin{array}{r} 239.32 \\ +\ 164.00 \\ \hline 403.32 \end{array}$$

UNIT 93. Deductions From Wages

The law requires that employers deduct money for two different taxes from the earnings of each employee. The employer then sends this money to the government. The exact amounts of these taxes may vary from year to year. The typical rates given in this unit are used to illustrate how the payroll deductions are computed.

One tax is the **federal income withholding tax,** commonly called the **income tax.** The amount deducted for this tax depends on the employee's earnings, and also on the number of **withholding allowances.**

Withholding allowances are not taxed. Thus, they are subtracted from earnings before the income tax is computed. The number of withholding allowances that an employee can claim depends on individual circumstances, such as the number of people the employee supports. The amount of each such allowance is set by the government. In a typical year, the amount might be $1,900 per year (or $36.54 per week).

Table XVII is one of the tables supplied by the government to tell an employer how much tax to withhold. There are separate tables for payrolls that are weekly, biweekly, monthly, etc. The table shown here is for a weekly payroll period. The wage amounts given are wages after withholding allowances have been subtracted.

Table XVII: Income Withholding Tax for Weekly Payroll Period

(a) SINGLE person—including head of household:				**(b) MARRIED person**—			
If the amount of wages is:		*The amount of income tax to be withheld shall be:*		*If the amount of wages is:*		*The amount of income tax to be withheld shall be:*	
Not over $12		0		Not over $36		0	
Over—	*But not over—*		*of excess over—*	*Over—*	*But not over—*		*of excess over—*
$12	—$47	11%	—$12	$36	—$93	11%	—$36
$47	—$335	$3.85 plus 15%	—$47	$93	—$574	$6.27 plus 15%	—$93
$335	—$532	$47.05 plus 28%	—$335	$574	—$901	$78.42 plus 28%	—$574
$532	—$1,051	$102.21 plus 35%	—$532	$901	—$1,767	$169.98 plus 35%	—$901
$1,051		$283.86 plus 38.5%	—$1,051	$1,767		$473.08 plus 38.5%	—$1,767

The other tax is the **Federal Insurance Contributions Act (F.I.C.A.) tax,** commonly known as the **social security tax.** In a typical year, the amount might be 7.15% of the first $43,800 earned that year. No tax would be withheld for earnings over $43,800. The F.I.C.A. tax is based on the full amount of an employee's earnings.

The salary that an employee gets after all deductions have been made is called the **net pay** or **take-home pay.** In this unit, we are considering only federal income tax and social security tax, but other deductions may also be made from an employee's salary. Some of these deductions are for state and city income taxes, union dues, and contributions to health insurance and pension funds.

EXAMPLE 1. Henry Sloan is single, earns $350 jer week, and claims one withholding allowance. Compute the amount to be withheld from his weekly salary for income tax.

Solution:

Step 1: Mr. Sloan is claiming one withholding allowance, or $36.54 per week. To find the amount to be taxed, subtract $36.54 from his weekly earnings of $350.

$$\begin{array}{r} 350.00 \\ -36.54 \\ \hline 313.46 \end{array}$$

Step 2: Find the appropriate line in the table for single persons. $313.46 is over $47 but not over $335. Thus, the tax is **$3.85 plus 15%** of the excess over $47. Subtract to determine the amount of the excess.

$$\begin{array}{r} 313.46 \\ -47.00 \\ \hline 266.46 \end{array}$$

Step 3: **Take 15% of the excess, $266.46.**

$$\begin{array}{r} 266.46 \\ \times .15 \\ \hline 13\ 3230 \\ 26\ 646 \\ \hline 39.9690 \end{array} \longrightarrow \$39.97$$

Step 4: **Add $3.85.**

$$\begin{array}{r} 39.97 \\ +3.85 \\ \hline 43.82 \end{array}$$

Answer: **The amount to be withheld for income tax is $43.82.**

EXAMPLE 2. Find the amount of social security tax to be deducted from Henry Sloan's weekly earnings of $350.

Solution: **Multiply the weekly salary of $350 by the social security tax rate of 7.15%.**

$$\begin{array}{r} 350 \\ \times .0715 \\ \hline 1750 \\ 350 \\ 24\ 50 \\ \hline 25.0250 \end{array} \longrightarrow \$25.03$$

Answer: **The social security tax is $25.03.**

EXAMPLE 3. Use the answers in Examples 1 and 2 to find Henry Sloan's net pay.

Solution: Find the total deductions by adding the income tax and the social security tax.

$$\begin{array}{r} 43.82 \\ +25.03 \\ \hline 68.85 \end{array}$$

Subtract the total deductions from the weekly salary of $350.

$$\begin{array}{r} 350.00 \\ -68.85 \\ \hline 281.15 \end{array}$$

Answer: Henry Sloan's net pay is $281.15.

EXAMPLE 4. Karen Smith is married and claims 5 withholding allowances. Her weekly earnings are $271.75. Find her net earnings.

Solution:

Step 1: Find the total for 5 withholding allowances of $36.54 each.

$$\begin{array}{r} 36.54 \\ \times 5 \\ \hline 182.70 \end{array}$$

Step 2: To find the amount taxable, subtract the total withholding allowances from the weekly pay.

$$\begin{array}{r} 271.75 \\ -182.70 \\ \hline 89.05 \end{array}$$

Step 3: Find the appropriate line in the table for married persons. The amount of $89.05 is over $36 but not over $93. Thus, the tax is 11% of the excess over $36. Find the amount of excess and multiply by 11%, to find income tax withheld.

$$\begin{array}{r} 89.05 \\ -36.00 \\ \hline 53.05 \end{array} \qquad \begin{array}{r} 53.05 \\ \times .11 \\ \hline 5305 \\ 5\,305 \\ \hline 5.8355 \end{array} \rightarrow \$5.84$$

Step 4: To find the social security tax, multiply the earnings by the tax rate of 7.15%.

$$\begin{array}{r} 271.75 \\ \times .0715 \\ \hline 135875 \\ 27175 \\ 19\,0225 \\ \hline 19.430125 \end{array} \rightarrow \$19.43$$

Step 5: Find the sum of both taxes.

$$\begin{array}{r} 19.43 \\ +5.84 \\ \hline 25.27 \end{array}$$

Step 6: Subtract the total taxes from the weekly earnings.

$$\begin{array}{r} 271.75 \\ -25.27 \\ \hline 246.48 \end{array}$$

Answer: The net earnings are $246.48.

APPLICATION PROBLEMS

In 1–4, use Table XVII (page 330), withholding allowances of $36.54 per week, and a social security tax rate of 7.15%.

1. William is married, earns $403.75 a week, and claims 3 withholding allowances.

   a. Find the amount of income tax withheld each week. ____________

   b. What is the social security deduction from his weekly pay? ____________

   c. What is his weekly net pay? ____________

2. Samantha is single, earns $19,526 a year (52 weeks), and claims 2 withholding allowances.

a. Find the amount of income tax withheld each week. ________

b. How much income tax is withheld for the entire year? ________

c. What is the social security deduction for the year? ________

d. What is the total of the income tax and social security deductions for the year? ________

e. What will her take-home pay be for the year? ________

3. Find the income tax, the social security tax, and the total of these deductions for each of the following employees. As a sample solution, this information has been computed for J. Baker.

	Name	Marital Status	With-holding Allow-ances	Weekly Wages	Income Tax	Social Security Tax	Total Deductions
	Baker, J	Single	2	$279.50	$27.76	$19.98	$47.74
a.	Bates, M.	Single	0	385.00			
b.	Cohen, J.	Married	4	294.75			
c.	Colon, M.	Single	1	396.20			
d.	Danzig, A.	Married	3	318.60			
e.	Dean, W.	Married	6	328.50			
f.	Dunbar, V.	Single	0	377.40			
g.	Engel, L.	Married	5	395.80			
h.	Fernandez, J.	Married	1	368.90			
i.	Gold, N.	Single	2	326.00			

Baker:

Income Tax:

$$\begin{array}{r} 36.54 \\ \times \quad 2 \\ \hline 73.08 \end{array} \qquad \begin{array}{r} 279.50 \\ -\ 73.08 \\ \hline 206.42 \end{array} \qquad \begin{array}{r} 206.42 \\ -\ 47.00 \\ \hline 159.42 \end{array} \qquad \begin{array}{r} 159.42 \\ \times\ .15 \\ \hline 79710 \\ 15942 \\ \hline 23.91(30) \end{array} \qquad \begin{array}{r} 23.91 \\ +\ 3.85 \\ \hline 27.76 \end{array}$$

Social Security Tax:

$$\begin{array}{r} 279.50 \\ \times\ .0715 \\ \hline 139750 \\ 27950 \\ 195650 \\ \hline 19.98(4250) \end{array}$$

Total:

$$\begin{array}{r} 27.76 \\ +\ 19.98 \\ \hline 47.74 \end{array}$$

4. Use a calculator for this problem.

Find the income tax, the social security tax, the total deductions, and the net wages for each of the employees listed in the following table. As a sample solution, this information has been computed for J. Abrams.

When this computation has been done, add the columns, and enter the sums in the Totals boxes at the bottom of the table. Use the Totals sums to check your work: The Income Tax sum plus the Social Security Tax sum should equal the Total Deductions sum. The Weekly Wages sum minus the Total Deductions sum should equal the total Net Wages.

	Name	Marital Status	Withholding Allowances	Weekly Wages	Income Tax	Social Security Tax	Total Deductions	Net Wages
	Abrams, J.	Single	3	$375.50	$36.68	$26.85	$63.53	$311.97
a.	Addison, F.	Married	1	298.65				
b.	Agnew, P.	Single	0	405.80				
c.	Brunillo, J.	Married	5	248.50				
d.	Budge, F.	Single	2	275.45				
e.	Curtis, A.	Single	1	367.90				
f.	DaMato, L.	Married	3	408.75				
g.	Delgado, G.	Single	0	298.70				
h.	Delio, J.	Single	1	393.65				
i.	Ferman, M.	Married	5	384.70				
j.	Finkelstein, C.	Married	6	403.60				
			Totals					

Abrams:

Income tax:

$36.54 \times 3 = 109.62$ $\quad 375.50 - 109.62 = 265.88$ $\quad 265.88 - 47.00 = 218.88$ $\quad 218.88 \times .15 = 32.83$ $\quad 32.83 + 3.85 = 36.68$

Social Security tax:

$375.50 \times .0715 = 26.85$

Deductions:

$36.68 + 26.85 = 63.53$

Net:

$375.50 - 63.53 = 311.97$

PART XXI. Spending Money

UNIT 94. Budgets

Paying bills is a necessary part of everyday living. Many people "pay as they go," paying bills when they have the money and putting off payments when they are short of cash. In addition to causing worry, this practice usually costs money in extra finance charges, and often results in a poor credit rating.

Many home economists feel that the best way to handle personal finances is to treat your annual income as one big sum and to decide how to spend it by making a **budget.**

To make a budget, go back through your records and receipts and find out how much you spent last year for such items as food, clothing, housing, doctors, and operating expenses. (Operating expenses include home repairs, car maintenance, vacations, recreation, etc.) Also include how much you saved last year. Then, estimate what your income will be for the entire year to come. Once you get all this information, you can figure out *in advance* how much you can afford to pay for the various expenses, and how much you can afford to save.

The chart shows how the Wilson family plans to spend its estimated annual income of $12,500. (The figure of $12,500 is Mr. Wilson's "take-home pay" from which all taxes have already been deducted.)

Note that the sum of all the expenses is the total income of $12,500.

Mr. Wilson wants to know what *percent* of his total income will be spent on each item. To find the percent of each expenditure, *divide the expenditure by the total income.*

Wilson Family Budget (Total Income: $12,500)

Food	Clothing	Housing	Operating	Medical	Savings
$3,000	$1,500	$3,500	$2,125	$1,125	$1,250

EXAMPLE. What percent of Mr. Wilson's total income will be spent on food?

Solution: Divide the expenditure, $3,000, by $12,500.

```
             .24 = 24%
12,500)3,000.00
       2 500 0
         500 00
         500 00
```

Answer: 24% of the total income will be spent on food.

Note: Recall that finding what percent one number is of another number can be calculated with the $\frac{\text{IS}}{\text{OF}}$ fraction. In the preceding example, the facts may be expressed as: "$3,000 is what % of $12,500?"

$$\frac{\text{IS}}{\text{OF}} = \frac{3,000}{12,500}$$

Change the fraction to a percent by dividing the denominator into the numerator. This division is shown in the preceding example.

The next chart is the Wilson family's budget with all percents figured out. Note that the sum of the percents must be 100%. (Recall that 100% means *all* of something, or the *total amount* of something.)

Wilson Family Budget (Total Income: $12,500)

Food	Clothing	Housing	Operating	Medical	Savings
$3,000	$1,500	$3,500	$2,125	$1,125	$1,250
24%	12%	28%	17%	9%	10%

APPLICATION PROBLEMS

Use a calculator for these problems.

1. If Mr. Wilson gets a raise, his estimated annual income will be $13,200. Using the same percentages as in the preceding example, find the amount available for each expenditure. (The sum of the expenditures should be $13,200.)

   Food ________ Clothing ________ Housing ________

   Operating ________ Medical ________ Savings ________

2. The Watts family's estimated income is $18,750. The budget allowances for the year are:

   Food: $4,688 Clothing: $2,436 Housing: $4,313
   Operating: $3,938 Medical: $1,875 Savings: $1,500

   Find the percent of income budgeted for each item.

   Food ________ Clothing ________ Housing ________

   Operating ________ Medical ________ Savings ________

3. Carl Davis has a take-home income of $14,845. His family budget allowances for the year are as follows:

   Food: 28% Clothing: 8% Housing: 25%
   Operating: 22% Medical: 6% Savings: the remaining percent

   a. What percent of the yearly income will the Davis family save? ________

   b. Find the amount of income budgeted for each item.

   Food ________ Clothing ________ Housing ________

   Operating ________ Medical ________ Savings ________

4. Next year, Mr. Davis's take-home income will be increased by $8\frac{1}{2}$% over his current income of $14,845.

   a. Find the amount of his new income. ________

   b. Using the budget percentages in problem **3**, find the amount of income to be budgeted for each item.

   Food ________ Clothing ________ Housing ________

   Operating ________ Medical ________ Savings ________

UNIT 95. Unit Pricing

Buying a Fractional Part of a Unit

There are times when you buy a fractional part of an item that is priced per whole unit. For example, what is the cost of $\frac{3}{4}$ of a yard of velvet that is priced at $5.75 a yard? What should you pay for 4 ounces of cheese priced at $3.85 a pound?

When finding the cost of a fractional part of a unit, the part bought must be of the same unit as the unit in the price. If you buy ounces and the price is per pound, change the ounces to a fraction of a pound. If you buy inches and the price is per yard, change the inches to a fraction of a yard. Then multiply the fraction by the price of a whole unit.

RULE

To find the cost of a fractional part when the unit price is known:

1. If necessary, change the units of the fractional part to a fraction of the whole unit.
2. Multiply the fractional part of the unit by the price per whole unit.

Example 1. If you buy $\frac{3}{4}$ of a yard of fabric at a price of $5.75 per yard, what is the price you pay?

Solution: The quantity bought, $\frac{3}{4}$ of a yard, is already a fraction of the same unit as the unit in the price. Go to step 2 and multiply the $\frac{3}{4}$ by $5.75, the price per yard. Use either the common fraction, $\frac{3}{4}$, or the decimal fraction, .75.

Common Fraction

$$\frac{3}{4} \times \frac{5.75}{1} = \frac{17.25}{4}$$
$$= 4.31(25)$$
$$= 4.32$$

Decimal Fraction

$$\begin{array}{r} 5.75 \\ \times .75 \\ \hline 28\ 75 \\ 4\ 02\ 5 \\ \hline 4.31(25) = 4.32 \end{array}$$

Answer: The cost of $\frac{3}{4}$ yard is $4.32.

Note: Generally, in making purchases, a fraction of a penny is considered an additional penny.

Example 2. If cheese sells at $4.80 per pound, how much should you pay for 6 ounces?

Solution:

Step 1. Change the 6 ounces to a fractional part of a pound.

$$6 \text{ ounces} = \frac{6}{16} = \frac{3}{8} \text{ pound}$$

Step 2: Multiply the fraction by the price per pound.

$$\frac{3}{8} \times \frac{4.80}{1} = \frac{14.40}{8} = 1.80$$

Answer: The cost of 6 ounces of cheese is $1.80.

Finding the Unit Price

A federal law requires most food stores to display the **unit price** of all food products. The unit price is the price of a standard unit of food, such as the price per gallon, the price per pound, or the price per dozen.

For example, if a certain sauce is priced at "35¢ for a 4-oz. can," the unit price of "$1.40 per pound" must also be displayed.

Small grocery stores, however, are not required to post the unit prices. You should be able to figure out unit prices in order to compare the cost of comparable items and get the best value for your money. For instance, which is a better buy, a 6-ounce box of puffed rice for $1.15 or a 7-ounce box for $1.29?

In Examples 1 and 2, the price of the whole unit was known. To find the cost of the fractional part, you *multiplied* the unit price by the fraction.

Now the problem is reversed; the cost of the fractional part is known. To find the price of the whole unit, you *divide* the cost of the fractional part by the fraction.

RULE

To find the unit price when the cost of a fractional part is known:

1. If necessary, change the unit of the fractional part to a fraction of the whole unit.
2. Divide the cost of the fractional part by the fraction.

EXAMPLE 3. Which is a better value, Brand A bacon at \$2.09 for 12 ounces or Brand B bacon at \$2.59 a pound?

Solution: Find the cost per pound of Brand A. 12 ounces is $\frac{12}{16}$ or $\frac{3}{4}$ of a pound.

Divide the cost of the fractional part, \$2.09, by the fraction, $\frac{3}{4}$.

$$2.09 \div \frac{3}{4} = \frac{2.09}{1} \times \frac{4}{3}$$

$$= \frac{8.36}{3} = 2.78\frac{2}{3} = 2.79$$

Answer: Brand B at \$2.59 a pound is a better value than Brand A at \$2.79 a pound.

Note: The $\frac{\text{IS}}{\text{OF}}$ fraction can be used to solve unit-price problems. The facts about Brand A can be reworded as follows: "(2.09 is) ($\frac{3}{4}$ of) what price?"

$$\frac{\text{IS}}{\text{OF}} = \frac{2.09}{\frac{3}{4}} = 2.09 \div \frac{3}{4}$$

Solve as before.

APPLICATION PROBLEMS

1. Use a calculator for this problem.
Mr. Ross made purchases at the supermarket and the hardware store. Find the cost of each purchase, and find the total amount that he spent.

	Item	Price per Unit	Fractional Part	Cost of Fractional Part
a.	steak	\$4.75 per lb.	12 oz.	
b.	coffee	\$11.45 per kg	.5 kg	
c.	milk	\$.69 per L	.75 L	
d.	eggs	\$.83 per doz.	6 eggs	
e.	shelf liner	\$1.75 per yd.	$\frac{2}{3}$ yd.	
f.	paint	\$7.45 per gal.	3 qt.	
g.			Total	

2. Mrs. Wilson bought 28 in. of velvet at \$8.65 a yard. What fraction of a yard did she buy? __________

How much did she pay? __________

3. Jerry bought 12 oz. of ham at $4.28 a pound. What fraction of a pound did he buy? ____________

How much did he pay? ____________

4. Mrs. Bianca bought the following items:

14 oz. of pork at $2.19 per pound
12 oz. of veal cutlets at $4.15 per pound
10 oz. of bacon at $1.79 per pound
8 oranges at $2.25 per dozen

How much did she spend? ____________

5. Use a calculator for this problem.

Hector is putting up unit prices of items in his grocery store. What is the unit price of each item listed below?

	Item	Fractional Weight or Measure	Cost of Fractional Unit	Price per Unit
a.	potato chips	8-oz. bag	$1.69	____ per lb.
b.	butter	4-oz. stick	47¢	____ per lb.
c.	coffee	.25 kg	$1.24	____ per kg
d.	laundry bleach	$\frac{1}{5}$ gal.	79¢	____ per gal.
e.	tomato juice	8-oz. can	63¢	____ per qt.
f.	cookies	200-g box	87¢	____ per kg

6. Mrs. Meyers bought 3 cucumbers for 31¢.

a. What fraction of a dozen did she buy? ____________

b. What is the price per dozen? ____________

7. *a.* Find the price per ounce of Brand X Puffed Rice, if a 6-ounce box costs $1.15. ____________

b. Find the price per ounce of Brand Y Puffed Rice, if a 7-ounce box costs $1.29. ____________

c. Which is the better buy, Brand X or Brand Y? ____________

8. *a.* Find the price per liter of Peppy Orange Drink, if a .75-liter bottle costs 89¢. ____________

b. Find the price per liter of Thirsty Orange Sip, if a 600-milliliter container costs 79¢. ____________

c. Which product costs less per liter? ____________

UNIT 96. Installment Buying

The **list price,** or *selling price,* of an expensive item such as a refrigerator or an automobile is usually greater than most people can pay all at once. Hence, such expensive items are often sold on the **installment plan.**

In an installment-plan purchase, the buyer agrees to pay a small part of the list price as a **down payment.** Subtracting this down payment from the list price gives the **balance due** on the purchase. The balance due is then divided into equal installments that are paid monthly over a specified period of time.

In allowing the buyer to pay off his purchase over a period of time, the retailer is actually lending the buyer money. Just as a bank charges interest for the use of money, the retailer adds **carrying charges,** or **finance charges,** to the monthly installments.

The carrying charges for installment buying are usually higher than the interest for a comparable bank loan. The higher rates are charged because, in addition to the interest, the retailer adds the clerical costs of maintaining the installment plan. Many buyers, however, are willing to pay the higher rates for the convenience of "buying now and paying later."

Example 1. You buy a ring for \$317, making a down payment of \$65. The balance due is to be paid in 12 monthly payments. How much do you pay each month?

Solution: To find the balance due, subtract the down payment, \$65, from the price, \$317.

$$\begin{array}{r} 317 \\ -65 \\ \hline 252 \end{array}$$

Divide the balance due, \$252, by the number of payments, 12.

$$\begin{array}{r} 21 \\ 12\overline{)252} \\ \underline{24} \\ 12 \\ \underline{12} \end{array}$$

Answer: You pay \$21 each month.

Example 2. A television set whose list price is \$122.50 may be purchased on the installment plan for \$25 down and 12 monthly payments of \$10.50 each. Find the carrying charges.

Solution: Multiply the amount of each payment by the number of payments to find the balance due.

$$\begin{array}{r} 10.50 \\ \times 12 \\ \hline 21\ 00 \\ 105\ 0 \\ \hline 126.00 \end{array}$$

Add to the balance due, \$126.00, the down payment of \$25.00.

$$\begin{array}{r} 126.00 \\ +25.00 \\ \hline 151.00 \end{array}$$

The total cost of the television set on the installment plan is \$151.00.

To find the carrying charges, subtract the list price, \$122.50, from the installment-plan price, \$151.00.

$$\begin{array}{r} 151.00 \\ -122.50 \\ \hline 28.50 \end{array}$$

Answer: Carrying charges are \$28.50.

APPLICATION PROBLEMS

1. At Sal's Furniture and Appliance Store, the following items were purchased on the installment plan one afternoon. Find the amount of each monthly payment. As a sample solution, the first answer has been calculated.

	Item	Installment Price	Down Payment	Number of Monthly Payments	Amount of Each Monthly Payment
	television	$473.80	$85	18	$21.60
a.	washer	362	50	24	
b.	dryer	297	45	12	
c.	furniture	867	75	36	
d.	refrigerator	589	165	18	

television

$$\begin{array}{r} 473.80 \\ -\ 85.00 \\ \hline 388.80 \end{array}$$

$$388.80 \div 18 = 21.60$$

$$\begin{array}{r} 36 \\ \hline 28 \\ 18 \\ \hline 108 \\ 108 \\ \hline \end{array}$$

2. Use a calculator for this problem.

The following installment-plan purchases were made at Martin's Mammoth Mart. Find the total installment-plan price and the carrying charges for each purchase. As a sample solution, the computation is shown for the first item.

	Item	List Price	Down Payment	Number of Monthly Payments	Amount of Each Monthly Payment	Total Installment-Plan Price	Carrying Charges
	stereo	$991	$115	24	$41.50	$1,109	$118
a.	jewelry	617	95	36	16.75		
b.	camera	324.80	50	12	26.40		
c.	sofa	1,250	375	24	43.50		
d.	boat	2,770	550	36	73.80		

stereo

$$\begin{array}{r} 41.50 \\ \times\ 24 \\ \hline 994.00 \end{array} \qquad \begin{array}{r} 994 \\ +\ 115 \\ \hline 1{,}109 \end{array} \qquad \begin{array}{r} 1{,}109 \\ -\ 991 \\ \hline 118 \end{array}$$

3. In buying a television set, which of the following installment purchase plans will cost you less?
Plan 1: No down payment and $9 a month for three years.
Plan 2: $60 down payment and $10 a month for two years. ____________

4. Stu is planning to buy a motorcycle whose list price is $1,050. The store will sell him the motorcycle for nothing down and monthly payments of $58 for two years. Stu's bank will lend him the $1,050, to be repaid in 2 years, at 12% interest per year. How much will Stu save by borrowing the money from the bank and then buying the motorcycle at its list price, compared to using the installment plan? ____________

5. In Problem 4, suppose that Stu has already saved the $1,050. How much will he save, compared to the bank loan? ____________

How much will he save, compared to the installment plan? ____________

PART XXII. Personal Banking

UNIT 97. Savings Accounts

TYPES OF SAVINGS ACCOUNTS

There are three basic types of savings accounts, the *day-of-deposit-to-day-of-withdrawal* account, the *time certificate* account, and the *money market* account.

The day-of-deposit-to-day-of-withdrawal account is the most convenient type because money may be deposited and withdrawn without any restriction. The interest rate for this type of account may be about 5%.

The time certificate, commonly called certificate of deposit or C.D., requires a minimum investment, for a specified period of time ranging from a number of months to a number of years. These restrictions are imposed because the certificate of deposit earns a substantially higher rate of interest than the day-of-deposit-to-day-of-withdrawal account.

The money market account also requires a minimum deposit, which may be as much as $2,500, but such an account is more flexible than the certificate of deposit, since it permits certain withdrawals and deposits. The interest rate is somewhat lower than the rate of a C.D.

On each of these types of accounts, interest rates and regulations vary at different times and at different banks.

CALCULATING INTEREST

The **principal,** which is money deposited as savings, earns **interest.** There are two types of interest, **simple interest** and **compound interest.** Simple interest is calculated on the principal only. Compound interest is calculated on all previously earned interest, as well as on the principal. Since finding compound interest is complicated, only simple interest is discussed here.

To find simple interest, multiply the *principal* (the amount deposited) by the *rate* (the annual interest rate) by the *time* (the period of time the money remains in the bank).

This procedure can be written as the formula

$$I = P \times R \times T$$

The rate, which is given as a percent, must be changed to a decimal. The time must be expressed in terms of years.

To find the new balance, add the principal to the interest earned.

$$\text{New balance} = P + I$$

EXAMPLE 1. Janet opened a savings account with a deposit of $975, and made no other deposits. If the interest rate is 5.8%, what will be her new balance at the end of a year?

Solution: First find the interest. Substitute the numbers into the formula. Change the rate to a decimal (5.8% = .058).

$$I = P \times R \times T$$
$$I = \$975 \times 5.8\% \times 1 \text{ year}$$
$$I = 975 \times .058 \times 1$$

$$\begin{array}{r} 975 \\ \times .058 \\ \hline 7\,800 \\ 48\,75 \\ \hline 56.550 \end{array}$$

$$I = \$56.55$$

Now add the interest to the principal, to get the new balance.

$$\begin{aligned} \text{New balance} &= P + I \\ &= \$975 + \$56.55 \\ &= \$1{,}031.55 \end{aligned}$$

Answer: The new balance is $1,031.55

EXAMPLE 2. Alex opens a 90-day money market account with a deposit of $3,500. If the interest rate is 9.75%, how much interest will he have earned at the end of 90 days?

Solution: Since a commercial year is considered to be twelve 30-day months or 360 days, the 90-day time period is $\frac{90}{360}$ or .25 of a year.

$$I = P \times R \times T$$

$$I = \$3{,}500 \times 9.75\% \times 90 \text{ days}$$

$$I = 3{,}500 \times .0975 \times .25$$

```
   3,500      341.25
  ×.0975        ×.25
  1 7500     17 0625
 24 500      68 250
315 00       85.3125
341.2500
```

$$I = \$85.31$$

Answer: At the end of 90 days, Alex will have earned $85.31 in interest.

EXAMPLE 3. Mrs. Mitchell earned $184.50 in interest for the year. If the amount of the principal was $1,800.00, what was the rate of interest?

Solution: Since $T = 1$, the basic formula $I = P \times R \times T$ becomes simply

$$I = P \times R$$

Substitute the given facts into this simplified formula.

$$\$184.50 = \$1{,}800 \times R$$

To find R, the unknown factor, use division to undo the multiplication.

$$\frac{184.50}{1{,}800} = \frac{\cancel{1{,}800} \times R}{\cancel{1{,}800}}$$

```
           .1025
1,800 )184.5000
       180 0
         4 500
         3 600
           9000
           9000
```

$$.1025 = R$$

$$R = .1025 \text{ or } 10.25\%$$

Answer: The rate of interest was 10.25%.

Example 3 could have been solved by using the $\frac{\text{IS}}{\text{OF}}$ formula. The problem can be rephrased as "($184.50 is) what percent (of $1,800)?"

$$\frac{\text{IS}}{\text{OF}} = \frac{184.50}{1{,}800} = .1025$$

APPLICATION PROBLEMS

1. Use a calculator for this problem.

Cory Wentworth, a bank teller, is calculating interest earnings. Find the amount of interest and the new balance for each account.

	Principal	Interest Rate	Period of Time	Amount of Interest	New Balance
a.	$1,280	5.6%	2 yr.		
b.	1,570	$5\frac{1}{4}$%	90 da.		
c.	1,600	6.6%	8 mo.		
d.	2,400	$5\frac{3}{4}$%	18 mo.		
e.	1,720	6.5%	180 da.		
f.	2,050	6.12%	45 da.		

2. Joseph deposited $6,875 in a 3-year certificate of deposit that pays an interest rate of 9.3%. What will be his new balance at the end of 3 years? ____________

3. Randy is saving money for her college education. In one year, she received $200.48 interest on her $3,580 principal. What was the rate of interest? ____________

4. William has a savings account of $4,375 that pays 5.9% interest. How much more interest can he earn in a year if he puts the $4,375 into a 1-year certificate of deposit that pays 10.5% interest? ____________

5. Sven opened a savings account that paid an interest rate of 5.7%. If he earned $133.95 in interest in one year, what was the amount of his deposit? ____________

UNIT 98. Checking Accounts

MAKING DEPOSITS

Edward J. Martin, a sales representative, has a **checking account** at the Newtown National Bank. When he receives payments for his salary and his commissions, he deposits the checks and cash in the bank. Then, when he has a bill to pay, he writes a check for the amount he owes and sends the check by mail.

By using his checking account, Mr. Martin does not have to spend time going from store to store and standing in line in order to pay his bills. Also, he does not have to worry about carrying large amounts of money with him. Finally, his cancelled checks are proof that his bills have been paid.

The bank gives Mr. Martin a supply of **deposit slips,** which he fills out whenever he deposits cash or checks. One day, Mr. Martin had the following items to deposit:

4 $20 bills
2 $10 bills
4 $1 bills
7 quarters
a salary check for $186.53
commission checks for $23.65, $19.67, $37.15, and $25.00

Here is how he filled out his deposit slip:

NEWTOWN NATIONAL BANK

DATE Dec. 29 19—

NAME Edward J. Martin

ADDRESS 106 West Washington St.
Newtown, N.Y. 10099

		DOLLARS	CENTS
Cash		105	75
Checks list separately	1	186	53
	2	23	65
	3	19	67
	4	37	15
	5	25	—
	6		
	7		
	8		
TOTAL		397	75

First, Mr. Martin counted all of his bills and coins and entered the sum of $105.75 after "Cash."

4 $20 bills	=	80.00
2 $10 bills	=	20.00
4 $1 bills	=	4.00
7 quarters	=	1.75
		105.75

Next, he listed each of his checks separately. Then, he added together all of his deposits and entered this sum, $397.75, after "Total." Notice that he also filled in the date and his name and address.

After the bank teller received the deposit slip and counted the cash and the checks, he returned a carbon copy of the deposit slip to Mr. Martin. This **customer's receipt,** which the teller dated, is proof that the bank received the deposit.

WRITING CHECKS

In addition to deposit slips, the Newtown National Bank gives Mr. Martin a supply of **checks.** These checks are in a checkbook that also contains either *check stubs* or pages called a **register** on which to keep track of his balance. Whenever Mr. Martin writes a check, he is careful to record the transaction and calculate the new balance.

To pay his December bill at Jay's Department Store, Mr. Martin filled out the following check:

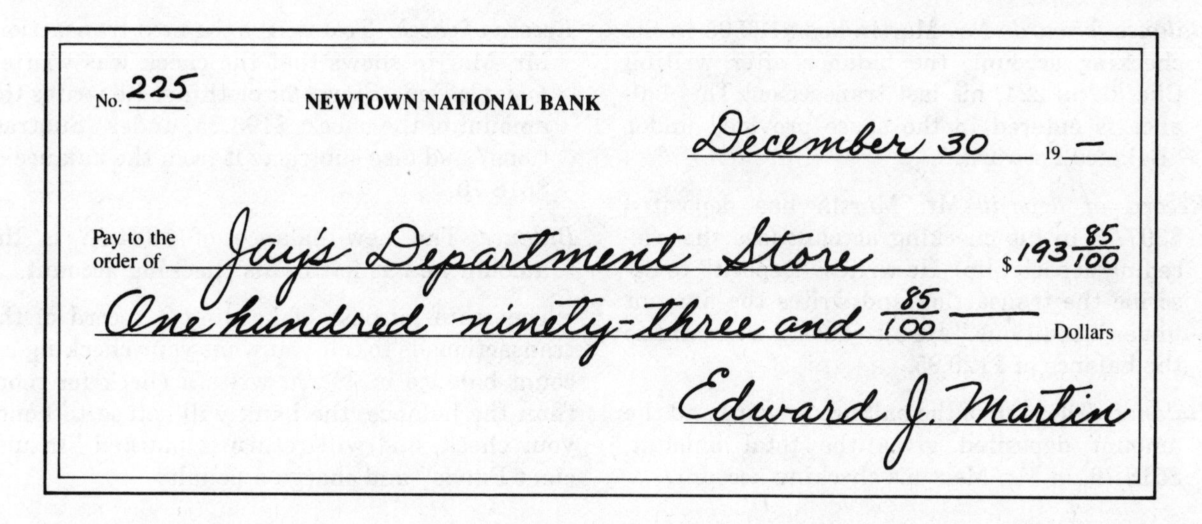

No. 225
NEWTOWN NATIONAL BANK
December 30 19—
Pay to the order of Jay's Department Store $ 193 85/100
One hundred ninety-three and 85/100 Dollars
Edward J. Martin

Check number: After "No." on both the register and the check, Mr. Martin writes "225." The last check he wrote was check No. 224; the next check he will write is check No. 226.

Date: The correct date is always written on both the register and the check.

To: With this check, Mr. Martin tells his bank to pay a certain amount to *Jay's Department Store.* He writes this name on the register and on the check.

For: When Mr. Martin goes over his records in a month or two, he may have forgotten what he bought at Jay's Department Store. Therefore, he writes on the register what the check pays for.

Face amount: The amount of the check (the face amount) is usually written on the register as a decimal. It must be written on the check in two different places, once in numbers and once in words and numbers. Amounts less than one dollar are written as a fraction, for example, $\frac{85}{100}$ or 85/100. Sometimes such amounts are written as $\frac{85}{xx}$ or as 85/. When the face amount is an even number of dollars, "no cents" is written as $\frac{no}{100}$ or as $\frac{xx}{100}$. Never write an amount as a decimal on a check.

Signature: Mr. Martin signs his *legal signature* to the check. He signs it in *exactly* the way he signed it when he opened his checking account. The bank has a record of his legal signature. If the check were signed "Ed Martin" or "E. J. Martin," the bank might return it to him for a proper signature.

Using the Register to Figure the Balance

ITEM NUMBER	DATE	DESCRIPTION OF TRANSACTION	SUBTRACTIONS AMOUNT OF PAYMENT OR WITHDRAWAL	ADDITIONS AMOUNT OF DEPOSIT OR INTEREST	BALANCE FORWARD
					120 95
	12/29	TO Deposit		397 75	397 75
		FOR			518 70
225	12/30	TO Jay's Dept. Store	193 85		193 85
		FOR clothing			324 85
		TO			
		FOR			
		TO			
		FOR			
		TO			
		FOR			

A Typical Register Page

Balance forward: Mr. Martin has $120.95 in his checking account, the balance after writing Check No. 224, his last transaction. This balance is entered in the space provided under "Balance Forward."

Record of deposit: Mr. Martin has deposited $397.75 in his checking account (see the preceding deposit slip). He writes "Deposit" to describe the transaction and writes the amount under "Additions." He also adds the amount to the balance of $120.95.

Balance: The sum of the balance forward and the amount deposited gives the total amount, $518.70, in Mr. Martin's checking account.

Record of check: To describe the next transaction, Mr. Martin shows that the check was written *to* Jay's Dept. Store, *for* clothing. He writes the amount of the check, $193.85, under "Subtractions" and also subtracts it from the balance of $518.70.

Balance: The new balance of $324.85 is the amount he has left in his checking account.

The main purpose of keeping a record of the transactions is to tell you what your checking account balance is. If you write a check for more than the balance, the bank will refuse to honor your check, and will return it marked "Insufficient Funds" and charge a penalty.

Application Problems

1. Use a calculator for this problem.

Below is a record of deposits made to the account of Edward J. Martin. Find the total of each deposit and the total of each column. The grand total of the deposits should be the same as the grand total of the columns.

Date	Cash	Checks 1	Checks 2	Checks 3	Checks 4	Total
Jan. 5	$115.60	$38.25	$19.63	$31.75	$34.75	*a.*
Jan. 20	120.75	25.85	20.15	43.85	31.60	*b.*
Feb. 5	125.80	37.50	21.20	32.60	27.80	*c.*
Feb. 20	120.75	35.65	19.70	31.10	47.63	*d.*
Mar. 5	117.80	30.15	20.75	29.68	39.15	*e.*
Total	*f.*	*g.*	*h.*	*i.*	*j.*	*k.* Grand Total

2. From the information provided, fill in the checks, and fill in the register that is shown after Check No. 29. As a sample solution, Check No. 26 has been filled in. Use your own name for the legal signature, and write the current year. Check No. 26 and the first deposit are shown in the register.

Balance forward: $152.85
Amount deposited: date, July 12; $85.16
Check No. 26: date, July 15; to Dr. S. Fontek for dental work; $36.50

No. 26 NEWTOWN NATIONAL BANK

July 15 19___

Pay to the order of Dr. S. Fontek $36 50/100

Thirty-Six and 50/100 Dollars

____

a. Check No. 27: date, July 25; to Ace Men's Shop for clothing; $63.71

No. ___ NEWTOWN NATIONAL BANK

_______________ 19___

Pay to the order of _______________ $___

_______________ Dollars

____

b. Check No. 28: date, July 25; to Rowe Stationers for office supplies; $23.53

No. ___ NEWTOWN NATIONAL BANK

_______________ 19___

Pay to the order of _______________ $___

_______________ Dollars

____

c. *Amount deposited:* date, July 27; $125.60

d. *Check No. 29:* date, July 30; to Home Appliance Services for washer repair; $73.85

No. ______ **NEWTOWN NATIONAL BANK**

______________ 19____

Pay to the order of ______________________ $________

______________________________________ Dollars

____

ITEM NUMBER	DATE	DESCRIPTION OF TRANSACTION	SUBTRACTIONS AMOUNT OF PAYMENT OR WITHDRAWAL		ADDITIONS AMOUNT OF DEPOSIT OR INTEREST		BALANCE FORWARD	
							152	85
	7/12	TO deposit					85	16
		FOR			85	16	BAL 238	01
26	7/15	TO Dr. S. Fontek					36	50
		FOR dental work	36	50			BAL 201	51
		TO						
		FOR					BAL	
		TO						
		FOR					BAL	
		TO						
		FOR					BAL	
		TO						
		FOR					BAL	
		TO						
		FOR					BAL	
		TO						
		FOR					BAL	

Note: If you have properly filled in the register, the final balance should be $166.02.

3. In the month of August, Joan Germaine made the deposits and wrote the checks indicated below. Complete the register. The starting Balance Forward is $325.60.

a. *Check No. 107:* date, August 1; to Quality Men's Shop for clothing; $136

b. *Deposit:* date, August 2; $116.70

c. *Check No. 108:* date, August 3; to Pioneer Supermarket for food; $53.65

d. *Check No. 109:* date, August 10; to Dr. Anna James for X-rays; $21.75

e. *Deposit:* date, August 13; $115.45

f. *Check No. 110:* date, August 15; to Dell Pharmacy for prescription; $21.70

g. *Check No. 111:* date, August 20; to Central Telephone Company for phone bill; $18.50

h. *Check No. 112:* date, August 25; to Ace Appliance Co. for TV set repair; $78.50

ITEM NUMBER	DATE	DESCRIPTION OF TRANSACTION	SUBTRACTIONS AMOUNT OF PAYMENT OR WITHDRAWAL	ADDITIONS AMOUNT OF DEPOSIT OR INTEREST	BALANCE FORWARD
		TO			
		FOR			BAL
		TO			
		FOR			BAL
		TO			
		FOR			BAL
		TO			
		FOR			BAL
		TO			
		FOR			BAL
		TO			
		FOR			BAL
		TO			
		FOR			BAL
		TO			
		FOR			BAL

Note: If you have properly filled in the register, the final balance should be $227.65.

PART XXIII. Algebra

UNIT 99. Algebraic Expressions

Algebra is a form of mathematics that uses number symbols and letter symbols to solve problems.

A **number symbol** is any word, numeral, combination of numerals, or picture that stands for a certain specific number. For example, all of the following number symbols stand for the number you know as "12":

XII	3×4	$15 - 3$
dozen	$24 \div 2$	$8 + 4$
$\frac{12}{1}$	$\frac{6 \times 6}{3}$	twelve
𝍸 𝍸 II	number of months in a year	

A **letter symbol** is a letter that stands for a number to be found or for the value of an unknown number. The letters x, y, and z in the following mathematical sentences are examples of letter symbols:

$x = 5 + 14 \quad y = 27 \div 3 \quad z = 9 \times 7$

By using number symbols, letter symbols, and the symbols of arithmetic that you already know (+, −, ×, ÷, =), you can write statements about numbers much more briefly than you can with words. The mathematical sentence, "When five is added to some unknown number, the sum is fourteen," may be written as follows:

$$n + 5 = 14$$

Just as you can change around the words in a sentence, so you can change around the symbols in a mathematical sentence. Here are three possible ways of rearranging the previous mathematical sentence, both in words and in symbols:

"When some unknown number is added to five, you get a total of fourteen." In symbols: $5 + \square = 14$ (Here, a "box" is used instead of a letter. In algebra, you can use any convenient symbol to stand for an unknown number.)

"Fourteen is the sum of five and some other number you want to find." In symbols:

$$14 = 5 + x$$

"Fourteen is what you get when you add an unknown number to five." In symbols:

$$14 = 5 + ?$$

Mathematical sentences that contain letters or other symbols in place of numbers are called **open sentences.**

In algebra, a letter is used to represent a number until the value of this number is discovered. In an algebra problem, you write an open sentence from the given facts and then find the value of the unknown number. Just as word sentences are made up of groups of words, so open sentences are made up of **algebraic expressions.** Therefore, in order to write an open sentence, you must learn to rewrite a word phrase as an algebraic expression.

Here are some examples: (In each case, the letter n stands for the unknown number.)

Word Phrase	*Algebraic Expression*
1. the number plus five	**1.** $n + 5$
2. the number minus 8	**2.** $n - 8$
3. the number increased by 6	**3.** $n + 6$
4. the number decreased by 9	**4.** $n - 9$
5. the number divided by 7	**5.** $n \div 7$ or $\frac{n}{7}$ or $\frac{1}{7}n$
6. the number multiplied by 3	**6.** $n \times 3$ or $3n$
7. the product of 8 times the number	**7.** $8 \times n$ or $8n$

Note: In algebra, the multiplication symbol ×, which might be confused with the letter x, is not usually used to show the product of a number and a letter symbol. Thus, $3 \times n$ and $8 \times n$ should be shown as $3n$ and $8n$. Also, the division sign, ÷, is seldom used in algebra. $n \div 7$ should be shown as $\frac{n}{7}$ or $\frac{1}{7}n$.

Here are some more examples. For practice, cover the right-hand column and try to guess what the algebraic expression should be. Then see if you guessed correctly.

Word Phrase	*Algebraic Expression*
1. 8 more than x	**1.** $x + 8$
2. r multiplied by 5	**2.** $5r$
3. 5 more than the product of y times 6	**3.** $6y + 5$
4. d times 3, decreased by 7	**4.** $3d - 7$
5. a boy's age 12 years from now (Use a for his age now.)	**5.** $a + 12$
6. one-half a given number n	**6.** $\frac{1}{2}n$ or $\frac{n}{2}$
7. s divided by 4 (Don't use the ÷ sign.)	**7.** $\frac{s}{4}$ or $\frac{1}{4}s$
8. ten more nickels than dimes (Use d for the number of dimes.)	**8.** $d + 10$
In examples 9 and 10, write a complete number sentence.	
9. Jane is twice as old as Mary. (Use J for Jane's age and M for Mary's age.)	**9.** $J = 2M$
10. The express train travels 25 miles per hour faster than the local train. (Use E for the speed of the express and L for the speed of the local.)	**10.** $E = L + 25$

EXERCISES

In 1–10, write an algebraic expression for each word phrase. Do not use the symbols × or ÷. Use n for the unknown number.

1. a number increased by 23

2. 16 less than a number

3. ten times a number

4. one-tenth of a number

5. the sum of a number and 13

6. 52 divided by some number

7. one-half the product of 7 and a number

8. one-third of the quotient you get when a number is divided by 17

9. twice the quotient of a number divided by 5

10. a number multiplied by three

In 11–20, write an algebraic expression for each word phrase. Use the letter symbols given.

11. x times y

12. x divided by y

13. the sum of n and m

14. p times 3, minus 5

15. 18 divided by x, plus 12

16. 5 times n, divided by 4

17. 3 times n, decreased by 15

18. 26 divided by x, decreased by 7

19. 5 more than the product of n and 7

20. 2 times y, divided by a plus 6

In 21–25, write a complete open sentence.

21. The area A equals the length L multiplied by the width W.

22. The distance D is the product of the time T and the rate R.

23. x is one-half of n, plus 10 more.

24. z is 15 more than 5 times n.

25. 6 times r, divided by 7, is the same as two-thirds of s.

APPLICATION PROBLEMS

1. Tom is x years old. Write John's age if he is twice as old as Tom. ______

2. The length of a room is 4 ft. longer than the width. Using w for the width, express algebraically the length of the room. ______

3. Jane is x years old. Mary is twice her age, and Sue is 5 years younger than Mary. Express Sue's age algebraically. ______

4. Scott has n dimes, half as many nickels, and 5 more quarters than nickels. Express algebraically the number of quarters Scott has. ______

5. Bill weighs x pounds and Jim weighs y pounds. If Jim gains 10 pounds and Bill loses 18 pounds, represent their total weight algebraically. ______

UNIT 100. Equations

In a numerical sentence such as $6 + 11 = 17$ or an open sentence such as $2n = 12$, the *quantity* to the left of the equals sign is equal to the *quantity* to the right of the equals sign. This expression of equality is called an **equation.** The quantity to the left of the equals sign is called the **left member** of the equation; the quantity to the right of the equals sign is called the **right member.**

An equation may be a numerical sentence, such as $3 \times 3 = \frac{18}{2}$, or it may be an open sentence that has one or more letter symbols, such as $3x = 18$ or $4S = P$.

When you find the value of the unknown number that is represented by the letter symbol in an open sentence, you **solve the equation.** To solve the equation $n + 15 = 20$, you must find some number that you can add to 15 to get 20. By counting ("sixteen, seventeen, eighteen, nineteen, twenty"), you see that the missing number is 5. In algebra, the solution would look like this:

$$n + 15 = 20$$
$$n = 5$$

You always *check* the found value of a letter symbol by putting its value back in the original equation.

Check:

$$n + 15 = 20$$
$$5 + 15 = 20$$
$$20 = 20 \checkmark$$

Since the left member equals the right member, you know that $n = 5$ is the correct value. The value of the letter that makes an equation true is called the **root** of the equation.

Algebra is often useful in solving word problems. To solve word problems algebraically, there are two basic steps to follow:

1. Write an equation that expresses the facts of the problem.
2. Solve the equation by finding its root.

The first step is often the more difficult of the two. By reading the problem carefully and by using a letter symbol in place of the unknown number, you will soon be able to rewrite word statements as equations.

EXAMPLE 1. Write each of the following sentences as an equation:

(*a*) Three times a certain number equals 36.

(*b*) A number increased by 12 equals 25.

(*c*) A number multiplied by 3, increased by 5, is equal to 38.

Solutions:

(*a*) $3n = 36$

(*b*) $x + 12 = 25$

(*c*) $3n + 5 = 38$

Often, you will have to read the problem carefully, rewrite the facts as *one complete sentence,* and then rewrite this sentence as the equation.

EXAMPLE 2. Jeri bought four cheeseburgers and paid for them with $4.00. If she got 20¢ in change, how much did *each* cheeseburger cost? Write an equation that can be used to solve this problem.

Solution: First, rewrite the above facts as one complete sentence. "The cost of 4 cheeseburgers plus $.20 change totals $4.00." Now, write this sentence as an equation, using C as the unknown cost.

Answer: $4C + .20 = 4.00$

EXAMPLE 3. When Tom started college, he was half his present age. When he graduated four years later, he was 22 years old. What is his present age? Write an equation that can be used to find the unknown age.

Solution: Rewrite the facts as one complete sentence. "Half of Tom's present age plus 4 years is 22." Use a for the unknown age and write an equation.

Answer: $\frac{1}{2}a + 4 = 22$

EXAMPLE 4. Henry is paid every two weeks, receiving a paycheck that is twice his weekly salary. Every payday, Henry puts $30, which is $\frac{1}{8}$ of his paycheck, in the bank. Write an equation that will enable you to find his *weekly* salary.

Solution: Rewrite the facts as one complete sentence. "When Henry's weekly salary is doubled, and then divided by 8, it is equal to $30." Write this sentence as an equation, using s as the unknown salary.

Answer: $\frac{2s}{8} = 30$

EXERCISES

In 1–12 write an equation for each word statement.

1. 8 taken from 4 times a number equals 64.

2. A number reduced by 15 is 36.

3. When 24 is subtracted from a number, the result is 15.

4. A number divided by 8, increased by 15, is equal to 47.

5. When 18 is added to a number, the result is 73.

6. One-quarter of a number added to 6 times the number is 53.

7. 8 added to 5 times a number results in 32.

8. If 3 times a given number is divided by 5, the result is 58.

9. When a number is decreased by 12, the result is 84.

10. The product of 8 and a number is 96.

11. A number multiplied by 5, increased by 10, results in 75.

12. When 10 is subtracted from 3 times a number, the result is 58.

APPLICATION PROBLEMS

1. Tom had x dollars. If he spent \$12, which was $\frac{1}{3}$ of what he had, write an equation to find how much money Tom had originally. __________
2. Jim weighs 146 pounds, which is 8 pounds more than twice as much as Harry. Write an equation to find Harry's weight. __________
3. Judy had x tennis balls. After buying 3 more, she had 11 tennis balls. Write an equation to find how many she had originally. __________
4. The price of a TV set is reduced by \$30. If this is $\frac{1}{3}$ of the original price, write an equation to find the original price. __________
5. 25% of an amount is equal to \$38. Write an equation to find the amount. __________
6. If 5 typewriters cost \$1,200, write an equation to find the cost of one typewriter. __________
7. A coat selling for \$68 was reduced to \$54. Write an equation to find the amount of the reduction. __________
8. Mary paid x dollars for a dress. Jane bought a dress for \$58, which was twice as much as Mary paid. Write an equation to find how much Mary paid. __________

UNIT 101. Solving Equations by Using Addition and Subtraction

In the preceding unit, you solved the equation

$$n + 15 = 20$$

You found the root to be $n = 5$. You started with an equation where the solution was not known; you ended up with an equation where the letter symbol was all alone in the left member.

To solve an equation, change it around so that *the letter symbol is in one member of the equation and all the number symbols are in the other member.*

In solving equations, you use addition, subtraction, multiplication, and division.

Consider the equation $n + 15 = 20$. You solved this equation by "common sense" in the preceding unit. Now, solve it by using algebra. The left member is $n + 15$. Since you want to get the n alone, you must separate the n from the 15. To do this, use the following fact:

The opposite of addition is subtraction.

The number 15 is *added* to the letter symbol n. Therefore, you must *subtract* 15 from the left member. When you subtract 15, however, you must follow this basic rule of solving equations:

RULE

Whatever you do to one member of an equation (add, subtract, multiply, or divide), you must do to the other member.

Here's how you solve the equation:

$$\begin{array}{rcrl} n + 15 & = & 20 & \\ -15 & & -15 & \text{Subtract 15 from both members.} \\ \hline n & = & 5 & \end{array}$$

(When you subtract 15 from 15, you get 0; but since $n + 0$ is the same as n, it is not necessary to write the 0 in the left member.)

Check: $n + 15 = 20$
$5 + 15 = 20$
$20 = 20$ ✓

The root of $n + 15 = 20$ is $n = 5$.

Here are more examples of solving equations by using subtraction. In each case, the unknown number has some other number added to it.

EXAMPLES. Solve each of the following equations for the unknown value of the letter symbol. Check each root in the original equation.

(*a*) $5 + n = 18$
(*b*) $20 = x + 9$
(*c*) $s + 20 = 9 \times 3$

Solutions:

(*a*)

$$\begin{array}{rcrl} 5 + n & = & 18 & \\ -5 & & -5 & \text{Subtract 5 from both members.} \\ \hline n & = & 13 & \end{array}$$

Check: $5 + n = 18$
$5 + 13 = 18$
$18 = 18$ ✓

Answer: $n = 13$

(*b*)

$$\begin{array}{rcrl} 20 & = & x + 9 & \\ -9 & & -9 & \text{Subtract 9 from both members.} \\ \hline 11 & = & x & \end{array}$$

Check: $20 = x + 9$
$20 = 11 + 9$
$20 = 20$ ✓

Answer: $x = 11$

(*c*) $s + 20 = 9 \times 3$

To simplify the equation, perform the indicated multiplication in the right member.

$$\begin{array}{rcrl} s + 20 & = & 27 & \\ -20 & & -20 & \text{Subtract 20 from both members.} \\ \hline s & = & 7 & \end{array}$$

Check: $s + 20 = 9 \times 3$
$7 + 20 = 9 \times 3$
$27 = 27$ ✓

Answer: $s = 7$

Remember

When you subtract any number from one member of an equation, you must subtract the same number from the other member.

You may think of an equation as a balanced scale, a scale with equal weights on both sides. Whenever you remove a weight from one side, you must remove the same weight from the other side in order to keep the scales balanced.

Consider the equation

$$x - 15 = 64$$

Here, the letter symbol has 15 subtracted from it. Since you want to get the x alone, you must separate it from the 15. To do this, use the following fact:

The opposite of subtraction is addition.

The 15 is *subtracted* from the x. Therefore, you must *add* 15 to both members.

$$\begin{array}{rcrl} x - 15 & = & 64 & \\ +15 & & +15 & \text{Add 15 to both members.} \\ \hline x & = & 79 & \end{array}$$

Check: $x - 15 = 64$
$79 - 15 = 64$
$64 = 64$ ✓

The root of $x - 15 = 64$ is $x = 79$.

EXAMPLES. Solve each of the following equations for the unknown value of the letter symbol. Check each root in the original equation.

(*d*) $x - 23 = 64$
(*e*) $27 = n - 13$
(*f*) $s - 4 = 12 - 2$

Solutions:

(*d*)
$$\begin{array}{rcl} x - 23 & = & 64 \\ +23 & & +23 \\ \hline x & = & 87 \end{array}$$
Add 23 to both members.

Check:
$$\begin{aligned} x - 23 &= 64 \\ 87 - 23 &= 64 \\ 64 &= 64 \ \checkmark \end{aligned}$$

Answer: $x = 87$

(*e*)
$$\begin{array}{rcl} 27 & = & n - 13 \\ +13 & & +13 \\ \hline 40 & = & n \end{array}$$
Add 13 to both members.

Check:
$$\begin{aligned} 27 &= n - 13 \\ 27 &= 40 - 13 \\ 27 &= 27 \ \checkmark \end{aligned}$$

Answer: $n = 40$

(*f*) $s - 4 = 12 - 2$

First, do the indicated subtraction in the right member.

$$\begin{array}{rcl} s - 4 & = & 10 \\ +4 & & +4 \\ \hline s & = & 14 \end{array}$$
Add 4 to both members.

Check:
$$\begin{aligned} s - 4 &= 12 - 2 \\ 14 - 4 &= 12 - 2 \\ 10 &= 10 \ \checkmark \end{aligned}$$

Answer: $s = 14$

Remember

When you add any number to one member of an equation, you must add this same number to the other member.

EXERCISES

In 1–9, solve each equation for the value of the letter symbol.

1. $n + 3 = 42$
2. $x + 13 = 27$
3. $y - 5 = 17$
4. $t - 3\frac{1}{2} = 15$
5. $n + 35 = 112$
6. $28 - x = 18$
7. $35 + y = 47$
8. $p - 15 = 120$
9. $t + 28 = 57$

In 10–20, write an equation and then solve it for the unknown number.

10. A number, decreased by 15, equals 52. Find the number.

11. What number, increased by 28, is equal to 75?

12. When 13 is added to a number, the result is 85. Find the number.

13. A number, reduced by 27, is equal to 85. Find the number.

14. A number, added to 15, results in 33. Find the number.

15. What number, less 24, is equal to 51?

16. When 27 is taken from a number, the result is 83. Find the number.

17. What number, decreased by 35, results in 47?

18. A number, increased by 17, is equal to 39. Find the number.

19. A number, plus 32, equals 51. Find the number.

20. If 13 is subtracted from a number, the result is 57. Find the number.

APPLICATION PROBLEMS

Write and solve an equation to answer each question.

1. Tom earns $35 more than Jim. If Tom earns $185 a week, what is Jim's salary? ______
2. Nathan sold his car for $565 less than he paid for it. If he paid $3,250 for the car, how much did he sell it for? ______
3. A class has 8 more girls than boys. If there are 28 students in the class, how many girls are there? ______

4. Juan bought a television set and made a down payment of \$58. If he still owes \$215 on the set, what was the full price of the television set? ____________

5. Jane is 5 years younger than Mary. How old is Jane if Mary is 23 years old? ____________

UNIT 102. Solving Equations by Using Division and Multiplication

The word statement, "Five times a certain number is equal to 60," can be rewritten as the equation:

$$5x = 60$$

In this equation, the left member is $5x$, which means that the letter symbol is *multiplied* by 5. Since you want to get the x alone, you must separate the x from the 5. To do this, use the following fact:

The opposite of multiplication is division.

The letter symbol x is *multiplied* by 5. Therefore, you must *divide* the left member by 5. Since whatever you do to one member of an equation you must do to the other member, you must divide the right member by 5 also.

$$5x = 60$$

$$\frac{\overset{1}{\cancel{5}}x}{\underset{1}{\cancel{5}}} = \frac{\overset{12}{\cancel{60}}}{\underset{1}{\cancel{5}}} \quad \text{Divide both members by 5.}$$

$$x = 12$$

Check to see if $x = 12$ is the root of the equation. In algebra, so as not to confuse the multiplication sign × with the letter x, it is usual to indicate multiplication by using parentheses ( ).

$$\begin{aligned} 5x &= 60 \\ 5(12) &= 60 \\ 60 &= 60 \checkmark \end{aligned}$$

The root of $5x = 60$ is $x = 12$.

Here are some examples of solving equations by using division. Notice that "algebraic division" is very similar to the arithmetic division that you have studied. In every case, the unknown number (the letter symbol) is multiplied by some other number.

EXAMPLES. Solve each of the following equations for the unknown value of the letter symbol. Check each root in the original equation.

(a) $9n = 36$
(b) $49 = 7x$
(c) $3x = 6(4)$

Solutions:

(a)

$$9n = 36$$

$$\frac{\overset{1}{\cancel{9}}n}{\underset{1}{\cancel{9}}} = \frac{\overset{4}{\cancel{36}}}{\underset{1}{\cancel{9}}} \quad \text{Divide both members by 9.}$$

$$n = 4$$

Check:
$$\begin{aligned} 9n &= 36 \\ 9(4) &= 36 \\ 36 &= 36 \checkmark \end{aligned}$$

Answer: $n = 4$

(b)

$$49 = 7x$$

$$\frac{\overset{7}{\cancel{49}}}{\underset{1}{\cancel{7}}} = \frac{\overset{1}{\cancel{7}}x}{\underset{1}{\cancel{7}}} \quad \text{Divide both members by 7.}$$

$$7 = x$$

Check:
$$\begin{aligned} 49 &= 7x \\ 49 &= 7(7) \\ 49 &= 49 \checkmark \end{aligned}$$

Answer: $x = 7$

(c) $3x = 6(4)$

To simplify the equation, perform the indicated multiplication in the right member.

$$3x = 24$$

$$\frac{\overset{1}{\cancel{3}}x}{\underset{1}{\cancel{3}}} = \frac{\overset{8}{\cancel{24}}}{\underset{1}{\cancel{3}}}$$ Divide both members by 3.

$$x = 8$$

Check: $3x = 6(4)$
$3(8) = 6(4)$
$24 = 24$ ✓

Answer: $x = 8$

The word statement, "A certain number, divided by 5, is equal to 80," can be rewritten as the equation:

$$\frac{n}{5} = 80$$

In this equation, the left member is $\frac{n}{5}$, which means that the letter symbol is *divided* by 5. Since you want to get the n alone, you must separate the n from the 5. Do this by using the following fact:

The opposite of division is multiplication.

The letter symbol n is divided by 5. Therefore, you *multiply* both members by 5.

$$\frac{n}{5} = 80$$

$$\frac{\overset{1}{\cancel{5}}}{1} \times \frac{n}{\underset{1}{\cancel{5}}} = 5(80)$$ Multiply both members by 5.

$$n = 400$$

Check to see if $n = 400$ is the root of the equation.

$\frac{n}{5} = 80$

$\frac{400}{5} = 80$

$80 = 80$ ✓

The root of $\frac{n}{5} = 80$ is $n = 400$.

Here are some examples of solving equations by using multiplication. In every case, the letter symbol is divided by some other number.

Examples. Solve each of the following equations for the unknown value of the letter symbol. Check each root in the original equation.

(d) $\frac{x}{8} = 6$

(e) $19 = \frac{n}{2}$

(f) $\frac{1}{4}x = 1\frac{3}{4}$

Solutions:

(d) $$\frac{x}{8} = 6$$

$$\frac{\overset{1}{\cancel{8}}}{1} \times \frac{x}{\underset{1}{\cancel{8}}} = 8 \times 6$$ Multiply both members by 8.

$$x = 48$$

Check: $\frac{x}{8} = 6$

$\frac{48}{8} = 6$

$6 = 6$ ✓

Answer: $x = 48$

(e) $$19 = \frac{n}{2}$$

$$2 \times 19 = \frac{\overset{1}{\cancel{2}}}{1} \times \frac{n}{\underset{1}{\cancel{2}}}$$ Multiply both members by 2.

$$38 = n$$

Check: $19 = \frac{n}{2}$

$19 = \frac{38}{2}$

$19 = 19$ ✓

Answer: $n = 38$

(*f*) $\frac{1}{4}x = 1\frac{3}{4}$

In dealing with mixed numbers, in algebra as well as in arithmetic, it is usual to change them to improper fractions when dividing or multiplying. Since $\frac{1}{4}x$ is the same as $\frac{x}{4}$ and since $1\frac{3}{4} = \frac{7}{4}$, you rewrite the original equation as follows:

$$\frac{1}{4}x = 1\frac{3}{4}$$

$$\frac{x}{4} = \frac{7}{4}$$

Now, solve by getting the x alone in one member.

$$\frac{\overset{1}{\cancel{4}}}{1} \times \frac{x}{\underset{1}{\cancel{4}}} = \frac{\overset{1}{\cancel{4}}}{1} \times \frac{7}{\underset{1}{\cancel{4}}}$$ **Multiply both members by 4.**

$$x = 7$$

Check: $\frac{1}{4}x = 1\frac{3}{4}$

$$\frac{1}{4}(7) = 1\frac{3}{4}$$

$$\frac{7}{4} = 1\frac{3}{4}$$

$$1\frac{3}{4} = 1\frac{3}{4} \checkmark$$

Answer: $x = 7$

EXERCISES

In 1–10, solve and check each equation.

1. $\frac{n}{12} = 2$
2. $\frac{1}{5}t = 25$
3. $9x = 27$
4. $46 = 23y$
5. $3x = 4\frac{1}{2}$
6. $11n = 110$
7. $15 = \frac{y}{4}$
8. $\frac{t}{3} = 3$
9. $\frac{x}{5} = 3.2$
10. $24 = \frac{1}{3}t$

In 11–20, write an equation from the facts given in each problem. Then solve for the unknown number.

11. Three times a number is 54. Find the number.

12. A number multiplied by .4 is 8. Find the number.

13. Two-fifths of a number is 65. What is the number?

14. Thirteen is equal to $\frac{1}{3}$ of a number. What is the number?

15. 5% of a number is equal to 25. Find the number.

16. A number divided by 8 equals 6. What is the number?

17. Three-quarters of a number is 150. Find the number.

18. When 120 is divided by a number, the result is 40. Find the number.

19. If 15 is equal to a number divided by 9, find the number.

20. A number divided by 50 is equal to $\frac{1}{2}$. Find the number.

APPLICATION PROBLEMS

In 1–5, write and solve an equation.

1. John got a 15% increase in salary. If the increase was \$25, what was the original salary? ________
2. Tony has a savings account that pays 6% interest per year. If he earned \$75 in interest one year, what was his principal? ________
3. Mary bought a coat at $\frac{1}{3}$ off the original price. If she saved \$36, what was the original price? ________
4. A salesperson sold a new car for \$13,490 and earned \$674.50 in commissions. What was the rate of commission? ________
5. Cesar saved $\frac{1}{5}$ of his weekly income. If he saved \$45, what was his weekly income? ________

UNIT 103. Solving Two-Step Equations

The equation $3x + 6 = 30$ states that, "When 3 times a certain number is increased by 6, the result is 30."

Since this equation indicates *two* mathematical operations (multiplying by 3 and adding 6), you must use *two* steps to solve it.

Since you must get the x alone in one member, the first step is to eliminate the "+6" by subtracting 6 from both sides of the equation.

$$\begin{array}{rcr} 3x + 6 &=& 30 \\ -6 && -6 \\ \hline 3x &=& 24 \end{array}$$

The left member of this new equation is $3x$. Therefore, the second step is to divide both members by 3.

$$3x = 24$$

$$\frac{3x}{3} = \frac{24}{3}$$

$$x = 8$$

Check to see if $x = 8$ is the root of the original equation.

$$\begin{aligned} 3x + 6 &= 30 \\ 3(8) + 6 &= 30 \\ 24 + 6 &= 30 \\ 30 &= 30 \ \checkmark \end{aligned}$$

The root of $3x + 6 = 30$ is $x = 8$.

RULE

When solving two-step equations, do the addition or subtraction first. Then do the multiplication or division.

Here are some examples of solving two-step equations.

Examples. Solve each of the following equations for the unknown value of the letter symbol. Check each root in the original equation.

(*a*) $4n - 20 = 4$ (*b*) $\frac{x}{3} + 9 = 13$

(*c*) $45 = 24 + 3s$

Solutions:

(*a*)
$$\begin{array}{rcr} 4n - 20 &=& 4 \\ +20 && +20 \\ \hline 4n &=& 24 \end{array}$$
Add 20 to both members.

$$\frac{4n}{4} = \frac{24}{4}$$
Divide both members by 4.

$$n = 6$$

Check:
$$\begin{aligned} 4n - 20 &= 4 \\ 4(6) - 20 &= 4 \\ 24 - 20 &= 4 \\ 4 &= 4 \ \checkmark \end{aligned}$$

Answer: $n = 6$

(*b*)

$$\frac{x}{3} + 9 = 13$$

$$-9 \quad -9$$ Subtract 9 from both members.

$$\frac{x}{3} = 4$$

$$3\left(\frac{x}{3}\right) = 3(4)$$ Multiply both members by 3.

$$x = 12$$

Check:

$$\frac{x}{3} + 9 = 13$$

$$\frac{12}{3} + 9 = 13$$

$$4 + 9 = 13$$

$$13 = 13 \checkmark$$

Answer: $x = 12$

(*c*)

$$45 = 24 + 3s$$

$$-24 \quad -24$$ Subtract 24 from both members.

$$21 = 3s$$

$$\frac{21}{3} = \frac{3s}{3}$$ Divide both members by 3.

$$7 = s$$

Check:

$$45 = 24 + 3s$$

$$45 = 24 + 3(7)$$

$$45 = 24 + 21$$

$$45 = 45 \checkmark$$

Answer: $s = 7$

EXERCISES

In 1–12, solve each equation for the value of the letter symbol.

1. $8x + 15 = 63$

2. $3n - 11 = 19$

3. $\frac{z}{5} + 5 = 12$

4. $\frac{1}{5}x - 3 = 6$

(*Hint:* $\frac{1}{5}x$ is the same as $\frac{x}{5}$.)

5. $40 = 25 + \frac{n}{6}$

6. $\frac{1}{4}x + 5 = 33$

7. $36 = 5x + 6$

8. $10 = \frac{x}{2} - 6$

9. $8 + 5x = 35 + 8$

10. $3n = 10\frac{1}{2}$

11. $9 + 5y = 24$

12. $8r - 7 = 25$

In 13–20, write an equation from the given facts. Then solve it for the unknown number.

13. Two times a certain number, decreased by 8, equals 64. Find the number.

14. When a number is multiplied by 5, and then the product is increased by 21, the result is 56. Find the number.

15. If 13 is subtracted from 8 times a certain number, the result is 83. Find the number.

16. When you triple a number and add 31, the result is 76. Find the number.

17. When a number is divided by 5 and the quotient is increased by 8, the result is 16. Find the number.

18. When 18 is added to a number that has been multiplied by 3, the result is 30. Find the number.

19. One-third of a number, increased by 10, is equal to 40. Find the number.
(*Hint:* $\frac{1}{3}x$ is the same as $\frac{x}{3}$.)

20. If 63 is equal to 6 times a certain number, increased by 15, find the number.

APPLICATION PROBLEMS

Write and solve an equation to answer each question.

1. George saved $25 more than three times the amount Harry saved. If George saved $85, how much did Harry save? ______

2. Jane spent $12 for two tickets to a concert, tax included. If the tax was 80¢, how much did *each* ticket cost? ______

3. Sean is 25 years old. He is 5 years less than three times the age of his sister. How old is his sister? ______

4. Eight more than one-third the number of boys in a class equals the number of girls. If there are 14 girls in the class, how many boys are there? ______

5. Jim earns $15 less than twice the amount earned by Pedro. If Jim earns $95 a week, how much does Pedro earn a week? ______

6. Mary spent $4 less than one-half her baby-sitting earnings. If she spent $13, how much did she earn? ______

7. The perimeter of a rectangle is 86 inches. If the width is 10 inches, what is the length? (*Hint:* Draw a figure, using the facts of the exercise and your knowledge of rectangles. Then write an equation for the perimeter, using a letter symbol for the length, and the given values for the width and the perimeter.) ______

8. On Saturday, Ken sold 10 more than twice the number of newspapers that he sold on Friday. If he sold 86 newspapers Saturday, how many did he sell Friday? ______

9. Geri traveled 5 miles less than one-fourth the distance Pat traveled. Geri traveled 17 miles. How far did Pat travel? ______

10. Before work one day, Chris had $15. After being paid for 6 hours of work, he then had a total of $40.50. How much did he earn per hour? ______

INDEX